S0-BLS-331

Ligand Exchange Chromatography

Authors

Vadim A. Davankov
Institute of Organo-Element Compounds
Academy of Sciences of the U.S.S.R.
Moscow, U.S.S.R.

James D. Navratil
Professor
University of New South Wales
Kensington, New South Wales
Australia

Harold F. Walton
Department of Chemistry
University of Colorado
Boulder, Colorado

CRC Press, Inc.
Boca Raton, Florida

Library of Congress Cataloging-in-Publication Data

Davankov, Vadim A.
Ligand exchange chromatography.

Includes bibliographies and index.
1. Ligand exchange chromatography. I. Navratil, James D., 1941- . II. Walton, Harold F., 1912- III. Title.
QD79.C4537D38 1988 543'.0892 87-31996
ISBN 0-8493-6775-1

This book represents information obtained from authentic and highly regarded sources. Reprinted material is quoted with permission, and sources are indicated. A wide variety of references are listed. Every reasonable effort has been made to give reliable data and information, but the author and the publisher cannot assume responsibility for the validity of all materials or for the consequences of their use.

Direct all inquiries to CRC Press, Inc., 2000 Corporate Blvd., N.W., Boca Raton, Florida, 33431.

International Standard Book Number 0-8493-6775-1

Library of Congress Card Number 87-31996
Printed in the United States

FOREWORD

About 20 years ago, a new type of chromatography was developed, namely, ligand exchange chromatography (LEC). It makes use of the formation of labile coordinate bonds between ligands and a metal cation, producing coordination compounds or complexes.

LEC, in its liquid and gas variations, has made it possible to resolve many of the outstanding problems in the separation, purification, and analysis of different substances which other types of chromatography had failed to do. The hallmark of the achievements of this technique is its ability to separate optical isomers. Several interesting modifications of LEC have also been developed.

This book presents a systematic and comprehensive review of the information on chromatographic processes that involve the formation of coordination compounds, aiming not only to demonstrate the achievements that have been made in the theory and praxis of chromatography, but also to point out, as far as possible, the future potential of LEC.

THE AUTHORS

Vadim A. Davankov, D.Sc., Professor of Chemistry, Institute of Organo-Element Compounds, Academy of Sciences of the U.S.S.R., started working at the above institute in 1962. There he received his Ph.D. degree and later, the degree of Doctor of Sciences in Chemistry. Since 1975 Dr. Davankov has been the head of the Laboratory for Stereochemistry of Adsorption Processes.

Dr. Davankov's main research field is separation of optical isomers by means of column liquid chromatography and immobilized enzyme techniques. In 1968, together with Rogozhin, he patented ligand exchange chromatography as a general approach to resolve racemates into enantiomeric pairs. Developing this idea has also included extensive studies on structure and enantioselectivity phenomena in copper(II) complexes with amino acids and diamines, as well as examining new types of polymeric networks, i.e., macronet isoporous polystyrene and hypercross-linked polystyrene ''Styrosorb''. Dr. Davankov has authored and co-authored more than 250 scientific publications.

Dr. Davankov serves as Vice Chairman of the Scientific Council on Chromatography of the Academy of Sciences of the U.S.S.R. and Chairman of its Liquid Chromatography Section. In 1978 he was awarded the Tswett Medal by the Academy of Sciences of the U.S.S.R.

James D. Navratil, Ph.D. Chemistry, University of Colorado, Boulder, started working at Rocky Flats, operated by Rockwell International for the Department of Energy, in 1961. He has held several positions in the Analytical Laboratories and Research and Development (present position, Manager of Chemical Research), and from 1978 to 1981 was on leave of absence to the International Atomic Energy Agency, Vienna. He is also an Adjunct Professor in the Department of Chemistry and Geochemistry, Colorado School of Mines, Golden.

He was named Rockwell International Scientist/Engineer of the Year in 1977. Dr. Navratil's research interests are mainly chemical separations and actinide chemistry.

He is founder and co-editor of the journal, *Solvent Extraction and Ion Exchange,* and serves on the editorial board of seven other journals. He is active in the American Chemical Society and is founder and past chairman of the ACS subdivision of Separation Science and Technology. Dr. Navratil was the recipient of the Colorado ACS award (1984) and two IR-100 awards (1983 and 1985). He has authored or co-authored more than 100 publications and co-edited nine books and co-authored the book, *Polyurethane Foam Sorbents in Separation Science.*

Harold F. Walton, Ph.D., Oxford University, joined the University of Colorado faculty as a member of the Department of Chemistry in 1947 and remained until his retirement in 1982, when he spent several months in Paris as the guest of Professor R. Rosset in the Ecole Superieure de Physique et de Chimie. He returned to the University of Colorado in an active capacity as Senior Research Associate of the Cooperative Institute for Research in Environmental Sciences and Professor Emeritus of Chemistry.

He was one of the first scientists to pursue a program of research in ligand exchange chromatography, which continued until his retirement. In 1976 he received the Colorado Section Award of the American Chemical Society.

He has taught in several countries of Latin America, especially Peru. He is a Corresponding Member of the Chemical Society of Peru and an Honorary Professor of the University of Trujillo and the University of San Marcos, Lima.

He is the author of seven books, including *Ion Exchange in Analytical Chemistry* (jointly with W. Rieman), several chapters in cooperative works, and some 120 research papers. These deal with ion exchange, chromatography, electrolyte solutions, electrochemistry, geochemistry, and water pollution.

TABLE OF CONTENTS

Chapter 1

INTRODUCTION: HISTORY, PRINCIPLES, TERMINOLOGY

H. F. Walton and V. A. Davankov

No method of chemical separation can equal chromatography in versatility and breadth of applications. Neutral molecules and ions can be separated, as can isotope species, small and large molecules, natural and artificial macromolecules, latex particles of different sizes, and even living cells. Chromatographic separations depend on differences in the distribution of substances between two phases, one stationary and the other moving. The moving phase can be a gas, a liquid, or a supercritical fluid. Because chromatography is a multistage process, even small differences in affinity can be exploited to produce useful separations. The forces that determine distribution ratios can be of many kinds, including dispersion forces, π-bonding, electrostatic forces, charge-transfer interactions, hydrogen bonding, and metal-ligand coordination. It is this last kind of interaction that is the basis of ligand-exchange chromatography.

The term "ligand-exchange chromatography" (LEC) dates from 1961 from a short report by Helfferich[1] entitled "Ligand Exchange: A Novel Separation Technique", which he later amplified in two longer papers dealing with the mechanism of LEC.[2,3] Helfferich had the problem of recovering a 1,3-diamine (1,3-diamino-2-hydroxypropane) from a dilute aqueous solution that also contained ammonia. He solved the problem by packing a glass column with a cation-exchange resin that was loaded with the blue copper(II) ammonia complex ions, mainly $Cu(NH_3)_2^{2+}$ under the conditions he was using, and passing the solution through this column. He saw a deep-blue band forming at the top of the column, which spread downward as the solution continued to flow; it contained the Cu-diamine complex. Each diamine molecule had displaced two molecules of ammonia:

$$Res_2Cu(NH_3)_2 + (\text{diamine}) = Res_2Cu(\text{diamine}) + 2NH_3$$

where Res denotes the functional group of the cation exchange resin, here the carboxyl ion.

After the deep-blue diamine band had spread through the length of the column, Helfferich displaced the absorbed diamine by passing a small volume of concentrated ammonia solution, reversing the reaction written above and restoring the column to its original condition. Displacement was efficient because two ammonia molecules replaced one diamine molecule, a process that is favored at high concentrations. (A similar condition exists in water softening, where one calcium ion displaces two sodium ions from the solid exchanger in dilute solution, and the column is regenerated by a concentrated solution of sodium chloride.) Little or no copper was removed from the resin, only the ligands; ammonia and the diamine changed places, and hence the name "ligand exchange".

The process of ligand exchange can be applied to any amines, and indeed to any compounds, that form labile coordination complexes with copper ions; moreover, the metal ions need not be those of copper, but could be any metal ions that form labile coordination complexes with the compounds to be separated. The process can be applied to analytical elution chromatography. If a mixture of amines, for example, is introduced into a column of metal-loaded ion exchanger and a solution of ammonia is passed, the amines will move along the column at different rates, with those forming more stable complexes remaining behind, while those that form weaker complexes move ahead (see Figure 1). Different selectivity orders can be expected with different metal ions, different ion exchangers, and different eluents. Helfferich wrote: "The method combines two fields of chemistry, namely,

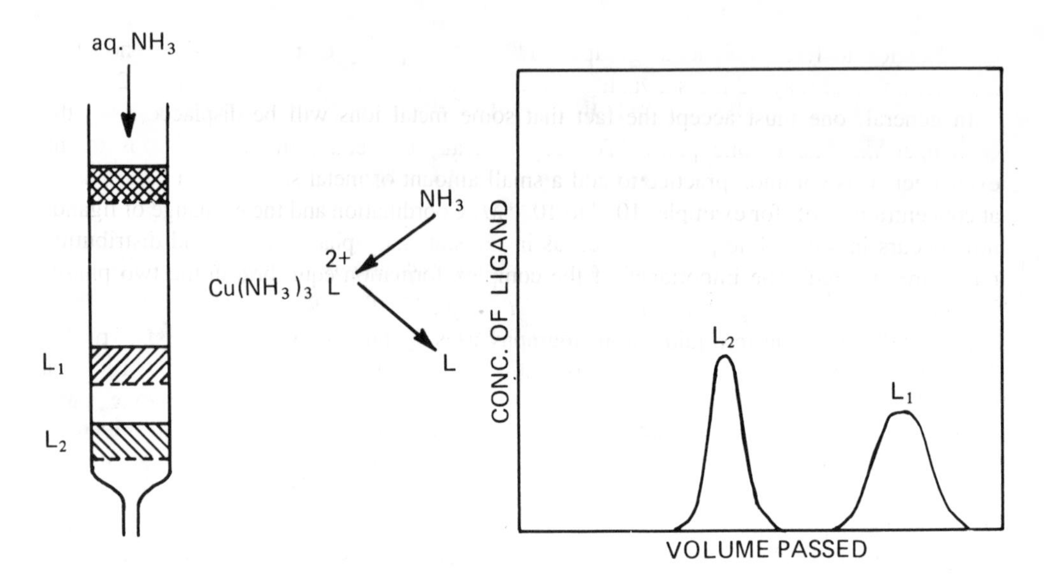

FIGURE 1. Ligand exchange chromatography, schematic.

ion exchange and coordination chemistry, in order to accomplish a task that neither could do alone."

Helfferich noted that the method could be applied to gas-liquid chromatography. One could use a carrier gas containing ammonia, for example, to analyze a mixture of volatile amines. To date, little use has been made of ligand exchange in gas chromatography, though labile metal-ligand complexes are exploited in gas chromatography of olefins on stationary phases carrying silver(I) ions. This topic will be treated later in the book (Chapter 7).

A "ligand" is something "tied on". In coordination chemistry, it is a neutral molecule or an anion attached by a coordinate link to a metal ion. By incorporating lone electron pairs of the donor atom of the ligand, the central metal cation completes its electron shell, building up to the stable shell structures characteristic of the noble gases. If the metal-ligand bond is labile (that is, easily formed and broken), one ligand can substitute for another. Water is a ligand, and coordination of water molecules is always important in aqueous solutions.

The ligand can be a negative ion. One of the most important uses of LEC is the separation of amino acids, which coordinate as their deprotonated/singly charged anions. The net charge on the metal ion must remain positive, however, or the metal will be stripped from the column.

We have considered that the ligands move in LEC while the metal ions remain stationary. Of course, this statement is an idealization. The metal ions do leave the exchanger and move through the column, because they enter into ion-exchange equilibrium with cations of the mobile phase. Even a dilute ammonia solution contains ammonium ions. Displacement of metal ions is minimized by choosing an exchanger, the functional groups of which themselves form coordinate bonds with the metal ions, like carboxylate or the iminodiacetate ions, which is the functional group of the chelating ion-exchange resin, Chelex®-100:

$$
\begin{array}{l}
| \\
CH{-}C_6H_4{-}CH_2N \begin{array}{c} \diagup CH_2COO \diagdown \\ {-}{-}{-}{-}{-}{\rightarrow} \\ \diagdown CH_2COO \diagup \end{array} Cu \\
| \\
CH_2
\end{array}
$$

Coordination to fixed functional groups reduces the capacity of metal ions to bind other ligands, and that may be a disadvantage. The matter will be discussed in Chapter 2.

In general, one must accept the fact that some metal ions will be displaced from the exchanger into the mobile phase. To keep a steady concentration of metal ions in the exchanger, it is common practice to add a small amount of metal salt to the mobile phase, at concentrations of, for example, 10^{-3} to 10^{-4} *M*. Coordination and the exchange of ligands now occurs in the mobile phase as well as in the stationary phase, and ligand distribution ratios are affected. The importance of the complex formation equilibria in the two phases to amino acid chromatography will be seen in Chapter 4.

A new development in liquid chromatography uses ligand exchange in the mobile phase. The stationary phase need not be an ion exchanger; it may be nonpolar, like porous silica bonded with octadecyl groups, the familiar reversed-phase packing. The metal ions are added to the mobile phase, and if the ligands are such that they form uncharged complexes, these uncharged metal-ligand complexes are distributed between the stationary and the mobile phase. Amino acids are examples of such ligands, and the technique of ligand exchange in the mobile phase is widely used today to analyze mixtures of enantiomers of amino acids, that is, of D- and L-optical isomers.

A characteristic of LEC is its sensitivity to the molecular structure of the ligands. Small differences in molecular shape or size often have a large effect on strengths of binding to a metal-loaded ion exchanger. In the restricted environment of an ion-exchange resin, substituents on a donor nitrogen atom may greatly hinder the formation of a coordinate link, and groups far removed from the donor atom may interact with the exchanger network. Hence, the great range of selectivity orders, which we have mentioned above and which will be seen in Chapter 3.

Nowhere is the selectivity of ligand exchange better seen than in the distinction between the D and L forms of optically active ligands. The most spectacular triumph of LEC is the separation of optical isomers of amino acids.

The use of a chiral ion exchanger for this purpose was first proposed by Rogozhin and Davankov in 1968.[4] They grafted an optically active amino acid, L-proline, on to a specially cross-linked polystyrene and treated the result with copper sulfate in aqueous ammonia to give a copper-loaded polymer having this structure:

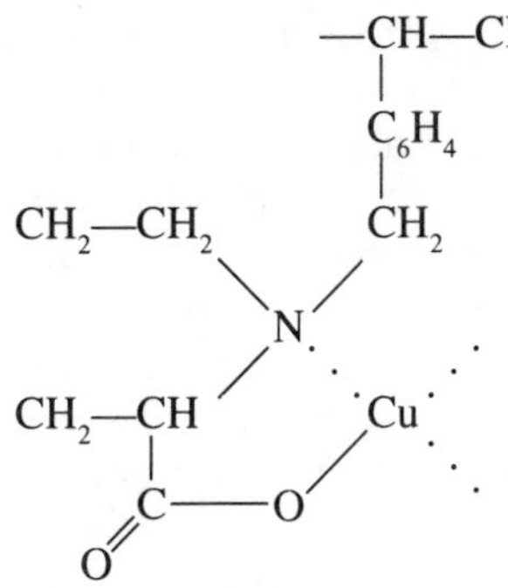

```
                      —CH—CH2—
                        |
                       C6H4
                        |
   CH2—CH2             CH2
          \           /
            N
          /   .
   CH2—CH        . Cu :
     |            /
     C——O
    //
   O
```

Each copper ion is coordinated with two fixed proline units and forms a bridge between them. The copper-loaded polymer can now absorb dissolved amino acids from an aqueous solution. In doing so, the copper bridges are broken, thus:

$$\overline{\text{RPro}}\text{–Cu–}\overline{\text{ProR}} + \text{HA} = \overline{\text{RPro}}\text{–Cu–A} + \overline{\text{HProR}}$$

where HA stands for a dissolved amino acid in its uncharged or dipolar-ion form, and $\overline{\text{RPro}}$ signifies a negatively charged proline anion bound to the polymer network. (Conventionally, the bar above formulas in ion-exchange chemistry indicates absorption in the exchanger phase.) This equilibrium, and others that accompany it, are discussed in detail in Chapter 5, Section V.

Once absorbed, the amino acid can be displaced by water if the attachment is weak, otherwise by ammonia:

$$\overline{RPro}\text{–Cu–A} + \overline{HProR} + NH_3 = \overline{RPro}\text{–Cu–}\overline{ProR} + NH_4^+A^-$$

Hence, the polystyrene-proline-copper resin can be used for the chromatography of amino acids, including proline itself.

In an early experiment, Rogozhin and Davankov packed a glass column, 9 mm wide and 50 cm long, with 12 g of this polymer (particle diameter, 30 to 50 μm), washed it with water, and introduced 0.5 g of racemic DL-proline dissolved in water. Then they passed water, followed by dilute ammonia. The first amino acid fractions to emerge from the column consisted entirely of L-proline; to get D-proline out of the column, they passed 1 *M* ammonia. In this way they recovered 0.25 g of L-proline and D-proline, each in 100% optical purity.[5-7] A small amount of copper was displaced from the column by ammonia along with the amino acid, but it was easily held back by interposing a short column of copper-free, proline-grafted polymer at the exit to the main column. The D- and L-proline were thus isolated in pure, copper-free form.

These results gave the impetus to an extended series of research by many investigators that will be described in detail in Chapter 5. At first, the emphasis was on the synthesis and use of chiral (optically active) polymers. These gave impressive separation factors and were good for preparative purposes, but bands were broad and the separation process was a bit slow. Then, chiral-bonded silicas were prepared and silica-based, reversed-phase supports were dynamically coated with optically active amino acids that carried long hydrocarbon chains to make them stick to the support. In every case, the stationary phases were loaded with copper ions. Another development, which we have noted, was to use an achiral support, sometimes a cation-exchange resin, sometimes a standard reversed-phase bonded silica, and add the chiral resolving agent to the mobile phase. Faster ligand exchange and more narrow chromatographic bands were achieved in this way, but the chiral mobile phase is obviously more suited to analytical than to preparative use. In every case, the chiral resolving agent was a metal complex, usually of copper(II), sometimes of zinc(II), or nickel(II), with a chiral amino acid, L-proline, L-hydroxyproline, or L-phenylalanine.

The function of the metal ion is to bring two amino-acid molecules close together in a fixed orientation. One molecule is of the optically active resolving agent, like L-proline. The other is the D or L form of the acid to be resolved. Each molecule is held in place by two coordinate-covalent bonds. The third interaction point, which is necessary for chiral discrimination, is provided by exchange forces or steric hindrance between the hydrocarbon side-chains of the two molecules. The result is that the free energies of formation of the L–Cu–L and L–Cu–D diastereoisomers may differ by 2 or 3 kJ, which is quite adequate to permit chromatographic resolution.

It is now time to formulate a definition of LEC. At first, it was taken for granted that ligands would be exchanged around metal ions that were immobilized in the stationary phase. We now see that the metal ions can reside in the mobile phase as well; indeed, there is no reason why they might not remain exclusively in the mobile phase. A working definition of LEC might read as follows:

> Ligand-exchange chromatography is a process in which complex-forming compounds are separated through the formation and breaking of labile coordinate bonds to a central metal atom, coupled with partition between a mobile and a

stationary phase. It separates ligands by causing them to change places around metal ions. The exchange can occur in either the stationary or the mobile phase.

In reading this definition we must again bear in mind that water is a ligand that can take the place of other ligands, and that nearly all LEC is performed in solutions that contain water.

An early use of metal-ligand complexing in chemical separations was made by Tsuji in 1960.[8] He absorbed isonicotinic acid hydrazide on a cation-exchange resin that was loaded with various metal ions. The strongest absorption was obtained with copper(II), but other cations, listed in decreasing order of absorption strength, were nickel(II), mercury(II), cobalt(II), cadmium(II), zinc(II), iron(II), lead(II), manganese(II), and aluminum (III). In every case, the compound of interest was displaced from the resin in concentrated form by aqueous ammonia. Both batch and column arrangements were used. Tsuji did not, however, call his process ligand exchange.

A more recent variation of ligand exchange is "metal chelate affinity chromatography," so called by Porath,[9] who has developed it as "a new approach to protein fractionation." To absorb and desorb proteins without denaturation, the stationary phase must be hydrophilic and the binding sites must be easily accessible by large molecules. Porath and co-workers have used agarose to which they attached long hydrophilic side chains terminating in an iminodiacetate group, which holds a metal ion, generally copper(II) or zinc(II). Proteins carrying histidine or cysteine are especially strongly held. The absorbed proteins can be selectively removed by passing aqueous buffers through the column, leaving the metal behind. Though the authors did not call it by that name, the process is clearly one of LEC. It is described in more detail in Chapter 4.

Yet another kind of LEC makes use of what is called "outer-sphere coordination". This is an interaction, a sort of ion-pair formation, between a stable coordination complex and another ion of opposite charge.

It should be reemphasized that very weak intermolecular interactions can serve as a base for a chromatographic process. Only formation of labile coordination compounds can be related to LEC. Numerous examples exist of chromatographic separations of stable, kinetically inert complexes of cobalt(III), chromium(III), platinum(II), platinum(IV), and some other metal ions, but they have little in common with LEC. These complexes can be separated according to an ion-exchange mechanism or reversed-phase technique, but there is no exchange of ligands in the inner coordination sphere of metal ions in these separations.

However, aside from the inner coordination sphere, the above complexes possess a highly organized solvation shell which can be regarded as a second, outer coordination sphere. Its ligands are bonded relatively weakly and can easily be exchanged under chromatographic conditions. And indeed, several interesting separations have been achieved using ligand exchange in the outer coordination sphere of inert complexes. They will be given consideration in Chapter 5, Section VI.

One of the first uses of outer-sphere coordination in LEC was made by Karger and associates in 1978,[10] though the complex ion was labile, not inert. Their mobile phase was a solution of the zinc complex of a hydrophobic triamine, $C_{12}H_{25}N(CH_2CH_2NH_2)_2$, in an acetonitrile-water mixed solvent; a molecule of acetonitrile coordinated with the zinc ion to give a fully coordinated complex ion, but nevertheless, this fully coordinated cation formed ion pairs with anions of sulfa drugs and permitted their chromatographic separation. The stationary phase was a C_8-bonded silica. This research extended to the use of asymmetric, optically active triamines and the separation of optical isomers of dansyl amino acids.[11,12]

Though mobile-phase interaction may have predominated in this system, it is quite likely that some of the hydrophobic complex cations were absorbed into the stationary phase and formed ion pairs there, as well as in the mobile phase. The chapters to follow will show many examples of mobile-phase interaction, ion-pair formation, and outer-sphere coordination.

REFERENCES

1. **Helfferich, F.,** Ligand exchange. A novel separation technique, *Nature (London),* 189, 1001, 1961.
2. **Helfferich, F.,** Ligand exchange. I. Equilibria, *J. Am. Chem. Soc.,* 84, 3237, 1962.
3. **Helfferich, F.,** Ligand exchange. II. Separation of ligands having different coordinative valencies, *J. Am. Chem. Soc.,* 84, 3242, 1962.
4. **Rogozhin, S. V. and Davankov, V. A.,** A chromatographic method for resolution of racemates of optically active compounds, *German Patent* 1,932,190, Jan. 8, 1970, Appl. U.S.S.R. July 1, 1968; *Chem. Abstr.,* 72, 90875c, 1970.
5. **Rogozhin, S. V. and Davankov, V. A.,** Ligand chromatography on dissymmetric complex-forming sorbents — a new principle of resolution of racemates, *Dokl. Akad. Nauk S.S.S.R.,* 192, 1288, 1970 (Engl. transl., p. 447).
6. **Davankov, V. A. and Rogozhin, S. V.,** Chromatography of ligands as a new method for studying mixed complexes. Stereoselective effects of α-amino acid complexes of copper(II), *Dokl. Akad. Nauk S.S.S.R.,* 193, 94, 1970 (Engl. transl., p. 460).
7. **Rogozhin, S. V. and Davankov, V. A.,** Ligand chromatography on asymmetric complex-forming sorbents as a new method for resolution of racemates, *Chem. Commun.,* p. 490, 1971.
8. **Tsuji, A. and Sekiguchi, K.,** Adsorption of nicotine acid hydrazide on cation exchanger of various metal forms, *Nippon Kagaku Zasshi,* 81, 847, 1960.
9. **Porath, J., Carlsson, J., Olsson, I., and Belfrage, G.,** Metal chelate affinity chromatography, a new approach to protein fractionation, *Nature (London),* 258, 598, 1975.
10. **Cooke, N. H. C., Viavattene, R. L., Eksteen, R., Wong, W. S., Davies, G., and Karger, B. L.,** Use of metal ions for selective separations in high-performance liquid chromatography, *J. Chromatogr.,* 149, 391, 1978.
11. **Lindner, W., LePage, J. N., Davies, G., Seitz, D. E., and Karger, B. L.,** Reversed-phase separation of optical isomers of Dns-amino acids and peptides using chiral metal chelate additives, *J. Chromatogr.,* 185, 323, 1979.
12. **LePage, J. N., Lindner, W., Davies, G., Seitz, D. E., and Karger, B. L.,** Resolution of the optical isomers of dansyl amino acids by reversed-phase liquid chromatography with optically active metal chelate additives, *Anal. Chem.,* 51, 433, 1979.

Chapter 2

GENERAL CONSIDERATIONS

H. F. Walton

I. EXPERIMENTAL CONDITIONS

A. Metal Ions

The only metal ions that do not form complexes in aqueous solutions are those of the alkali metals, and these ions, of course, are hydrated. The other metallic cations may be conveniently classified into "hard" and "soft" acids, with word "acid" understood to mean an electron acceptor. Hard acid cations are those having a noble-gas electronic structure or a large charge:radius ratio. These associate preferably with hard bases like F^-, OH^-, and ligands containing oxygen. Soft acid cations are those of transition and posttransition metals; they do not have a noble-gas electronic structure. They associate preferably with ligands having nitrogen or sulfur as electron donors.

Ligand exchange chromatography (LEC) is most often performed with metal ions of the soft acid type, and primarily with copper(II). Copper(II) is the ion of choice because it forms very stable complexes. The Irving-Williams series states that as one proceeds along the first transition series, the divalent ions form complexes, the stability of which rises in the order Mn $<$ Fe $<$ Co $<$ Ni $\ll$ Cu with a sharp peak at Cu, then falls going from Cu to Zn. This stability order is the same for all soft-acid ligands. The copper(II) ion has a square planar distribution of coordinate valences, which sets it apart from the others; nickel(II) favors octahedral coordination, and zinc(II) favors tetrahedral. As we know, copper(II) complexes can also be octahedral, but the octahedron is a distorted one with the axial bonds being much weaker than the square planar or equatorial bonds. The axial coordination of copper(II), however, is vitally important in ligand exchange selectivity (see Chapter 5).

In choosing a cation for LEC, we must consider how strongly it is bound to the ion exchanger which is the stationary phase. The sulfonated polystyrene resins that are commonly used do not bind copper(II) very strongly; it is too easily displaced by NH_4^+ and other cations. For this reason, much of the earlier work on LEC was done with nickel ions attached to a sulfonated polystyrene resin. Copper(II) ions were supported on an acrylic or chelating resin, which holds them strongly enough so that they are virtually not displaced by aqueous ammonia. Another circumstance that favors the use of nickel(II) or zinc(II) rather than copper(II) is the exchange of ligands that are bound very strongly, like 1,2-diamines. These compounds are held so strongly by copper(II)-loaded exchangers that the only practical way to get them off the column is to strip the metal ions off with dilute acid.

Yet another circumstance that favors the use of zinc(II) is the speed of the exchange. In the LEC of amino acids on a sulfonated polystyrene exchanger, zinc(II) gives sharper bands than copper(II), indicating faster exchange.[1,2] The extreme case of slow exchange, of course, is in kinetically stable complexes formed by cobalt(III), chromium(III), and platinum(IV). These complexes have their uses in LEC, but only through "outer-sphere" complexing or ion-pair association. This topic will be mentioned later. Meanwhile, note that cobalt(II) is seldom used in LEC, because it is easily oxidized by air to cobalt(III) in the alkaline media commonly used.

After copper(II), zinc(II), and nickel(II), the soft-acid cations that have been most used in LEC are cadmium(II), silver(I), and mercury(II). Cadmium ions have been used in the column chromatography of sulfur ligands, including thiourea,[3] but their main use has been in thin-layer chromatography, where they have been incorporated into silica and used to analyze mixtures of aromatic amines.[4-7] In effectiveness, there is little to choose between cadmium(II) and zinc(II). Silver ions incorporated in ion-exchanging polymers, as well as in silica, have been used for the chromatography of heterocyclic nitrogen bases[8] as well as various olefinic compounds,[9,10] but since the solvents used were nonaqueous, it is questionable whether the processes should be called LEC. However, separations were due to differences in stability of metal ligand complexes. Mercury(II) has a great affinity for ligands containing sulfur, and a macroporous cation exchanger loaded with mercury(II) may be used to remove sulfur compounds as a class from petroleum;[11] a mercury-loaded resin served for the chromatography of aromatic hydroxy acids.[12] However, mercury(II) salts react irreversibly with polystyrene-based ion-exchange resins with covalent attachment of mercury to the aromatic ring.[13]

Turning now to the hard-acid cations, the one most used is probably calcium(II), which forms complexes with polyhydric alcohols and certain sugars. Calcium-loaded columns of cation-exchange resins are used routinely for the analysis of sugar mixtures, with water as the eluent. The calcium complexes are weak, but their formation constants have been measured in some cases, and the mechanism of chromatographic retention is mainly ligand exchange (see Chapter 6).

Several publications describe the use of iron(III)-loaded ion exchangers for recovery and chromatography of phenols,[14,15] aromatic acids and hydroxy acids,[16] beta-diketones,[17] and even aromatic diamines.[18] Titanium(IV) has been used to separate hydroxy acids.[19] Aluminum(III)-loaded cation-exchange resins were used for fractionation of DNA and RNA, using alkaline glycine buffers as eluents.[20]

A complication in using iron(III) and aluminum(III) is the easy hydrolysis of these ions, which in aqueous solution starts above pH 3 to 4. The lanthanum ion, lanthanum(III), being larger, is much less hydrolyzed — only about 1% at pH 6. A resin loaded with lanthanum(III) selectively retains anions of carboxylic acids and hydroxy acids and can be used for their chromatography[21] using an acetate buffer as eluent. A problem here, which is quite general in LEC, is the poor chromatographic efficiency, i.e., the band broadening which is associated with slow exchange. The rate of ion exchange is limited, as a rule, by the rate at which ions and molecules can diffuse in and out of the exchanger. Conventional gel-type resins loaded with trivalent ions are more compact and have a smaller water content than those loaded with univalent ions, and diffusion in such resins is correspondingly slow.

B. Exchangers

The exchanger used to retain the cations in LEC can be either organic or inorganic, or a combination of both, namely, a bonded silica. Most work has been done with organic exchangers, and principally those derived from cross-linked polystyrene because these are easily available and their particles are sufficiently rigid that they can be used in closed columns under pressure.

The most common organic cation exchanger is sulfonated polystyrene, which can be obtained in various degrees of cross-linking and various particle sizes. For analytical chromatography, small and uniform particles are desired, about 10 μm in diameter. A cross-linking of 8% is sufficient to give enough rigidity, and 6% cross-linked resins — even 4% cross-linked — can be used with divalent counter-ions. (Doubly charged ions act electrostatically to pull the polymer chains together, and cause the resin particles to contract and become more rigid and more tolerant of high-pressure gradients; these are the ions used in LEC.)

The common gel-type resins, which are internally homogeneous, are better for analytical chromatography than are macroporous resins. In our experience, they give narrower and more symmetrical chromatographic peaks. Macroporous resins are aggregates of very small, highly cross-linked microspheres, and it appears that the environment of the ionic functional groups is not uniform; the peaks show much tailing. Where good chromatographic performance is not needed, as in the recovery of traces of ligands from large volumes of water, macroporous resins can be used to advantage, as they can and must be when treating nonaqueous solutions. Macroporous sulfonated polystyrene, in our experience, retains metal ions more poorly than do the gel-type resins, but this problem is less acute in nonaqueous solvents.

None of these objections apply to the "macronet" and "isoporous" polymers developed by Davankov and co-workers.[22,23] These resins are solvent-modified polymers made by using long rod-like molecules as cross-linking agents. They are distinguished by high internal porosity and uniformity of cross-linking. Their properties are described elsewhere in this book.

When gel-type sulfonated polystyrene resins are used in LEC, it is necessary to add metal salt to the mobile phase, for copper(II) and zinc(II) are not retained with sufficient strength to prevent loss from the column. As we have noted, nickel(II) is held more strongly.

Metal ions are held much more strongly if the resin has other functional groups. Some authors have used phosphonate groups attached to cross-linked polystyrene.[24] More commonly used is the iminodiacetate chelating resin, Chelex®-100, but diffusion in and out of this resin is slow and chromatographic performance is poor; further, the ligand-binding capacity is limited by the coordination of the metal ion to the resin functional group. Nevertheless, a nickel(II)-loaded chelating resin has been used for the column chromatography of organic acids.[25] The most important use of chelating resins, both in ligand exchange and in the exchange of inorganic ions, is the recovery and concentration of trace substances from large volumes of water. A nickel-loaded iminodiacetate chelating resin has been used to recover amino acids from waste water; 11 samples were passed through a bed containing 10 mℓ of nickel-loaded resin, and the absorbed amino acids were later eluted with concentrated aqueous ammonia.[26] Similarly, a copper-loaded chelating resin recovered dissolved amino acids from sea water[17] and from urine.[28,29] Phenols were absorbed from industrial waste water by a chelating resin loaded with iron(III) and stripped from the resin by dilute sodium hydroxide.[30] For applications like these, high affinity is necessary, but high resolution is not.

Functional carboxyl groups retain copper(II) and other metals more strongly than do sulfonate groups, and carboxyl is the functional group of cross-linked polyacrylate-methacrylate resins, sold commercially as Bio-Rex® 70. Acrylic resins whose structure is aliphatic, have the advantage that there is little π-bonding between the resin and solutes of aromatic character. We saw this difference in the LEC of amphetamine drugs;[31] an acrylic resin gave symmetrical and fairly narrow peaks, while a polystyrene-based resin gave broad, asymmetrical peaks with much tailing, indicative of a mixed retention mechanism. An acrylic resin is better for the LEC of alkaloids.[32] However, acrylic resins are less uniform and less reproducible than polystyrene-type resins. We found that an experimental batch of acrylate resin was much better at retaining alkaloids than the regular commercial material. The ligand-binding behavior of the copper in the two resins was correlated with the electron-spin resonance of Cu^{2+}.[33] Subtle differences in polymer structure may make large differences in metal-ligand binding.

A great disadvantage of acrylic resins in high-resolution chromatography is their softness. They are easily deformed under pressure and must be treated with great care in closed columns.

Even softer than the acrylic resins are the exchangers derived from natural polymers like cellulose and dextran. Such exchangers are used in LEC primarily to retain large biological molecules like those of proteins and peptides, but a DEAE cellulose loaded with antimony served to separate aliphatic and aromatic amines, including diamines.[34] Copper-loaded Sephadex®G-25, a dextran carrying carboxyl groups, served to separate amino acids as a group from peptides.[35] Dextran and Sepharose® (a polysaccharide) have been treated to introduce iminodiacetate chelating groups, and the products, loaded with copper(II) and used in gravity feed columns, served to collect and separate proteins and peptides.[36,37] Loaded with mercury(II) instead of copper(II), these materials were selective sorbents for proteins having sulfhydryl groups, like papain.[38]

In high-performance liquid chromatography (HPLC), it is a great advantage to use packings based on silica. Porous silica is hard and will stand high flow rates and high pressure gradients; it can be obtained in well-defined particle sizes and porosities. A great variety of organic groups can be attached chemically to silica; the techniques for doing this are well known.[39-49] Much high-resolution chromatography is done with silica coated with chemically bonded organic groups, primarily hydrocarbon chains like $C_{18}H_{37}$. Naturally, bonded silicas have been developed for use in LEC.

One of the early attempts to use bonded silica for this purpose was made in 1977 by Chow and Grushka.[39] They took silica carrying amino groups as $-CH_2CH_2CH_2NH_2$, the common amino bonded phase, and impregnated it with copper(II) from a solution of copper sulfate in dry methanol. Aqueous solutions stripped copper from the exchanger. To hold copper and other metal ions more tightly, the next step was to bind diamines, polyamines, and iminodiacetate groups to silica. Masters and Leyden[46] attached diamine groups not to silica, but to controlled pore glass, by refluxing with *N*-β-aminoethyl-γ-aminopropyltrimethoxysilane dissolved in dry toluene. The product carried the groupings $-Si-CH_2-CH(NHC_2H_5)CH_2NH_2$. It was loaded with copper(II) by contact with a copper(II)-ammonia solution in water, taking up about 0.4 μmol copper per gram. Chromatography of amino acids and amino sugars was accomplished with an eluent 0.1 *M* in ammonia and ammonium ions, pH 9.5, which was 10^{-4} *M* in $CuCl_2$. Chromatographic efficiency was poor, probably because the particle size was too great, but it established that the diamine function held copper(II) sufficiently and strongly enough that aqueous eluents could be used. Later, Gimpel and Unger[40] bonded silica with several aminosilanes, introducing the groups $Si-(CH_2)_3NH(CH)_2$, $Si-(CH_2)_3N(CH_2COOH)_2$, $Si-(CH_2)_3N(CH_2COOH)CH_2CH_2N-(CH_2COOH)_2$. Loaded with copper(II), these materials were used to separate amino acids and mixtures of aliphatic carboxylic acids. Various buffer solutions were used as eluents, generally in the pH range of 3 to 6, each containing 8×10^{-5} *M* copper(II).

Other functional groups can be grafted to silica; there are many possibilities. Starting with γ-aminopropyl-bonded silica, the common commercial "amino bonded phase", Chow and Grushka[47] attached dithiocarbamate ligands by reaction with carbon disulfide, and diketo groups by reaction with ethyl benzoylacetate; the bonded ligands coordinated copper(II). By first attaching a diamine ligand, they produced a bonded silica that carried the cobalt(III)-*tris*-ethylenediamine complex, which was effective for chromatography of nucleotides with buffered phosphate eluents;[48] this separation makes use of "outer-sphere coordination".

Tartaric acid has been attached to silica via the bonded γ-aminopropyl group.[42] Optically active D-tartaric acid was used in the synthesis, giving an asymmetric stationary phase that allowed separation of optical isomers by chromatography. Catecholamines and related compounds, including amino acids like "dopa", were separated on the copper-loaded stationary phase, as was the hydroxy-acid mandelic acid. Eluents were phosphate buffers made 3×10^{-4} *M* in copper(II). Optical isomers were separated with good resolution in running times of 15 min.

Starting again with the amino bonded phase, $Si-(CH_2)_3NH_2$, Shahwan and Jezorek[41] attached 8-quinolinol to silica in this form:

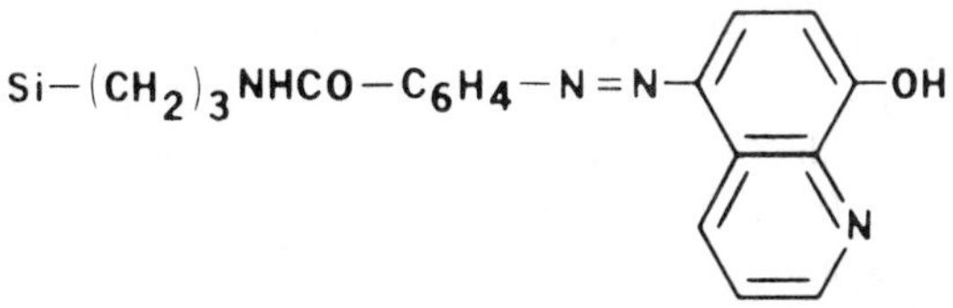

$$Si-(CH_2)_3NHCO-C_6H_4-N=N-\text{(8-hydroxyquinolin-5-yl)}-OH$$

The functional group combines with iron(III) in the ratio of 1:1, leaving iron(III) with four coordinate valences available for ligand exchange. The material was used for chromatography of phenols, chloro-, and nitrophenols.

Amino acids may be attached to silica in various ways. Of special interest are optically active amino acids bonded to silica, for these stationary phases, carrying copper(II), enable the separation of D- and L-isomers of amino acids injected into the mobile phase. Chapter 5 of this book deals at length with the separation of amino-acid enantiomers. Recently, hydroxy acids, including mandelic and phenyllactic acid, have been separated on silica bonded via glycidoxypropyl groups to L-proline and L-hydroxyproline.[49]

Silica itself is an ion exchanger by virtue of its surface silanol groups, –Si–OH. Silica gel can be made to carry metal cations and serve as a substrate for ligand exchange. Metal ions may be bound hydrothermally; Vogt[7] heated silica gel with saturated aqueous cadmium chloride in a closed vessel to 350°C for 2 hr. Cadmium was bound strongly and the surface area of the silica was reduced, causing a high surface concentration of cadmium ions. The solid was used for the chromatography of aromatic and heterocyclic amines, with a hexane-acetonitrile mixture as the mobile phase.

Copper ions, and perhaps other metal ions, can be attached firmly to silica gel by simple ion exchange in aqueous solution at room temperature. Foucault, Caude, and Rosset[43,44] have made supports for ligand exchange by simply packing columns with 7- μm silica particles and passing solutions of copper sulfate in 1 *M* aqueous ammonia, then washing with water. The surface silanol groups exchange their protons for copper ions. As the copper-ammonia solution is passed, it first displaces the solvent used to pack the silica column (ethanol), then pure water emerges as copper(II) and NH_4^+ are adsorbed by the silica. Then aqueous ammonia, copper free, emerges, as copper(II) in the effluent displaces ammonium ions from the silica. Finally, the deep-blue $Cu(NH_3)_4SO_4$ breaks through.[44] After washing with water, the silica carries about 0.75 mmol copper per gram and is free from sulfate ions.

Because the copper ions are on the surface, diffusion paths are short, ligand exchange is rapid, and sharp peaks are obtained (see Figure 1). The investigators have used copper-loaded silica for the chromatography of amino acids and peptides, using as eluents solutions of ammonia in 1:1 water-acetonitrile mixtures. Eight amino acids were separated in 15 min. The resolution of di- and tripeptides was particularly impressive, suggesting the use of this kind of chromatography in medical diagnosis. Tripeptides are eluted first, then dipeptides, then amino acids.[44]

Silica gel dissolves in alkaline solutions. In liquid chromatography on bonded silica with aqueous eluents, one is advised not to use solutions of pH above 8, and preferably not above 7. Yet Foucault and co-workers[43,44] report the use of ammonia solutions of 0.5 *M* and more. The use of these highly alkaline eluents was possible because (1) copper-loaded silica is somewhat less soluble in alkaline solutions than bare silica; (2) silica is much less soluble in organic solvents than it is in water; and (3) the eluents are presaturated with silica by passing them through a guard column packed with copper-loaded silica gel. This guard column performs the dual function of saturating the eluent with silica and maintaining a copper concentration high enough to prevent loss of copper from the analytical column. With this precaution, the analytical column may be used for months, and when it shows signs of deterioration, it is easily and cheaply repacked.

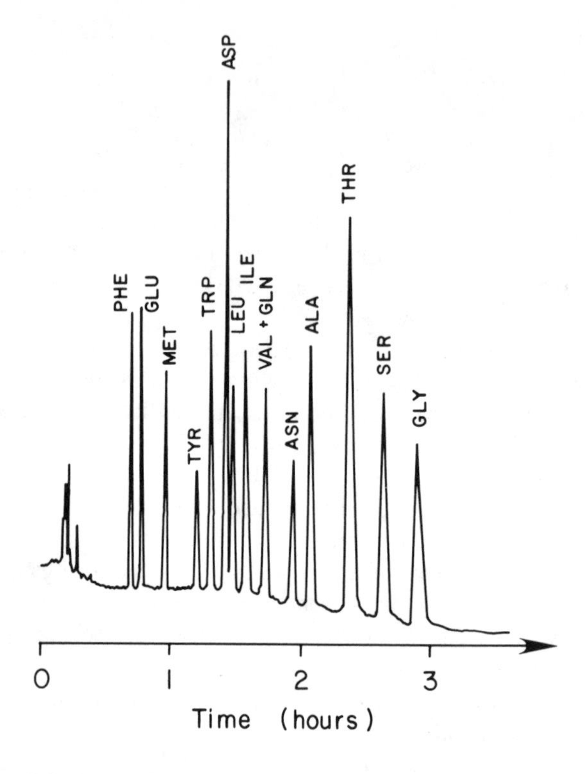

FIGURE 1. Chromatogram of amino acids on a Cu(II)-loaded silica gel. Column, 1000 × 1 mm; particle size 7μm; temperature, 50°C; eluent, acetonitrile-water, 1:1 by volume, 0.5 *M* in ammonia. (From Foucault, A. and Rosset, R., *J. Chromatogr.*, 317, 41, 1984. With permission.).

Guyon et al.[45] measured the effects of copper(II) and solvent composition on the solubility of silica in 0.5 *M* ammonia and interpreted the results in terms of two reactions: the hydration of silica and the dissociation of silicic acid. The effect of solvent composition is shown in Figure 2. Silica is far less attacked by ammonia in 1:1 methanol-water or acetonitrile-water than it is in water alone.

C. Eluents and Detection

These topics must be considered together, for the choice of one affects the choice of the other. In the early days of LEC, the emphasis was on the chromatography of amines, using aqueous ammonia as the eluent. Ultraviolet (UV) absorbance monitoring could be used for aromatic amines that absorbed in the UV, but even then, leakage of metal ions from the column contributed to the UV absorbance and reduced the sensitivity. Aliphatic amines do not absorb in the UV. The alternative was refractive index detection, which responds to any change in effluent composition, but is generally less sensitive than UV absorption. One complication with refractive index detection is that it shows displacement peaks, like the ammonia concentration pulse that is produced when a sample containing amines is injected into an aqueous ammonia eluent. The amines, on entering the column, displace ammonia from the coordinated metal ions. The pulse of increased ammonia concentration travels down the column at a rate that depends on the slope of the adsorption isotherm; if the column is fully saturated with ammonia, but only then, the ammonia pulse (or refractive index peak) travels at the void volume. More often, the pulse appears after the void volume and may interfere with the sample peaks.[50]

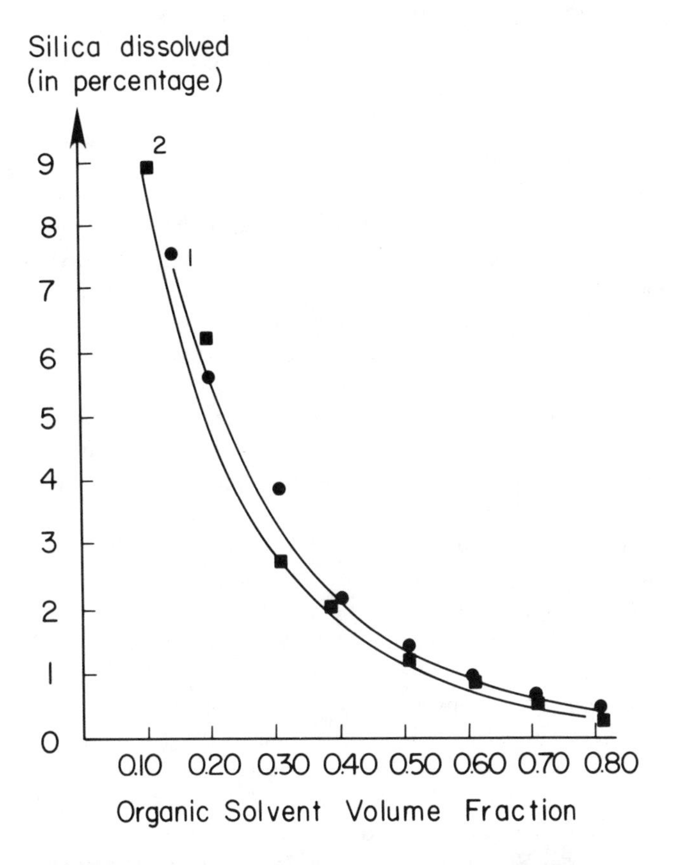

FIGURE 2. Solubility of silica gel vs. fraction of organic solvent. Curve 1, methanol and water; Curve 2, acetonitrile and water. (From Guyon, F., Chardonnet, L., Caude, M., and Rosset, R., *Chromatographia*, 20, 30, 1985. With permission.)

For research purposes, one may use isotopically labeled amines as test samples, preferably labled with carbon-14, then collect fractions and measure their radioactivity.[51] In an important study of LEC of amino acids, Doury-Berthod[52] used carbon-14 amino acids and devised a detector that measured the radioactivity of the effluent as it left the column.

Detection and measurement of amino acids in ion-exchange chromatography is customarily done by postcolumn reaction with ninhydrin, with a several-minute delay time in a reaction coil heated to 100°C. Red- or violet-colored solutions are produced and their light absorbance is measured. Ninhydrin reacts indiscriminately with amines, amino acids, and ammonia, and is therefore useless for detection with ammonia eluents. Moreover, traces of metal ions affect the color. Davankov[53] overcame these difficulties by using pyridine as the eluent, rather than ammonia; ninhydrin does not react with pyridine. He added EDTA to the ninhydrin reagent to form a complex with the copper ions that bled out of the column.

Ninhydrin was used for detection in the commercial amino-acid analyzer of Arikawa, which used ligand exchange on a cation-exchange resin loaded with zinc ions. The eluent was not ammonia, but a series of acetate buffers of pH 4 to 5 made 10^{-3} M in zinc ions.[2] Zinc does not interfere with the ninhydrin reaction.

A reagent that reacts with amines, amino acids, and amino sugars, but does not react with ammonia, was developed by Maeda et al.[54] It is pyridoxal:

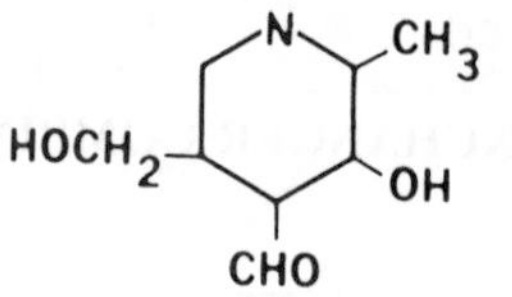

N CH_3
$HOCH_2$ OH
CHO

Its reaction with column effluents takes 10 min at 75°C. It produces a fluorescent derivative, observed by excitation at 390 nm, with emission at 470 nm, and, like most fluorescence reactions, it is extremely sensitive. It was used in LEC on a zinc-loaded, ion-exchange resin column, with sodium acetate-acetic acid eluents of pH 4.1 and 5.1, 10^{-3} *M* in zinc(II).[1] Zinc ions do not interfere, and ammonia may be added to the eluent to hasten the elution of strongly basic amino acids.

A fluorescence-producing reagent which has become very popular is *o*-phthalaldehyde. In the presence of a strong reducing agent, this compound reacts rapidly at room temperature with primary amines and amino acids, with the exception of proline and hydroxy-proline, which are secondary amines. It does not react with ammonia, or only to a small extent. The reagent was first described by Roth[55] and applied to amino-acid chromatography by Benson and Hare,[56] who compared its performance with that of ninhydrin and fluorescamine. A typical reagent formulation[56] has 0.8 g *o*-phthalaldehyde and 2 g 2-mercaptoethanol (the reducing agent) in 1 ℓ of 0.4 *M* borate buffer, pH 9.7. A nonionic surfactant is added, which facilitates the reaction with lysine. Different amino acids give different fluorescent intensities; sensitivities are in the picomole range. The excitation wavelength is at 340 nm, with emission at 455 mn. The reagent was used in LEC by Hare and Gil-Av,[57] who resolved optical isomers of amino acids on a column of copper-loaded cation-exchange resin, with the eluent, an acetate buffer containing copper(II) and L-proline. The chiral resolving agent was in the mobile phase. To avoid the precipitation of copper when the alkaline *o*-phthalaldehyde was added, the authors added EDTA to the reagent. As we have mentioned, this reagent does not react with proline.

We now turn to ultraviolet (UV) absorption as a means of detection. Amino acids and carboxylic acids generally absorb appreciably below 220 nm and, provided the eluents do not contain enough copper(II) to interfere (that is, not more than 0.5 mg/ℓ), amino acids can be monitored by their absorption at 206,[23] 210,[58] or 214 nm, the wavelength of the strong resonance line in a zinc vapor lamp. Most UV detectors in general use have mercury vapor lamps, however, and their working wavelength is 254 nm.

Now it turns out that the presence of copper ions, which is a disadvantage for observation at low UV wavelengths, is an advantage at higher wavelengths, like 254 nm. In 1975, Navratil et al.,[59] separating amino sugars and amino acids on a copper-loaded carboxylate ion-exchange resin with 1 *M* ammonia as the eluent, noticed that the elution peaks were accompanied by strong absorption at 254 nm. These compounds were stripped from the resin in the form of their copper complexes, and these complexes absorb UV light more strongly and at higher wavelengths than do the copper(II)-ammonia complexes. The fact that copper(II)-amino acid complexes show very high molar absorptivities between 230 and 240 nm was known some time ago,[60] but had not been applied to chromatography. It is advantageous, therefore, to use eluents that contain a low concentration of copper(II) — between 10^{-3} and 10^{-4} *M* — so as to ensure that the effluents contain enough copper(II) and to monitor the UV absorbance at 230 nm. This detection method was used by Masters and Leyden,[46] who measured the wavelengths of maximum absorbance of several amino acid-copper complexes, and by Foucault,[44] who, as we have mentioned above, used a precolumn of copper-loaded silica and thus not only suppressed the attack of the alkaline ammonia eluent on the copper-silica analytical column, but also provided a convenient concentration of copper(II) for the detection. It was also employed by Grushka et al.,[61] who

Table 1
ELUTION ORDERS OF SELECTED AMINES ON DIFFERENT METAL ION-EXCHANGER COMBINATIONS

	Nickel				Copper		
Exchanger	**Sulfonic**	**Carboxylic**	**Chelating**	**Zr Phosphate**	**Carboxylic**	**Chelating**	**Cellulose PO_4**
Diethanol-amine	1	1	1	4	1	1	1
Ethanol-amine	2	4	2	3	(2)	2	2
Dimethyl-amine	3	3	3	2	3	3	4
n-Butyl-amine	4	2	4	1	(2)	4	3

From Shimomura, K., Dickson, L., and Walton, H. F., *Anal. Chim. Acta*, 37, 102, 1967. With permission.

used ligand exchange in the mobile phase to separate amino acids on a bonded C_{18}-silica column. His eluent was 3×10^{-4} *M* $CuCl_2$, sometimes containing an acetate buffer. Grushka has tabulated the molar absorptivities of copper complexes of several amino acids at 230 nm; they range from 1100 ℓ/mol-cm for valine to 4400 ℓ/mol-cm for glutamic acid, and permit detection limits of a few nanograms. This method of detection is now routine for amino acids and amino sugars, and can probably be used for diamines and polyamines. Of course, detection can be done at 254 nm, the wavelength of the strong mercury line, with some loss of sensitivity.

D. Elution Orders

Selectivity and elution orders in LEC reflect the free energies of ligand substitution in the exchanger, and the interaction of the ligands with the exchanger matrix play an important part. Stability constants of metal-ligand complexes in aqueous solution are only a rough guide. For example, the 1:1 complexes of nickel(II) with ethanolamine and dimethylamine in water have formation constants of 950 and 50 ℓ/mol-cm, respectively, yet a nickel-loaded sulfonated polystyrene resin binds dimethylamine more strongly than ethanolamine.[62] By manipulating the exchanger matrix, the fixed ions, and the metallic counter-ions, one can get a great variety of elution orders, as is shown in Tables 1 and 2. The ability to change elution orders is one of the strengths of LEC.

From the observed elution orders for different compound classes, one may offer these three generalizations concerning the elution of amines from metal-loaded polystyrene-based resins.

First, ligand binding in the resin is very sensitive to steric hindrance. Primary amines are held much more strongly than secondary amines, which in turn are more strongly held than tertiary amines. Alkyl substituents on the carbon atom next to the amino group, as in $-CHR{\cdot}NH_2$, are held more weakly than the corresponding unsubstituted amines.

Second, among isomeric primary aliphatic amines, the more highly branched is the carbon chain; the more weakly bound is the amine to the metal-loaded resin. Among the primary butylamines, the following elution order was found on nickel-loaded sulfonic and chelating polystyrene resins:[63]

Table 2
ELUTION ORDERS OF DIAMINES AND POLYAMINES

Exchanger type:	Sulfonated polystyrene		Polyacrylic (carboxylic)		Cellulose (carboxylic)		
Metal ion:	Cu	Zn	Cu	Zn	Cu	Zn	Ni
Diamines							
C_2	5	6	7	—	—	—	—
C_3	4	1	6	2	6	1	5
C_4	—	2	3	4	1	2	1
C_5	—	5	4	5	2	5	3
C_6	3	7	5	6	4	6	4
Polyamines							
Spermidine	1	3	1	1	3	3	2
Spermine	2	4	2	3	5	4	3
Ammonia concentration *M*	7		4		1.1	1.7	1.1

From Walton, H. F. and Navratil, J. D., in *Recent Developments in Separation Science,* Vol. 6, Li, N. N., Ed., CRC Press, Boca Raton, Fla., 1981, chap. 5.

```
  C
  |
C-C-NH2
  |
  C
```

(most weakly held)

```
    C
    |
C-C-C-NH2
```

```
  C
  |
C-C-C-NH2
```

$C{-}C{-}C{-}C{-}NH_2$

(most strongly held)

Isopropylamine was held more weakly than *n*-propylamine. A factor here is the interaction between the amine molecule and the water of the mobile phase. The more extended the molecule, the more it disrupts the hydrogen-bonded structure of water, and the more it is displaced from the water phase into the resin phase. This generalization holds for partition of isomeric alcohols between water and sulfonated polystyrene resins, regardless of the resin counter-ion.

Third, hydroxyl groups in the molecule may strengthen or weaken the retention of amines by a metal-loaded resin, depending on whether the formation of bidentate chelate rings with the metal outweighs the increase of hydrophilic character. Thus, ethanolamine is held more weakly than ethylamine on a nickel-loaded resin, but norephedrine, $C_6H_5CHOH{\cdot}CHCH_3NH_2$ is held more strongly on resins loaded with nickel(II) or copper(II) than amphetamine, $C_6H_5CH_2{\cdot}CHCH_3NH_2$.[31]

II. METAL-LIGAND EQUILIBRIA IN EXCHANGERS AND SOLUTIONS

We have seen that stability constants of metal-ligand complexes measured in aqueous solution are a poor guide to ligand selectivities and elution orders in metal-loaded ion-exchange resins. Inside a gel-type ion exchanger, the volume of water is limited, and interactions between ligands and the resin matrix contribute significantly to the energy of metal-ligand association.

The easiest uncharged ligand to study is ammonia. Metal-ammonia complexes in cation exchange resins were studied by Stokes and Walton[64] in 1953. They took known weights of dry hydrogen-form resins and shook them with solutions containing copper or silver nitrate, ammonium nitrate, and enough ammonia to neutralize all the hydrogen ions in the resin and in addition, to coordinate with the metal ions. After equilibrium was reached, they measured the amounts of metal ions (which were small), total titratable ammonia, and the pH. The ammonium-ion concentration in the solution was easily found from the quantities of salt added, and the activity of free ammonia was calculated from the ammonium-ion concentration and the pH. Graphs were then drawn of the ratio of bound ammonia to total metal ions in the resin vs. pNH_3, the negative logarithm of the free ammonia activity in solution. Curves for copper(II) and silver(I) are shown in Figure 3. They are drawn for two different ion-exchange resins: one a sulfonated polystyrene with 10% cross-linking, the other an acrylic polymer with functional carboxylate ions. The graphs also show the curves obtained by Bjerrum for the metal ion-ammonia association in aqueous solution.

Figure 3 shows that in the sulfonated polystyrene resin the metal-ammonia complexes are just as stable (no more and no less) as they are in concentrated aqueous solutions. In the acrylate resin, this is not the case. Here, the complexes are much less stable than in water or concentrated salt solutions. An obvious reason is that the metal ions are covalently bound, forming complexes with the carboxyl ions of the resin and thus reducing their capacity to bind ammonia molecules. Helfferich,[65] attempting a theoretical description of ligand exchange of ammonia for diamines in a copper-loaded carboxylic resin, considered that two of the coordination sites of copper(II) were blocked, leaving only two sites free for coordination with ammonia and diamines. Figure 3A shows that this description is not quite correct, for the copper ions hold three ammonia molecules and possibly more. It seems that the carboxylate ions cause a general destabilization of the copper(II)-NH_3 complexes. We must remember, however, that copper(II) has six coordination sites, not four. The geometry of coordinated copper(II) is a distorted octahedron, where four coordinate bonds in a single plane are strong, and the two axial bonds relatively weak. The axial bonds are nevertheless important to coordination within ion-exchanging polymers, especially in discriminating between optical isomers.[49] Loewenschuss and Schmuckler[66] found that copper ions in an iminodiacetate chelating resin held three water molecules, implying that they were octahedrally coordinated. The ability of silver(I) to bind two ammonia molecules in the acrylate resin (Figure 3B) is not so easily explained.

Walton and collaborators[62,64,67] made similar experiments to study the association of aliphatic and aromatic amines with copper(II), silver(I), zinc(II), and nickel(II). Graphs were drawn to compare the stabilities of the complexes in the resin with those in solution (see Figure 4).

From Figure 4, it is apparent that the complexes in the resin do not have the same stabilities as those in solution. Aromatic amines form more stable complexes in polystyrene-type resins than in solution, undoubtedly due to π-bonding with the aromatic rings of the resin. Piperidine, a nonaromatic, secondary amine, forms weaker complexes in the resin, perhaps because of steric hindrance (see the discussion of elution orders above). In general, the complexes are more stable at low ligand:metal ratios than at high, perhaps because of an "overcrowding" effect at high ligand concentrations. In chromatography, one operates at

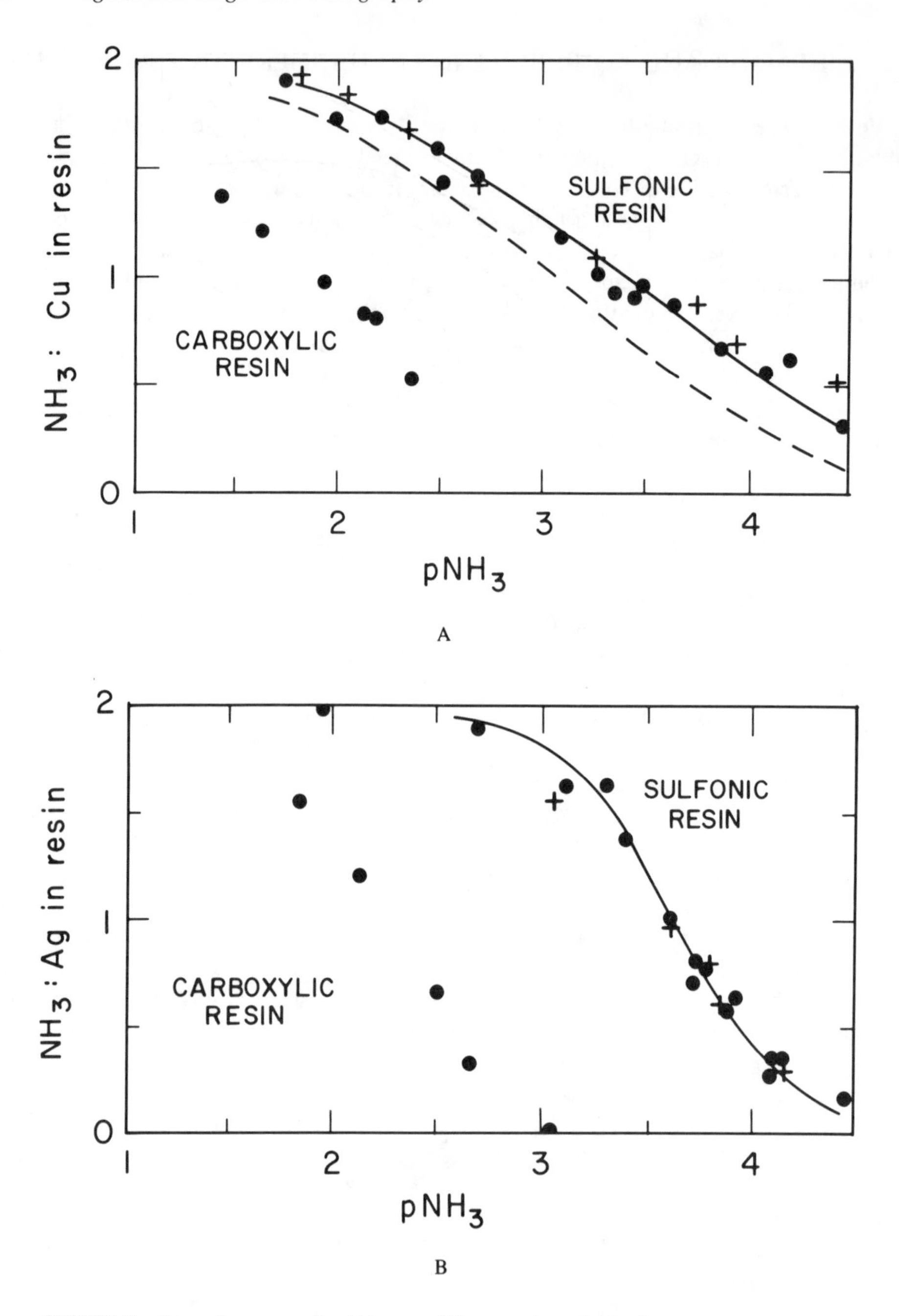

FIGURE 3. Formation curver for (A) copper(II)-ammonia and (B) silver-ammonia complexes. Points, observed in ion-exchange resins; solid lines, Bjerrum's data for formation in aqueous ammonium nitrate — 5 *M* in (A), 2 *M* in (B); dashed line, formation curve in 0.5 *M* NH_4NO_3. (From Stokes, R. H. and Walton, H. F., *J. Am. Chem. Soc.*, 76, 3327, 1954. With permision.)

low ligand:metal ratios, and here the complexes are likely to be more stable in the resin than in solution, as appears for *n*-butylamine in Figure 4.

Mono-, di-, and triethanolamine are bound more weakly in the resin than a solution;[62] a reason may be the greater opportunities for hydrogen bonding in the aqueous phase.

The case of diamines is very interesting. Figure 5 shows formation curves in the resin and in solution for silver(I) and ethylenediamine. At low ligand numbers, the complex is

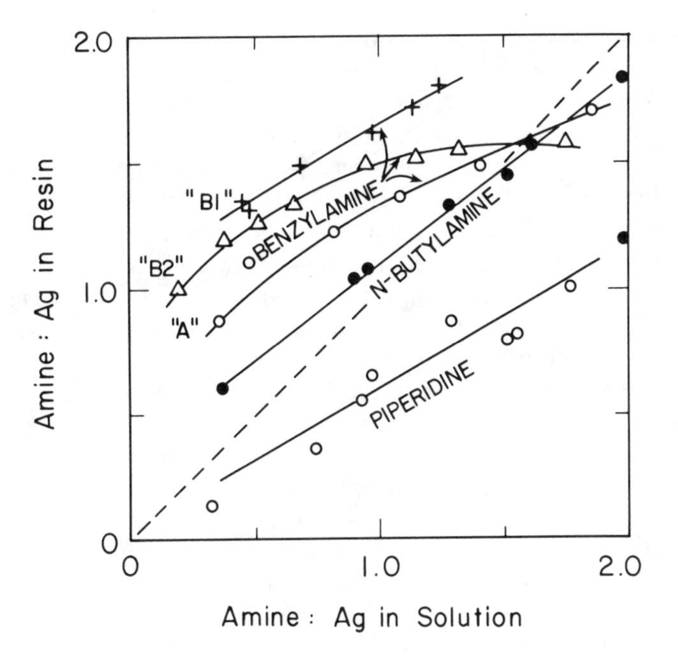

FIGURE 4. Comparative stabilities of complexes in solution and in an ion-exchange resin. Ordinates are bound amine: total silver in an 8% cross-linked sulfonated polystyrene resin; abscissae are bound ammonia: total silver in aqueous solution. For benzylamine, curve "A" is for a resin carrying 2 meq Ag per gram of resin; curve "B1" and "B2" are for a different resin carrying 2 meq Ag and 5 meg Ag per gram of resin, respectively. (From Stokes, R. H. and Walton, H. F., *J. Am. Chem. Soc.*, 76, 3327, 1954. With permission.)

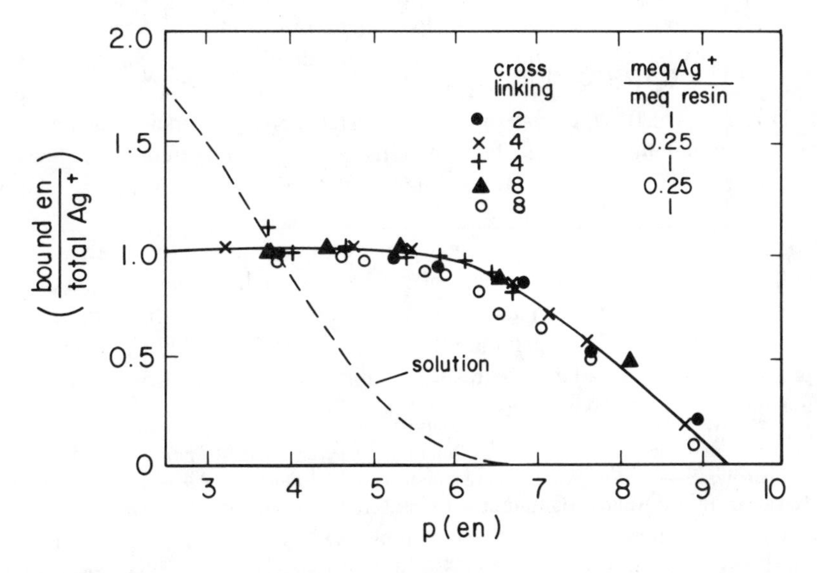

FIGURE 5. Formation curves for silver-ethylenediamine complexes in solution and in a cation-exchange resin. (From Suryaraman, M. G. and Walton, H. F., *J. Phys. Chem.*, 66, 78, 1962. With permission.)

more stable in the resin by three powers of ten. However, coordination stops at one ligand molecule per silver ion, whereas in solution it proceeds to 2:1. The suggestion was made that ethylenediamine acts as a bridging ligand connecting the two silver ions.

A series of papers by Skorokhod and co-workers describes the association of metal ions with uncharged ligands in sulfonated polystyrene cation-exchange resins. The first studies[3,68] deal with thiourea and the cations of cadmium(II), zinc(II), and lead(II). Metal-loaded exchangers were shaken with aqueous solutions of thiourea, and after reaction, the thiourea remaining in the solution was measured. Bjerrum-type formation curves, like those in Figure 3, were obtained by plotting the ratio of bound ligand to total metal, n, in the exchanger against p(TU), the negative logarithm of the thiourea concentration in solution, which ranged between 0 and 2; in other words, the complexes were not highly stable. The curves showed that the complexes were more stable in the resin than in the aqueous solution, and moreover, that association continued to higher ligand ratios in the resin; the complex $Cd(TU)_4^+$ was fully stabilized in 1 *M* aqueous thiourea. When the effect of resin cross-linking was examined, however, it was seen that the complexes with higher ligand numbers were destabilized on going from 8% cross-linking to 12%. The cadmium-thiourea complexes in the resin were much more stable than those of zinc, and it was very easy to separate zinc from cadmium on a column of cation-exchange resin by eluting the former with a dilute thiourea solution (containing sodium nitrate), leaving cadmium behind.

Commenting on these results, Skorokhod suggests that that association occurs by ion-dipole interaction which is accentuated in the resin because of the lower dielectric constant within the resin.

Other ligands studied by Skorokhod were pyridine and methyl-substituted pyridines (picolines)[69] bipyridyl isomers, 1,10-phenanthroline, and ethylenediamine.[70] Alpha-picoline (2-methylpyridine) was much less strongly held by copper(II) and cadmium(II) in the resin than were β- and γ-picoline, undoubtedly due to the steric hindrance effect noted above. The complexes were more stable in the resin than in solution; stabilization increasing in the order ethylenediamine (least), bipyridyl, phenanthroline (most); aromatic solutes are attracted to the polystyrene matrix of the resin. At low ligand ratios, the stabilization increases with the resin cross-linking, but at higher ratios, the ''crowding'' effect was apparent. The higher complexes were less stable in the resin or not formed at all. For ethylenediamine with copper(II) and nickel(II), the first formation constant, K_1, was about ten times greater in the resin than in solution; K_2 was about three times greater. The trend to high stabilization at low ligand ratios and destabilization at high ligand ratios is seen in Figure 6. It was very marked indeed, with α,α'-bipyridyl.

Comparing complex stabilities in exchangers with those in solution, the tacit assumption is made that the thermodynamic activity of the uncombined, uncharged ligand is the same in the exchanger as in solution. This statement is probably true, but it is not true to say that the concentrations of the free ligand are the same in both phases. For ammonia, the molar concentrations inside and outside the resin probably are nearly equal, and the same may be true for small molecules like methylamine and ethylenediamine, but we may expect that most organic ligands, and particularly those having aromatic character, will be more concentrated in the resin, i.e., the ratio of uncombined ligand to water will be greater in the resin. Skorokhod[3] tried to evaluate this effect for thiourea by measuring the sorption of thiourea by a resin carrying noncomplexing ions, Na^+, K^+, and H^+. The sorption was not zero, but it was very small — only 2 to 3% of that which occurred when complexing metal ions were present.

The concentration of free or uncombined ligand in the systems we have been describing is very hard to measure experimentally and is, therefore, generally ignored. The chemical potential of an uncharged, free ligand must, however, be the same in both phases.

A new approach to complex stabilization was made by Maes, Cremers and co-work-

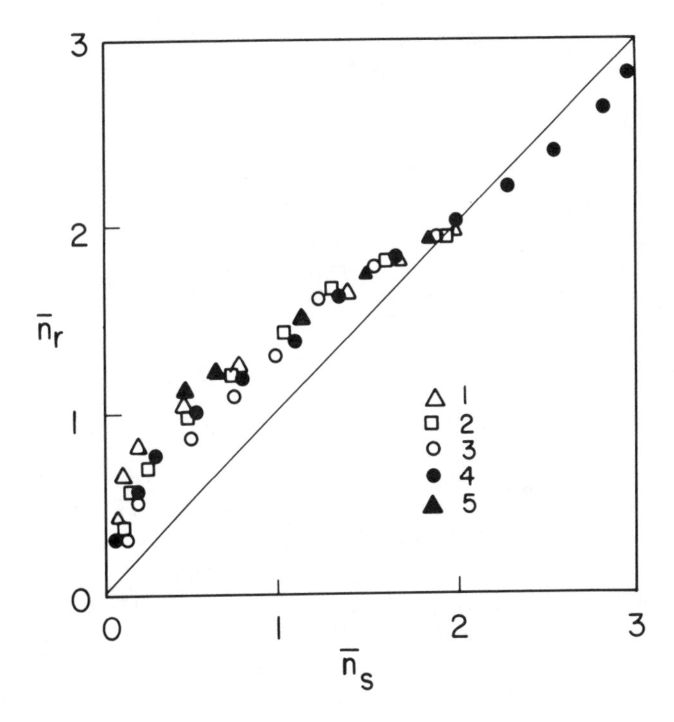

FIGURE 6. Distribution of ethylenediamine between cation-exchange resins and aqueous $Cu(NO_3)_2$, Curve 3, or $Ni(NO_3)_2$, Curve 4. Curve 5 is calculated from Reference 62. (From Skorokhod, O. R. and Kalinina, A. A., *Zh. Fiz. Khim.*, 49, 317, 1975. With permission.)

ers.[71-74] Suppose that two cations are distributed between an exchanger and a solution, and suppose that a ligand is added that forms a complex with one of the cations, but not the other. If the complex is more stable in the exchanger than in solution, adding the ligand will cause the complexing cation to move from the solution to the exchanger; if the complex is less stable in the exchanger, then adding the ligand will bring the complexing cation out of the exchanger and into the solution.

That addition of an uncharged ligand could cause changes in ion-exchange distribution ratios had been noted by Inczedy[75,76] and used to improve separations of metal ions by ion-exchange chromatography; thus, addition of neutral pyridine to a suspension of pyridium-form cation-exchange resin in a solution of pyridinium chloride containing a small amount of copper(II) caused displacement of copper(II) from the solution into the resin. Zinc(II) was so displaced, but to a much smaller degree. Zinc and copper ions were effectively separated on a column of pyridium-form cation-exchange resin with an eluent of pyridinium chloride containing a small excess of pyridine, with zinc emerging first. More recently, Weinert and Strelow[77] have shown that thiourea, an uncharged ligand, increases the distribution coefficient of certain metals, like copper(II), between a cation-exchange resin and dilute hydrochloric acid solutions, but decreases the distribution coefficient of others, like zinc(II).

Maes, Cremers, and co-workers[71-74] chose the copper(II)-ethylenediamine system for study, with calcium as the noncomplexing competing cation. Ethylenediamine coordinates with copper(II) in two stages, giving the cations $Cu(en)^{2+}$ and $Cu(en)_2^{2+}$, with step-wise formation constants (as logarithms) $\log K_1 = 10.72$, $\log K_2 = 9.31$. These constants are high enough that a small excess of ethylenediamine stabilizes the 2:1 complex, $Cu(en)_2$, so that only this complex needs to be taken into account. The equilibrium constants for the exchanges of Cu^{2+} (aqueous) and $Cu(en)_2^{2+}$ with Ca^{2+} are as follows:

$$\frac{[\overline{Cu^{2+}}]\,[Ca^{2+}]}{[Cu^{2+}]\,[\overline{Ca^{2+}}]} = K, \quad \frac{[\overline{Cu(en)_2^{2+}}]\,[Ca^{2+}]}{[Cu(en)_2^{2+}]\,[\overline{Ca^{2+}}]} = K^L$$

The barred symbols refer to the exchanger phase.

The ratio

$$\frac{K^L}{K} = \frac{[\overline{Cu(en)_2^{2+}}]\,[Cu^{2+}]}{[Cu(en)_2^{2+}]\,[\overline{Cu^{2+}}]} = \frac{\overline{\beta_2}}{\beta_2} \times \frac{[\overline{en^2}]}{[en^2]}$$

where β_2 is the cumulative formation constant, K_2K_2, for the $Cu(en)_2{}^{2+}$ complex. Maes and Cremers called the ratio $\frac{\overline{\beta_2}}{\beta_2}$ the "excess stability constant" or K_{syn}. If the activities of ethylenediamine are the same in both phases, as we have every reason to expect they are, then K_{syn} is simply the ratio of the two equilibrium constants, K^L (measured with excess ligand) and K (measured in the absence of ligand). These are true equilibrium constants, for they were found by measuring the concentration quotients, K_c, over the whole range of exchanger compositions and integrating by the Gaines-Thomas method.

Measurements were made with two ion exchangers, a macroporous sulfonated polystyrene resin, and a montmorillonite clay. The values found for K_{syn} were 22 and 850, respectively.

To find the step-wise formation constants K_1 and K_2, two methods were used. One was the Bjerrum method, in which the metal-loaded exchanger is brought to equilibrium with solutions containing varied concentrations of free ligand, and the ratio (bound ligand-to-total metal) in the exchanger is plotted against the negative logarithm of the free ligand concentration, giving graphs like the ones in Figure 3, which are analyzed numerically to find K_1 and K_2. From such curves, K_1 can be found with fair certainty, but the computational errors in K_2 may be large. The other method was to see how the ion-exchange distribution of the complexing metal changed as the ratio of ligand to metal changed. This second method is especially useful where the first complex is destabilized. In such cases, the distribution ratio of the metal ion shows a maximum at a free ligand concentration close to $\beta_2{}^{-0.5}$. The maximum distribution ratio is sensitive to the ratio of K_2:K_1, and allows K_2 to be calculated with good accuracy.

Applying this method to the silver-thiourea complex, Maes and Cremers found that K_1 in the macroporous resin was more than 10^4 times what it is in water; however, K_2 was considerably smaller in the resin.

The data for the copper-ethylenediamine complexes are summarized in Table 3. Both K_1 and K_2 are greater in the two ion exchangers than they are in water, and the stablization of K_2 in montmorillonite clay is especially great. The investigators[74] later showed that the stabilization of K_2 depended on the charge density of the clay, and vanished when the charge density was extrapolated to zero.

The reasons for the stabilization of complexes in ion exchangers are still not clear. Stabilization occurs with monodentate ligands like thiourea and pyridine, as well as with bidentate ligands, and the internal cross-linking postulated for silver(I) and ethylenediamine[67] cannot be a general explanation. Any attractive force between the ligand and the exchanger matrix supplements the force of the metal-ligand bond and stabilizes the complex. Organic ligands of aromatic character are attracted to the aromatic matrix of a polystyrene-based ion exchanger, and the stabilization of pyridine complexes, for example, is to be expected. Such interactions do not, however, explain stabilization in ion-exchanging clays.

Table 3
STABILITY CONSTANTS FOR THE COPPER ETHYLENEDIAMINE COMPLEXES FROM BJERRUM FORMATION CURVES

Exchanger	log K_1	log K_2	log β_2	Overall stabilization
Montmorillonite	11.60	11.50	23.10	$10^{3.1}$
Macroporous resin	11.65	9.70	21.35	$10^{1.32}$
In bulk solution	10.72	9.31	20.03	—

From Maes, A., Peigneur, P., and Cremers, A., *J. Chem. Soc. Faraday Trans.* 1, 74, 182, 1978. With permission.

III. PARTITION OF NONIONIC ORGANIC COMPOUNDS BETWEEN EXCHANGERS AND SOLUTIONS

Retention of organic solutes by ion exchangers is a complex phenomenon that involves several kinds of interaction. In ligand exchange the primary effect is metal-ligand coordination, but this is not the only effect. We have seen that metal-ligand complexes may be many times more stable in an ion exchanger than in solution, and they may also be less stable. Metal-ligand coordination may be modified by other interactions.

To understand these interactions, it is helpful to examine the sorption of uncharged organic compounds by ion exchangers in cases where metal-ligand coordination is absent, or at least of minor importance.

A good starting point is the sorption of aromatic hydrocarbons by ion-exchange resins of the polystyrene type. The hydrocarbon molecules are bound to the polymer by aromatic π-electron overlap. Molecules having aromatic character (and this class includes xanthines and other nonbenzenoid aromatic compounds) are always bound more strongly to polystyrene-based exchangers than comparable aliphatic compounds, whether the primary mechanism is ion exchange, ligand exchange, or simply the solvent action of the polymer. The partition of aromatic hydrocarbons between cation exchangers and mixed aqueous-organic solvents has been studied by Walton and co-workers.[21,78] It is found to depend on the nature of the inorganic cation in the exchanger. The higher the charge on the ion, the more the hydrocarbon is retained. Phenanthrene, for example, had the following capacity factors for the counter-ions indicated in a 4% cross-linked sulfonated polystyrene cation exchanger, with 37% 2-propanol in water as the mobile phase: Na,3.7; NH_4,4.4; Ca,7.9; Fe(III), 8.2. Nonionic, polar aromatic compounds like caffeine and acetanilide show a similar progression. Triply charged La^{3+} gives about the same retention as Fe^{3+} — roughly double the retention with the Na^+-form exchanger. The exact relations depend on the resin cross-linking and its physical form, whether gel-type or macroporous.

To interpret these results, we may first consider the ion-exchange resin to be a solid organic solvent the properties of which are modified by the ionic groups and by the imbibed water. Placed in a solvent containing water, the resin is like a mixed aqueous-nonaqueous solvent itself, and will retain nonpolar or weakly polar organic solutes more strongly, the less water it contains.

Now, the water content of the swollen resin depends on the inorganic counter-ion. The degree of hydration of this ion is important, but more important is the ionic charge. When a doubly charged cation is substituted for two singly charged cations, the resin shrinks; water is expelled.We visualize this effect as caused by electrostatic attraction, with the doubly charged ion attracting fixed sulfonate ions on different polymer chains, and so drawing the polymer chains together.

One more concept is invoked here. If we consider a linear polymer chain bearing negative ion at intervals along its length, and then bring up to it a doubly charged mobile ion, the inverse square law of force predicts that the most stable position for this doubly charged ion is near one of the fixed ions, not between them. Of course, the position of the mobile ion is modified by thermal disorder, but experimental data, namely, the self-diffusion rates of cations of different charges in a cross-linked polystyrene cation exchanger,[79] suggest strongly that the counter-ions have to surmount an energy barrier when they move from one position to another in the resin, and that this barrier is higher, the higher the ionic charge. In other words, ion pairing exists, and it is easy to see how doubly and triply charged cations can produce electrostatic cross-linking. Whatever the reason, the resins shrink and expel water when these ions are introduced. At the same time, they become better solvents for aromatic hydrocarbons and for aromatic compounds of moderate polarity.

An interesting effect is shown by the dipolar ions of amino acids, including trigonelline and homarine:

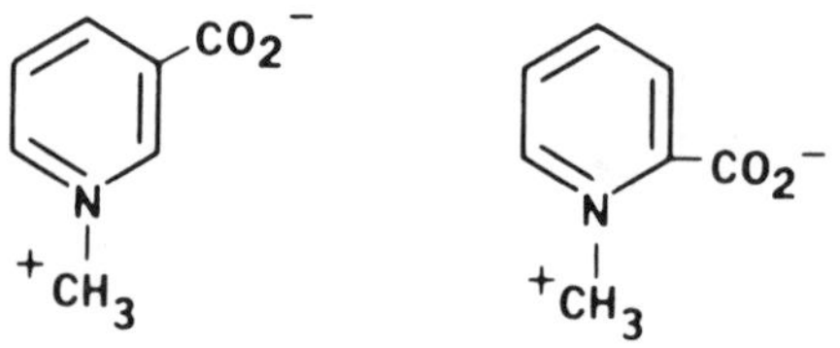

trigonelline **homarine**

These compounds are held much more strongly by calcium-form resins than by sodium-form resins, and still more strongly by lanthanum-form resins. The same is true of simple amino acids like glycine and α- and β-alanine, and the effect is greater, the larger the dipole moment; thus, trigonelline is bound more strongly than homarine by resins with Ca^{2+} or La^{3+} as counter-ions, and β-alanine is bound more strongly than α-alanine.[80] The effect may simply be one of ion-dipole attraction, such as occurs in aqueous solutions, or it may be due to the strong electrostatic field existing in the exchanger between a di- or trivalent ion and a neighboring, unoccupied fixed ion (see Figure 7).

The effect of ionic hydration may be seen by comparing retentions on an ion-exchange resin carrying different counter-ions of equal charge, such as the alkali-metal cations. This was done by Dieter and Walton[81] for a group of water-soluble, polar aromatic compounds, with the results summarized in Figure 8. The counter-ions chosen were Li^+, Na^+, K^+, Mg^{2+}, and Ca^{2+}. Among the solutes, two separate trends were noted: for the xanthines, caffeine, theobromine, and theophylline for the dipolar ion trigonelline, for the amides acetanilide and phenacetin, and perhaps for ethyl and methyl benzoate, retention increases with counter-ion in the order $K^+ < Na^+ < Li^+$ and increases from K^+ to Mg^{2+}. Compounds having a free phenolic group, methyl and ethyl paraben, and phenol itself show retentions increasing in the order $Li^+ < Na^+ < K^+$ and falling from K^+ to Mg^{2+}. In the first group, the more the counter-ion is hydrated, the greater the retention. In the second group, the reverse is true.

At first sight, the higher the water content of the swollen resin, the weaker should be its solvent action for organic compounds. Water that is bound up with ions as water of hydration, however, is not available to modify the solvent character of the resin polymer. The well known phenomenon of "salting out" organic solutes from water depends on the binding of water by the added salt. A quantitative measure of the water of hydration attached to ions

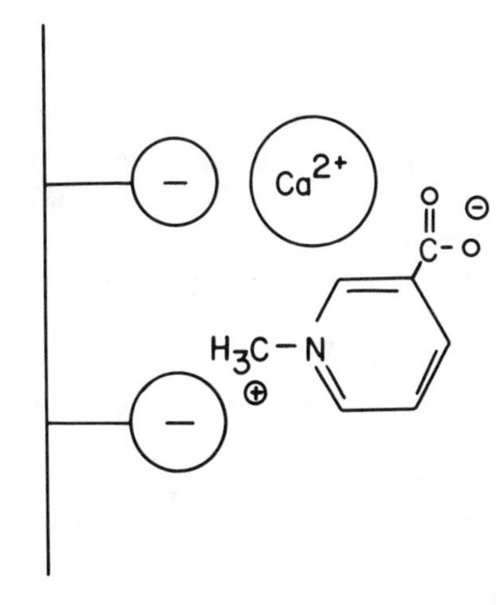

FIGURE 7. Postulated retention of a dipolar ion, trigonelline, by a resin carrying divalent ions (Ca^{2+}). (From Otto, J., deHernandez, C. M., and Walton, H. F., *J. Chromatogr.*, 247, 91, 1982. With permission.)

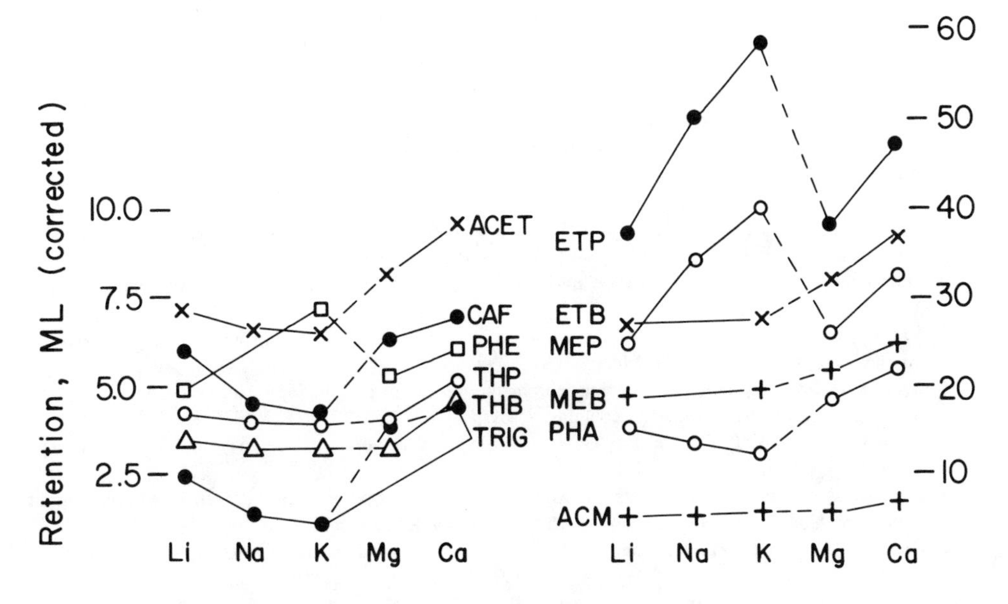

FIGURE 8. Retention volumes of organic compounds on an 8% cross-linked cation-exchange resin as functions of the inorganic counter-ion. Abbreviations for compounds are as follows:

ACET:	acetanilide, *N*-acetylphenylamine
CAF:	caffeine, 1,3,7-trimethylxanthine
ETB:	ethyl benzoate
ETP:	ethyl paraben, ethyl *p*-hydroxybenzoate
MEB:	methyl benzoate
MEP:	methyl paraben, methyl *p*-hydroxybenzoate
PHA:	phenacetin, 4′-ethoxyacetanilide
PHE:	phenol
THB:	theobromine, 3,7-dimethylxanthine
THP:	theophylline, 1,3-dimethylxanthine
TRIG:	trigonelline or *N*-methylpyridinium-3-carboxylate

can be made from activity coefficients and the theory of strong electrolytes; the water in the swollen ion exchanger can be measured by centrifugation and weighing. Combining these two measurements, Dieter and Walton[81] concluded that although the total water content per equivalent of resin was greatest with lithium as the counter-ion, the "free" water (water not bound as water of hydration) was least (among the alkali metals) with Li^+. It was still less with Mg^{2+} and even smaller with Ca^{2+}, and as we have already noted, doubly charged counter-ions cause stronger sorption of aromatic hydrocarbons.

With phenols as solutes, another mechanism comes into play. Here, the free water favors sorption by the resin instead of suppressing it. The most likely mechanism is hydrogen bonding between the free water and the phenolic hydroxyl groups.

The partition of sugars between solutions and ion-exchange resins is governed by hydrogen bonding. The classic work of Rückert and Samuelson[82] showed that the sorption of glucose by cation-exchange resins carrying alkali-metal ions was least with Li^+, greater with Na^+, and still greater with K^+. The same trend has been noticed by other workers.[80] It is the free water, not bound as water of hydration, that draws the sugar into the resin. More evidence on this subject will be presented in Chapter 6.

REFERENCES

1. **Maeda, M., Tsuji, A., Ganno, S., and Onishi, Y.,** Fluorophotometric assay of amino acids by using automated ligand-exchange chromatography and pyridoxal-zinc(II) reagent, *J. Chromatogr.*, 77, 434, 1973.
2. **Wagner, F. W. and Shepherd, S. L.,** Ligand exchange amino acid analysis: resolution of some amino sugars and cysteine derivatives, *Anal. Biochem.*, 41, 314, 1971.
3. **Skorokhod, O. R. and Varavva, A. G.,** Thiourea complexes of cadmium in the KU-2 sulfonic cation exchanger, *Zh. Fiz. Khim.*, 46, 1708, 1972 (Engl. transl., p. 980).
4. **Shimomura, K. and Walton, H. F.,** Thin-layer chromatography of amines by ligand exchange, *Sep. Sci.*, 3, 493, 1968.
5. **Yasuda, K.,** Thin-layer chromatography of aromatic amines on cadmium sulphate-impregnated silica gel thin layers, *J. Chromatogr.*, 60, 144, 1971.
6. **Kunzru, D. and Frei, R. W.,** Separation of aromatic amine isomers by high-pressure liquid chromatography on cadmium-impregnated silica gel columns, *J. Chromatogr. Sci.*, 12, 191, 1974.
7. **Vogt, C. R., Ryan, T. R., and Baxter, J. S.,** High-speed liquid chromatography on cadmium-modified silica gel, *J. Chromatogr.*, 136, 221, 1977.
8. **Frei, R. W., Beall, K., and Cassidy, R. M.,** Determination of aromatic nitrogen heterocycles in air samples by high-speed liquid chromatography, *Mikrochim. Acta*, p. 859, 1974.
9. **Guha, O. K. and Janak, J.,** Charge transfer complexes of metals in the chromatographic separation of organic compounds, *J. Chromatogr.*, 68, 325, 1972.
10. **Houx, N. W. H., Voerman, S., and Jongen, W. M. F.,** Purification and analysis of synthetic insect sex attractants by liquid chromatography on a silver-loaded resin, *J. Chromatogr.*, 96, 25, 1974.
11. **Snyder, L. R.,** Nitrogen and oxygen compound types in petroleum, *Anal .Chem.*, 41, 314, 1969.
12. **Funasaka, W., Hanai, T., Fujimura, K., and Ando, T.,** Nonaqueous solvent chromatography, *J. Chromatogr.*, 78, 424, 1973.
13. **Walton, H. F. and Martinez, J. M.,** Reactions of Hg(II) with a cation-exchange resin, *J. Phys. Chem.*, 63, 1318, 1959.
14. **Petronio, B. M., DeCaris, E., and Iannuzzi, L.,** Some applications of ligand-exchange. II, *Talanta*, 29, 691, 1982.
15. **Shahwan, G. J. and Jezorek, J. R.,** Liquid chromatography of phenols on an 8-quinolinol silica gel-iron(III) stationary phase, *J. Chromatogr.*, 256, 39, 1983.
16. **Maslowska, J. and Pietek, W.,** The use of ligand exchange for the separation of aromatic acids on synthetic cation exchangers, *Chromatographia*, 20, 46, 1985.
17. **Webster, P. V., Wilson, J. N., and Franks, M. C.,** Macroreticular ion-exchange resins: some analytical applications to petroleum products, *Anal. Chim. Acta*, 38, 193, 1967.
18. **Funasaka, W., Fujimura, K., and Kuriyama, S.,** Ligand exchange chromatography. I. Separation of phenylaminediamine isomers, *Bunseki Kagaku*, 18, 19, 1969.

19. **Fujimura, K., Koyama, T., Tanigawa, T., and Funasaka, W.,** Studies in ligand-exchange chromatography. IV. Separation of hydroxybenzoic and hydroxynaphthoic acid isomers, *J. Chromatogr.*, 85, 101, 1973.
20. **Kothari, R. M.,** Some aspects of the fractionation of DNA on an IR-120-Al(III) column, *J. Chromatogr.*, 57, 83, 1971.
21. **Otto, J., de Hernandez, C. M., and Walton, H. F.,** Chromatography of aromatic acids on La-loaded ion-exchange resins, *J. Chromatogr.*, 247, 91, 1982.
22. **Davankov, V. A., Rogozhin, S. V., and Tsyurupa, M. P.,** Factors determining the swelling power of crosslinked polymers, *Angew. Makromol. Chem.*, 32, 145, 1973.
23. **Davankov, V. A., Rogozhin, S. V., and Tsyurupa, M. P.,** Influence of polymeric matrix structure on performance of ion-exchange resins, in *Ion Exchange and Solvent Extraction*, Vol. 7, Marinsky, J. A. and Yizhok, M., Eds., Marcel Dekker, New York, 1977, chap 2.
24. **Del Rosario Maury, M., Poitrenaud, C., and Tremillon, B.,** Sorption of ammonia from an aqueous solution by complexation with ion exchanger containing phosphonic groups, *Chem. Anal. (Warsaw)*, 17, 1059, 1972.
25. **Bedetti, R., Carunchio, V., and Marino, A.,** An application of the ligand exchange method to the separation of some aliphatic carboxylic acids, *J. Chromatogr.*, 95, 127, 1974.
26. **Hemmasi, B. and Bayer, E.,** Ligand-exchange chromatography of amino acids on copper, cobalt, and zinc Chelex 100, *J. Chromatogr.*, 109, 43, 1975.
27. **Siegel, A. and Degens, E. T.,** Concentration of dissovled amino acids from saline waters by ligand-exchange chromatography, *Science*, 151, 1098, 1966.
28. **Buist, N. R. M. and O'Brien, D.,** The separation of peptides from amino acids in urine by ligand exchange chromatography, *J. Chromatogr.*, 29, 398, 1967.
29. **Bellinger, J. F. and Buist, N. R. M.,** The separation of peptides from amino acids by ligand-exchange chromatography, *J. Chromatogr.*, 87, 513, 1973.
30. **Petronio, B. M., Lagana, A., and Russo, M. V.,** Some applications of ligand-exchange. I. Recovery of phenolic compounds from water, *Talanta*, 28, 215, 1981.
31. **de Hernandez, C. M. and Walton, H. F.,** Ligand exchange chromatography of amphetamine drugs, *Anal. Chem.*, 44, 890, 1972.
32. **Murgia, E. and Walton, H. F.,** Ligand-exchange chromatography of alkaloids, *J. Chromatogr.*, 104, 417, 1975.
33. **Warren, D. C. and Fitzgerald, J. M.,** Parameters influencing the electron spin resonance signal intensity of metal ion complex-exchanged resins, *Anal. Chem.*, 49, 1840, 1977.
34. **Muzzarelli, R. A. A., Martelli, A. F., and Tubertini, O.,** Ligand-exchange chromatography on thin layers and columns of natural and substituted celluloses, *Analyst*, 94, 616, 1969.
35. **Rothenbühler, E., Waibel, R., and Solms, J.,** An improved method for the separation of peptides and amino acids on copper-Sephadex, *Anal. Biochem.*, 97, 367, 1979.
36. **Gozdzicka-Josefiak, A. and Augustyniak, J.,** Preparation of chelating exchangers with a polysaccharide network and low cross-linkage, *J. Chromatogr.*, 131, 91, 1977.
37. **Tsuji, A. and Sekiguchi, K.,** Adsorption of nicotine acid hydrazide on cation exchanger of various metal forms, *Nippon Kagaku Zasshi*, 81, 847, 1960.
38. **Sluyterman, L. A. E. and Wijdenes, J.,** An agarose mercurial column for the separation of mercaptopapain and nonmercaptopapain, *Biochim. Biophys. Acta*, 200, 593, 1970.
39. **Chow, F. K. and Grushka, E.,** Separation of aromatic amine isomers by high pressure liquid chromatography with a copper(II)-bonded phase, *Anal. Chem.*, 49, 1756, 1977.
40. **Gimpel, M. and Unger, K.,** Monomeric vs. polymeric bonded iminodiacetate silica supports in high-performance ligand-exchange chromatogrphy, *Chromatographia*, 17, 200, 1983.
41. **Shahwan, G. J. and Jezorek, J. R.,** Liquid chromatography of phenols on an 8-quinolinol silica gel-iron(III) stationary phase, *J. Chromatogr.*, 256, 39, 1983.
42. **Kicinski, H. G. and Kettrup, A.,** Determination of enantiomeric catecholamines by ligand-exhcange chromatography using chemically modified tartaric acid silica gel, *Fresenius Z. Anal. Chem.*, 320, 51, 1985.
43. **Caude, M. and Foucault, A.,** Ligand exchange chromatography of amino acids on copper(II) modified silica gel with ultraviolet spectrophotometric detection at 210 nanometers, *Anal. Chem.*, 51, 459, 1979.
44. **Foucault, A. and Rosset, R.,** Ligand-exchange chromatography on copper(II)-modified silica gel: improvements and use for screening of protein hydrolyzate and quantitation of dipeptides and amino acid fractions, *J. Chromatogr.*, 317, 41, 1984.
45. **Guyon, F., Chardonnet, L., Caude, M., and Rosset, R.,** Study of silica gel and copper silicate gel solubilities in high performance liquid chromatography, *Chromatographia*, 20, 30, 1985.
46. **Masters, R. G. and Leyden, D. E.,** Ligand-exchange chromatography of amino sugars and amino acids on copper-loaded silylated controlled-pore glass, *Anal. Chim. Acta*, 98, 9, 1978.

47. **Chow, F. K. and Grushka, E.,** High performance liquid chromatography with metal-solute complexes, *Anal. Chem.*, 50, 1346, 1978.
48. **Chow, F. K. and Grushka, E.,** High performance liquid chromatography of nucleotides and nucleosides using outersphere and innersphere metal-solute complexes, *J. Chromatogr.*, 185, 361, 1979.
49. **Gubitz, G. and Mihellyes, S.,** Direct separation of 2-hydroxy acid enantiomers by high-performance liquid chromatography on chemically bonded chiral phases, *Chromatographia,* 19, 257, 1984.
50. **Shimomura, K., Hsu, T.-J., and Walton, H. F.,** Ligand-exchange chromatography of aziridines and ethanolamines, *Anal. Chem.*, 45, 501, 1973.
51. **Hill, (Sister) A. G., Sedgley, R., and Walton, H. F.,** Separation of amines by ligand exchange. III. A comparison of different cation exchangers, *Anal. Chim. Acta,* 33, 84, 1965.
52. **Doury-Berthod, M. and Poitrenaud, C.,** Continuous detection of molecules labeled with C-14, *Analusis,* 5, 270, 1977.
53. **Davankov, V. A., Rogozhin, S. V., Semechkin, A. V., Baranov, V. A., and Sannikova, G. S.,** Ligand-exchange chromatography of racemates; influence of temperature and concentration of eluent on ligand-exchange chromatogrphy, *J. Chromatogr.*, 93, 363, 1974.
54. **Maeda, M., Kinoshita, T., and Tsuji, A.,** A novel fluorimetric method for determination of hexosamines using pyridoxal and zinc(II), *Anal. Biochem.*, 38, 121, 1970.
55. **Roth, M.,** Fluorescence reaction for amino acids, *Anal. Chem.*, 43, 880, 1971.
56. **Benson, J. R. and Hare, P. W.,** σ-phthalaldehyde: fluorogenic detection of primary amines in the picomole range. Comparison with fluorescamine and ninhydrin, *Proc. Natl. Acad. Sci. U.S.A.*, 72, 619, 1975.
57. **Hare, P. E. and Gil-Av, E.,** Separation of D and L amino acids by liquid chromatography: use of chiral eluants, *Science,* 204, 1226, 1979.
58. **Foucault, A., and Caude, M., and Oliveros, L.,** Ligand-exchange chromatography of enantiomeric amino acids on copper-loaded chiral bonded silica gel and of amino acids on copper(II)-modified silica gel, *J. Chromatogr.*, 185, 345, 1979.
59. **Navratil, J. D., Murgia, E., and Walton, H. F.,** Ligand-exchange chromatography of amino sugars, *Anal. Chem.*, 47, 122, 1975.
60. **Spies, J. R.,** An ultraviolet spectrophotometric micromethod for studying protein hydrolysis, *J. Biol. Chem.*, 195, 65, 1952.
61. **Grushka, E., Levin, S., and Gilon, C.,** Separation of amino acids on reversed-phase columns as their copper(II) complexes, *J. Chromatogr.*, 235, 401, 1982.
62. **Cockerell, L. and Walton, H. F.,** Metal amine complexes in ion exchange. II. 2-Aminoethanol and ethylenediamine complexes, *J. Phys. Chem.*, 66, 75, 1962.
63. **Shimomura, K., Dickson, L., and Walton, H. F.,** Separation of amines by ligand exchange. IV. Ligand exchange with chelating resins and cellulosic exchangers, *Anal. Chim. Acta,* 37, 102, 1967.
64. **Stokes, R. H. and Walton, H. F.,** Metal-amine complexes in ion exchange, *J. Am. Chem. Soc.*, 76, 3327, 1954.
65. **Helfferich, F. G.,** Ligand exchange. I. Equilibria, *J. Am. Chem. Soc.*, 84, 3237, 1962.
66. **Loewenschuss, H. and Schmuckler, G.,** Chelating properties of the ion exchanger Dowex®-A-1, *Talanta,* 11, 1399, 1964.
67. **Suryaraman, M. G. and Walton, H. F.,** Metal-amine complexes in ion exchange. III. Diamine complexes of silver(I) and nickel(II), *J. Phys. Chem.*, 66, 78, 1962.
68. **Skorokhod, O. R. and Baravva, A. G.,** The ligand sorption of thiourea by salt-forms of a sulphonic cation-exchange resin, *Zh. Fiz. Khim.*, 48, 429, 1974 (Engl. transl., p. 247).
69. **Skorokhod, O. R. and Kalinina, A. A.,** Ligand chromatography of isomeric organic bases, *Zh. Fiz. Khim.*, 48, 2830, 1974 (Engl. transl., p. 1663).
70. **Skorokhod, O. R. and Kalinina, A. A.,** The stability of complexes with the counterions in the sulphonic acid cation exchanger, *Zh. Fiz. Khim.*, 49, 317, 1975 (Engl. transl. p. 187).
71. **Maes, A., Marynen, P., and Cremers, A.,** Stability of metal uncharged ligand complexes in ion exchangers. I. Quantitative characterization and thermodynamic basis, *J. Chem. Soc. Faraday Trans.* 1, 73, 1297, 1977.
72. **Maes, A., Peigneur, P., and Cremers, A.,** Stability of metal uncharged ligand complexes in ion exchangers. II. The copper + ethylenediamine complex in montmorillonite and sulphonic acid resin, *J. Chem. Soc., Faraday Trans. 1,* 74, 182, 1978.
73. **Maes, A. and Cremers, A.,** Stability of metal uncharged ligand complexes in ion exchangers. III. Complex ion stability and stepwise stability constants, *J. Chem. Soc. Faraday Trans.* 1, 74, 2470, 1978.
74. **Maes, A. and Cremers, A.,** Stability of metal uncharged ligand complexes in ion exchangers. IV. Hydration effects and stability changes of copper-ethylenediamine complexes in montmorillonite, *J. Chem. Soc. Faraday Trans.* 1, 75, 513, 1979.
75. **Gabor-Klatsmanyi, P. and Inczedy, J.,** *3rd Proc. Anal. Chem. Conf.*, 1970, 15; *Chem. Abstr.*, 74, 57627s, 1971.

76. **Inczedy, J., Gabor-Klatsmanyl, P., and Erdey, L.,** The use of complex forming agents in ion exchange chromatography. VI, *Acta Chim. Acad. Sci. Hung.*, 69, 137, 1971; *Anal. Abstr.*, 19, 4959, 1972.
77. **Weinert, C. H. S. W., Strelow, F. W. E., and Bohmer, R. G.,** Cation exchange in thiourea-HCl solutions, *Talanta,* 30, 413, 1983.
78. **Ordemann, D. M. and Walton, H. F.,** Liquid chromatography of aromatic hydrocarbons on ion-exchange resins, *Anal. Chem.*, 48, 1728, 1976.
79. **Helfferich, F.,** *Ion Exchange,* McGraw-Hill, New York, 1962, chap. 6.
80. **Walton, H. F.,** Counter-ion effects in partition chromatography, *J. Chromatogr.*, 332, 203, 1985.
81. **Dieter, D. S. and Walton, H. F.,** Counter-ion effects in ion-exchange partition chromatography, *Anal. Chem.*, 55, 2109, 1983.
82. **Rückert, H. and Samuelson, O.,** Distribution of glucose between ion-exchange resins and alcohol-water mixtures, *Acta Chem. Scand.*, 11, 315, 1957.

Chapter 3

SEPARATION OF AMINES

J. D. Navratil

I. INTRODUCTION

The separation of amines on a column of cation-exchange resin loaded with copper(II) ions is the archetype of LEC. The copper ions are in the form of their ammonia complexes and aqueous ammonia is used as the eluent. The copper ions remain sorbed on the column while ammonia and amine molecules move down the column, exchanging places with each other in the coordination sphere of the copper ion. The more stable the copper-amine complex, the more slowly the amine moves.

Other metal ions that have been used to separate amines include silver(I), cobalt(II), cadmium(II), nickel(II), and zinc(II).[1-6] Although silver(I) is mainly used for olefins and unsaturated compounds (see Chapter 7), it has been used for amines.[7,8] An antimony-loaded column of cellulose was also used for the chromatography of aliphatic and aromatic amines.[9]

The first use of LEC was the recovery of a diamine, 1,3-diamino-2-hydroxy propane, from dilute aqueous solution that contained ammonia.[10-12] The solution was passed through a column of copper-loaded cation exchanger to concentrate the amines, and later concentrated ammonia was passed to elute and recover the amines.

Afterwards, Walton and co-workers, as well as other research groups, published a series of papers dealing with practical applications of LEC for the separation of mixtures of aliphatic and aromatic amines and diamines, hydrazines, aziridines, alkanolamines, amino sugars, nucleic bases, etc.[1-6] Several analytical problems have been elegantly solved; among them, analysis of aziridines, alkaloids, and amino sugars.

II. ALIPHATIC AMINES

The LEC of simple aliphatic amines is of little practical interest, for these compounds can be analyzed very effectively by gas chromatography, but the elution orders show clearly the effect of steric hindrance about the nitrogen atom.

Walton and co-workers[13,14] studied the elution behavior of various aliphatic amines on different exchangers loaded with copper and nickel ions. The results (Table 1) show that by choosing the proper combination of exchanger and metal ion, almost any desired elution sequence can be obtained. The elution orders corresponded to the decreasing polar or hydrophilic character of the amine and the increasing hydrocarbon character when polystyrene-type exchangers were used, but were reversed when zirconium phosphate was used.

Data in Table 2 show that copper(II) ions give better discrimination between amines than do nickel or zinc. The base strength of the amines do not correlate with their sorption. Primary amines are bound more strongly than secondary, and secondary amines more strongly than tertiary. Substitution on the amine nitrogen weakens the binding. With isomeric primary amines, the binding is weaker the more branched the chain or compact the molecule.

The effect of straight-chain primary aliphatic amines being retarded more than branched-chain amines was also observed in 1959 by Barber et al.,[15] using liquid films of metal stearates supported on Celite® as stationary phases in gas chromatography. They used an ammonia-hydrogen mixture as a carrier gas and found amines to be well retarded compared to retardation on Apiezon oil (Table 3).

Table 1
ELUTION ORDERS OF SELECTED AMINES ON DIFFERENT METAL ION-EXCHANGER COMBINATIONS

	Elution order					
	Polystyrene sulfonic	Polystyrene chelating	Polyacrylic (carboxylic)		Cellulose phosphate	Zirconium phosphate
Compound	Ni	Ni, Cu	Ni	Cu	Cu	Ni
Diethanolamine	1	1	1	1	1	4
Ethanolamine	2	2	4	2	2	3
Dimethylamine	3	3	3	3	4	2
n-Butylamine	4	4	2	2	3	1

From Shimonura, K., Dickson, L., and Walton, H. F., *Anal. Chim. Acta*, 37, 102, 1967. With permission.

Table 2
COLUMN DISTRIBUTION RATIOS, ALIPHATIC AMINES IN 1.4 *M* AMMONIA

		Distribution ratios[a] in			
Amine	pK_b	Dowex®-50 Ni	Chelex® Ni	Chelex® Cu	Chelex® Zn
Methylamine	3.38	—	3.8	5.9	8.9
Dimethylamine	3.23	—	2.5	3.0	9.5
Trimethylamine	4.20	—	1.1	1.3	—
Ethylamine	3.37	—	3.8	7.1	9.8
Diethylamine	3.07	—	3.4	4.1	6.8
Triethylamine	3.13	—	2.9	3.0	5.7
n-Propylamine	3.47	12.5	3.8	7.7	9.6
iso-Propylamine	3.37	8.2	2.7	4.5	7.5
n-Butylamine	3.40	21.6	6.6	13.4	18.9
iso-Butylamine	3.51	13.4	4.1	9.1	12.9
sec-Butylamine	3.44	11.6	3.4	5.9	10.2
tert-Butylamine	3.55	9.1	2.9	3.4	7.1

[a] Distribution ratios from Reference 35. Base ionization constants from literature, especially A. E. Martell and L. G. Sillén, *Stability Constants*.

Table 3
RELATIVE RETENTION TIMES FOR AMINES OR APIEZON L AND ON MANGANESE STEARATE COLUMNS AT 156°C

	Primary			Secondary			Tertiary	
	Apiezon	Manganese stearate		Apiezon	Manganese stearate		Apiezon	Manganese stearate
Pr^i	66	2304	Et_2	99	1610	Et_3	187	328
Pr_n	173	3655	Pr_2^n	276	2134	Pr_3^n	619	649
Bu^n	215	6032	Bu_2^n	758	4871	Bu_3^n	2382	3342
Bu^t	79	1550						

From Barber, D. W., Phillips, C. S. G., Tusa, G. F., and Verdin, A., *J. Chem. Soc.*, p. 18, 1959. With permission.

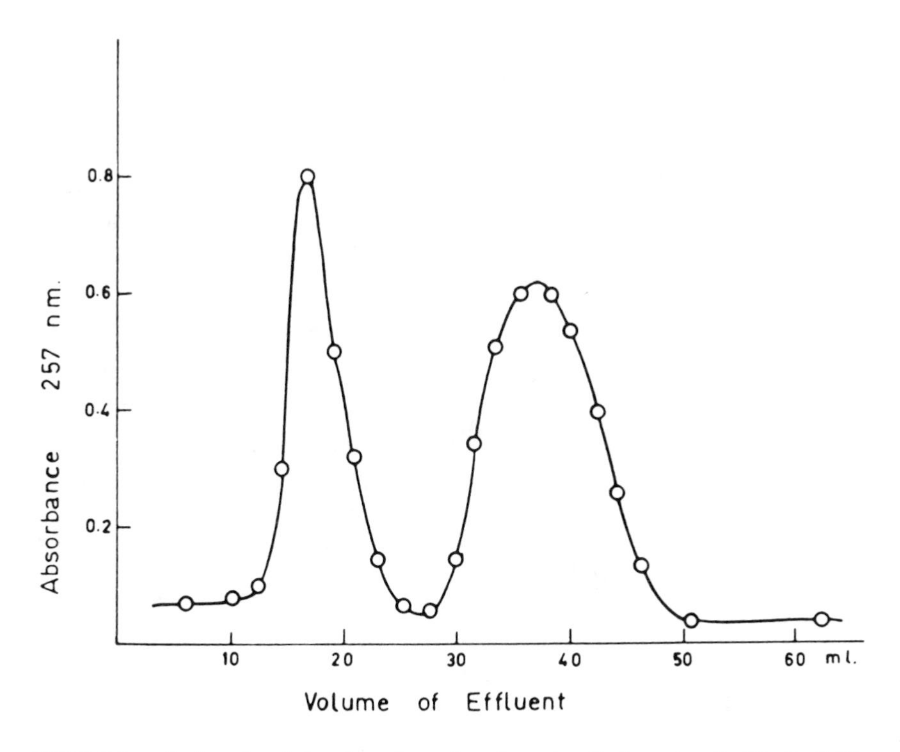

FIGURE 1. Elution of methylamphetamine and amphetamine. Resin, carboxylic (Bio-Rex® 70), 100-2 mesh, Cu form; column 60 × 0.8 cm^2; eluent, 0.1 *M* ammonia in 50% aqueous methanol. Methylamphetamine is eluted first. (From deHernadez, C. M. and Walton, H. F., *Anal. Chem.*, 44, 890, 1972. With permission.)

III. AROMATIC AMINES

Aromatic amines have been separated on several metal-exchanger combinations. Aniline, pyridine, and benzylamine were eluted in that order by ammonia from a nickel-chelating resin (Dowex® A-1) using 2 *M* ammonia;[16] Inczédy[17] found the same elution order on a nickel-sulfonated polystyrene resin with an eluent of aqueous-alcoholic nickel chloride solution. Benzylamine and phenethylamine are held very tightly by polystyrene-based resins, both in the LEC mode and in simple cation exchange. Their molecules are similar to the monomer units of the resin, and this may be the reason for the strong attachment. However, on an inorganic zirconium-base exchanger, benzylamine is weakly held.[14]

Using DEAE-cellulose, Muzzarelli[9] observed the elution order: aniline, trimethylamine, and dimethylamine. Aniline and the methylamines could be displaced with diethyl ether and ethyl alcohol, respectively. Ethylene diamine, which was also present, stayed on the column. Muzzarelli performed similar separations on cobalt(III)- and antimony(III)-loaded cellulose phosphate using aqueous ammonia, ether, and alcohol as eluents. In all cases, aniline eluted first, and he found that amines do interact with celluloses, unlike resins, since aniline was sorbed by nonloaded cellulose, but was more strongly sorbed by metal-loaded celluloses.

Polystyrene resins give broad, asymmetrical bands with aromatic compounds in LEC, perhaps by the superposition of π-bonding on the metal-ligand interaction. Hernandez and Walton,[18] studying the chromatography of amphetamine drugs by LEC, found that the acrylic resin Bio-Rex 70, which has an aliphatic matrix, gave much sharper bands than a polystyrene resin, while at the same time binding the metal more tightly (Figure 1). With copper, nickel, and cadmium the order of elution was

Table 4
THIN-LAYER CHROMATOGRAPHY ON ABSORBENTS IMPREGNATED WITH METAL SALTS (R_f VALUES)

	Solvent					
	Benzene-methanol, 5:1		Benzene-methyl ethyl ketone, 3:1		Benzene-methanol, 20:1	
	Absorbent					
	Silica gel		Silica gel		Alumina	
Compound	Alone	With Zn	Alone	With Cd	Alone	With Zn
Aniline	0.53	0.28	0.47	0.21	0.83	0.40
o-Toluidine	0.60	0.45	0.51	0.29	0.86	0.70
m-Toluidine	0.55	0.27	0.47	0.15	0.83	0.46
p-Toluidine	0.50	0.17	0.43	0.10	0.82	0.22
2,4-Xylidine	0.58	0.31	0.48	0.25	0.86	0.48
N-Methylaniline	0.68	0.56	0.62	0.57	—	—
N,N'-dimethylaniline	0.74	0.67	0.69	0.61	—	—
N,N'-diethylaniline	0.78	0.28	0.73	0.35	—	—
α-Naphthylamine	0.60	0.59	0.52	0.44	0.85	0.79
β-Naphthylamine	0.55	0.44	0.47	0.21	0.85	0.55
N-Methylnaphthylamine	0.72	0.82	0.67	0.72	—	—
p-Phenetidine	0.49	0.11	0.31	0.06	0.70	0.12
Methyl *p*-aminobenzoate	0.49	0.50	0.46	0.45	0.70	0.82

Adapted from Shimomura, K. and Walton, H. F., *Sep. Sci.*, 3, 493, 1968.

1. Metamphetamine, $C_6H_5CH_2CHCH_3NHCH_3$ (eluted first)
2. Ephedrine, $C_6H_5CHOH{\cdot}CHCH_3NHCH_3$
3. Amphetamine, $C_6H_5CH_2CHCH_3{\cdot}NH_2$
4. Norephedrine, $C_6N_5CHOH{\cdot}CHCH_3{\cdot}NH_2$
5. Phenethylamine, $C_6H_5CH_2CH_2NH_2$ (eluted last)

The effects of steric hindrance and of cooperative coordination by the hydroxyl group are clearly seen.

By current standards the plate numbers of these columns were low, but the separation factors were high. High separation factors are characteristic of LEC. They were observed by Funasaka[19-21] for isomeric aminobenzoic acids and aromatic diamines.

Recent work with metal-loaded exchangers bonded to porous silica has given higher plate numbers, better resolution, and faster separations. Vogt,[22] using silica hydrothermally loaded with cadmium, with 2% CH_3CN in hexane as the eluent, eluted 2-, 3-, and 4-picoline in that order, α-naphthylamine before the β-isomer, and separated methyl- and ethylanilines. Chow and Grushka,[23,24] using the copper-loaded bonded phases we have described, separated aminonaphthalenes, chloroanilines, nitroanilines, toluidines, anisidines, and some nonnitrogenous ligands, such as fluorene and fluorenone, with the same solvent system. High-performance copper(II)-silica LEC has also been applied for the determination of *N*-methylpyridinium 2-aldoxime in biological fluids.[25]

Thin-layer chromatography by ligand exchange has been very successful in separating aromatic amines (Table 4). The stationary phase is silica or alumina impregnated with metal salts.[8,26-31] Isomeric toluidines, chloro- and nitroanilines, naphthylamines, anisidines, xylidines, and aminophenols were separated with very good resolution. Generally, the mobile phase was hexane or benzene modified by a small proportion of a polar solvent-like methanol or acetic acid. Once more, steric hindrance around the amino nitrogen (ortho-substitution, $-NHCH_3$) loosens the attachment to the metal ions and causes the substance to migrate faster.

Aromatic amines have also been separated in the gas phase. Liquid films of metal stearates supported on Celite® as stationary phases in gas chromatography (GC) was tested to separate aromatic amines;[15] ammonia and hydrogen mixture was tested as a carrier gas, and the amines were retained well compared to Apiezon oil (Table 5). Chromosorb® coated with manganese(II) stearate was found to be more suitable for the GC separation of aniline bases than zirconium phosphate gels in the manganese(II) form.[32]

IV. ALIPHATIC DIAMINES AND POLYAMINES

We recall that 1,2-diamines are held so strongly by copper-loaded resins that they can hardly be displaced by ammonia. Other diamines are less strongly held and can be separated by LEC. As was noted above, Helfferich[10-12] was the first to use LEC for the separation and concentration of 1,3-diamino-2-propanol from dilute solutions containing ammonia on a nickel-loaded and ammonia-conditioned column of carboxylic cation exchanger (Amberlite® IRC-50). The amine displaced ammonia and concentrated it on the resin; later, the diamine was eluted from the column with concentrated ammonia solution.

Ethylene diamine is also held very strongly on nickel-loaded resins and can be filtered out from dilute solutions by short columns, with chelating as well as sulfonic resins.[16]

Walton and Latterell[7,33,34] found that other diamines are bound less strongly with elution orders of 1,6-hexanediamine, 1,4-butanediamine, and 1,3-propanediamine. Shimomura[35] found that 1,4-butanediamine eluted before 1,6-hexanediamine on a nickel-loaded sulfonic resin using 4 *M* ammonia, and also on a copper-loaded chelating resin. She found that propanediamine bound with two coordination sites on copper, whereas hexanediamine bound only at one site.

Certain diamines and polyamines occur widely in plant and animal tissue and are of medical interest. They are

- Putrescine, or 1,4-butanediamine
- Cadaverine, or 1,5-pentanediamine
- Spermine, $H_2N(CH_2)_3NH(CH_2)_4NH(CH_2)_3NH_2$
- Spermidine, $H_2N(CH_2)_3NH(CH_2)_4NH_2$

Navratil and Walton[36] made an extensive study of these compounds, as well as other diamines and amino acids, on three different exchangers with three different metal ions (copper, zinc, and nickel). Several elution orders were observed (see Table 6).

The concentrations of ammonia and metal ions affected the retention differently for different compounds, and thus it was possible to move the elution peaks back and forth and optimize the separations by varying these parameters. For instance, on a copper-loaded exchanger, the retention volume of 1,3-propanediamine depended inversely as the square of the ammonia concentration, suggesting that the compound was a bidentate ligand; two molecules of ammonia would displace one molecule of diamine. For 1,5-pentanediamine, the retention varied inversely as the first power of the ammonia concentration, suggesting that this compound acted as a monodentate ligand. Spermine and spermidine acted as bidentate ligands.

Copper-loaded carboxylic resin, Bio-Rex® 70, gave the best separation factors, but the softness of this resin is a disadvantage. Zinc-loaded sulfonated polystyrene resin (Aminex

Table 5
RELATIVE RETENTION TIMES FOR AMINES ON APIEZON L AND ON MANGANESE STEARATE COLUMNS AT 156°C

	Apiezon	Manganese stearate		Apiezon	Manganese stearate		Apiezon	Manganese stearate
Aniline	855	2094	α-Picoline	416	1230	2,6-Lutidine	571	661
NN-Dimethylaniline	1835	1989	β-Picoline	535	7460	Pyrrole	202	301
Pyridine	296	3016	γ-Picoline	540	9670			

From Barber, D. W., Phillips, C. S. G., Tusa, G. F., and Verdin, A., *J. Chem Soc.*, p. 18, 1959. With permission.

Table 6
ELUTION ORDERS OF DIAMINES AND POLYAMINES

	Exchanger type						
	Sulfonated polystyrene		Polyacrylic (carboxylic)		Cellulose (carboxylic)		
Metal ion:	**Cu**	**Zn**	**Cu**	**Zn**	**Cu**	**Zn**	**Ni**
Diamines							
C_2	5	6	7	—	—	—	—
C_3	4	1	6	2	6	1	5
C_4	—	2	3	4	1	2	1
C_5	—	5	4	5	2	5	3
C_6	3	7	5	6	4	6	4
Polyamines							
Spermidine	1	3	1	1	3	3	2
Spermine	2	4	2	3	5	4	3
Ammonia concentration *M*	7		4		1.1	1.7	1.1

From Walton, H. F. and Navratil, J. D., in *Recent Developments in Separation Science*, Vol. 6, Li, N. N., Ed., CRC Press, Boca Raton, Fla., 1981, chap. 5.

A-7, particle size 7 to 11 μm, cross-linking 8%) gave the best all-around performance; the eluent was 5.5 *M* ammonia, 0.002 *M* in Zn^{2+}, and the temperature was 55°C. Changing the zinc concentration changed the relative position of the peaks, and close control was necessary to get the best separation of the peaks of interest, i.e., those of the four compounds mentioned above. Analysis took about 90 min.

Refractive index was used for detection. This was convenient, but the sensitivity was quite poor. About 10 to 15 μg of spermine or spermidine could be detected. Better detection methods are needed if LEC is to compete with other chromatographic methods, such as orthodox amino acid analysis by cation exchange with nitrate eluents.[37] When such detection methods are forthcoming, however, LEC will have the advantage of simplicity. All amino acids, even the most basic, are eluted near the void volume, well before the diamines and polyamines, and only one single eluent is needed.

Various basic and acidic components of petroleum products were separated on copper-, nickel-, and iron(III)-loaded macroporous (porous cross-linked polystyrene sulfonic acid) resin (Amberlyst®-15).[38] Long-chain aliphatic diamines were absorbed (along with napthenic acids, salicylic acid, benzotriazole, and other compounds), and could be displaced with methanol.

V. AZIRIDINES, ALKANOLAMINES, AND HYDRAZINES

Aziridines, derived from ethyleneimine, CH_2–CH_2 (ring) NH, are good subjects for LEC because they are very reactive and hydrolyzed in neutral- and acid-aqueous solutions, but stable in alkaline solutions. Chromatography was performed on a nickel-sulfonated polystyrene resin with 1 *M* ammonia eluent.[39] The separation of three representative compounds,

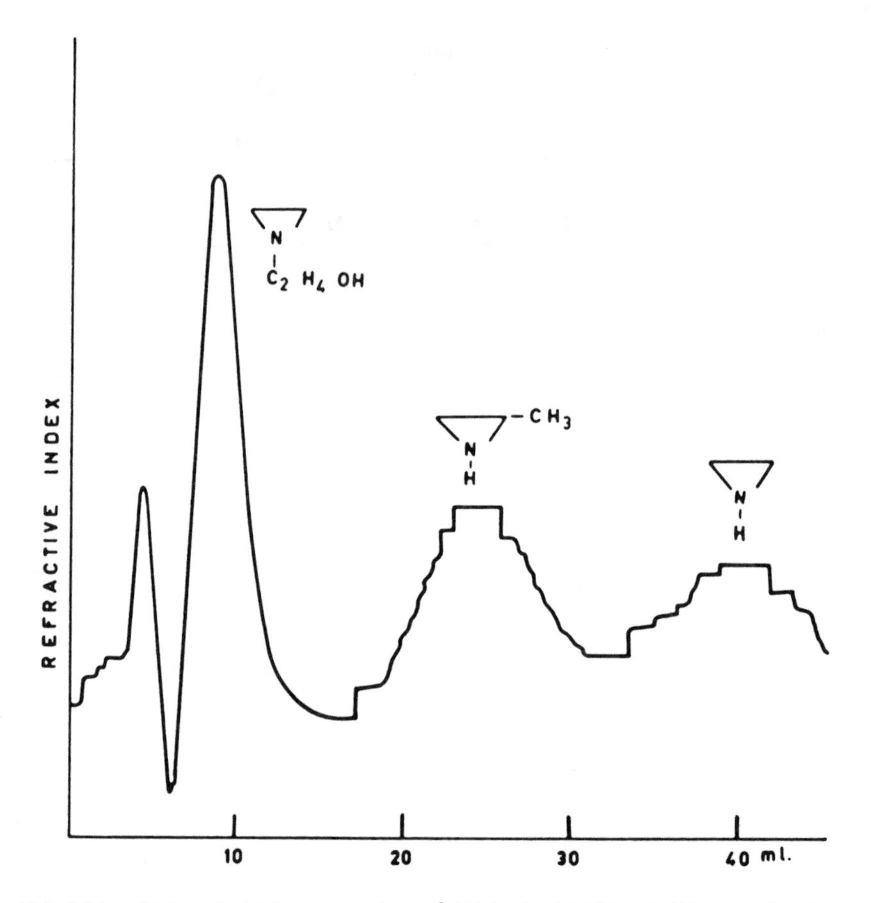

FIGURE 2. Elution of aziridines. Resin, Dowex®-50W × 8, 200-400 mesh, Ni form; bulk column volume, 30 mℓ; eluent, 1.0 *M* ammonia; flow rate 11 mℓ/hr. Loading, 5 mg *N*-(2-hydroxy-ethyl) aziridine, 10 mg each propylenemine and ethyleneimine. (Courtesy of Shimomura, K., Ligand Exchange Chromatography, Ph.D. thesis, University of Colorado, Boulder, 1968. See also Reference 39.)

N-(2-hydroxyethyl)aziridine, propyleneimine, and ethyleneimine, is shown in Figure 2 and elution data are shown in Table 7. As expected, the parent compound, ethyleneimine, eluted last. Substitution on the nitrogen or on one of the carbon atoms reduced the retention.

Distribution ratios for mono- and diethanolamine, which are hydrolysis products of ethyleneimine and *N*-(2-hydroxyethyl)aziridine, respectively, are also shown in Table 7. It was interesting to note that a copper-loaded chelating resin can separate aziridines from alkanolamines, while a nickel-loaded resin cannot. Figure 3 illustrates the important separation of aziridines from their hydrolysis products.

Alkanolamines are polar and hydrophilic with low volatility, and hard to analyze by G.C. By LEC they are easily separated.[35] The increasing base strengths are the same as their elution orders; triethanolamine emerges first, diethanolamine second, then monoethanolamine (Figure 4). It is best to raise the ammonia concentration after the diethanolamine peak, for monoethanolamine is much more strongly bound.

The three ethanolamines have been separated by conventional ion exchange using 1.5 *M* hydrochloric acid[40] and borate buffer of pH 9.2,[41] and the elution order is again the order of increasing base strengths. However, LEC gives the best separation of mono- and diethanolamine. Monoethanolamine has been separated from triethanolamine on a zirconium phosphate column using hydrochloric acid elution, with monoethanolamine being eluted first, but the separation was poor.[42]

Table 7
COLUMN DISTRIBUTION RATIOS, AZIRIDINES, AND ETHANOLAMINES[a]

	Sulfonic resin		Chelating resin			
	Ni		Ni		Cu	
Amine	NH_3 conc	k	NH_3 conc	k	NH_3 conc	k
Ethylenimine	1.0	11	0.25	14	0.1	14
Propyleneimine	1.0	6	0.25	4.5	0.1	12
N-Ethylaziridine	1.0	0.67	0.05	2.3	0.1	1.8
N-(2-Hydroxyethyl) aziridine	1.0	1.4	0.05	5.5	0.1	1.8
Ethanolamine	1.0	10	0.25	10.7	0.1	17
Diethanolamine	—	—	0.05	7.2	—	—

[a] Data of Shimomura.[35] See also Reference 39.

Hydrazine and substituted hydrazines are separated similarly to alkanolamines, except that copper ions catalyze the decomposition of these compounds and cannot be used. On nickel-loaded resins, hydrazine can be absorbed, left overnight, and eluted the next day with only 10% loss by decomposition. The unsymmetrical compounds dimethylhydrazine and monomethylhydrazine can be separated on nickel-loaded sulfonated polystyrene and were eluted by 0.4 *M* ammonia in that order, then hydrazine was removed with 5*M* NH_3 (Figure 5).[13]

Again, these compounds illustrate the fact that methyl groups substituted on nitrogen impede metal sorption on the resin as well as destabilize the metal complexes in aqueous solutions.[13] The hydrazines seem to be bound as singly coordinated ligands in the resin.

In 1960, Tsuji[43,44] studied the adsorption of isoniazide or isonicotinic hydrazide using sulfonated polystyrene resin saturated with various metal ions; the strength of isoniazide binding decreased in the order Cu > Ni > Hg > H > Co > Cd > Zn > Fe(II) > Pb > Mr > Al; no adsorption was found on Na-, Ca-, Mg-, or Ba-loaded resins.

VI. AMINO SUGARS

The sensitivity of LEC to molecular geometry is nowhere better seen than in the amino sugars. The isomeric hexosamines, glucosamine, galactosamine, and mannosamine are eluted in that order from the copper(II)-loaded carboxylic resin Bio-Rex® 70 by 1 *M* aqueous ammonia at room temperature, with retention volumes in the ratio 1:1.4:2.5. Copper-loaded sulfonated polystyrene resin gave even higher separation factors, 1:1.7:8.0, but here the retention of mannosamine is too great for convenient isocratic operation.[45,46]

The amino sugars are carried as copper complexes in the mobile phase and are detected by UV absorbance. Figure 6 shows a chromatogram. The mannosamine peak width corresponds to a plate height of 0.3 mm, which is very satisfactory. Peak heights are proportional to the quantities injected to within 2 to 3%.

Amino acids also travel as light-absorbing copper complexes and appear in the chromatogram. Most of them emerge close to the void volume and do not interfere with the amino sugars. Only the basic amino acids elute near the amino sugars, and their peaks can be kept from overlapping by the fact that the retention volumes of the amino acids depend (roughly) on the inverse square of the ammonia concentration while those of the amino sugars depend

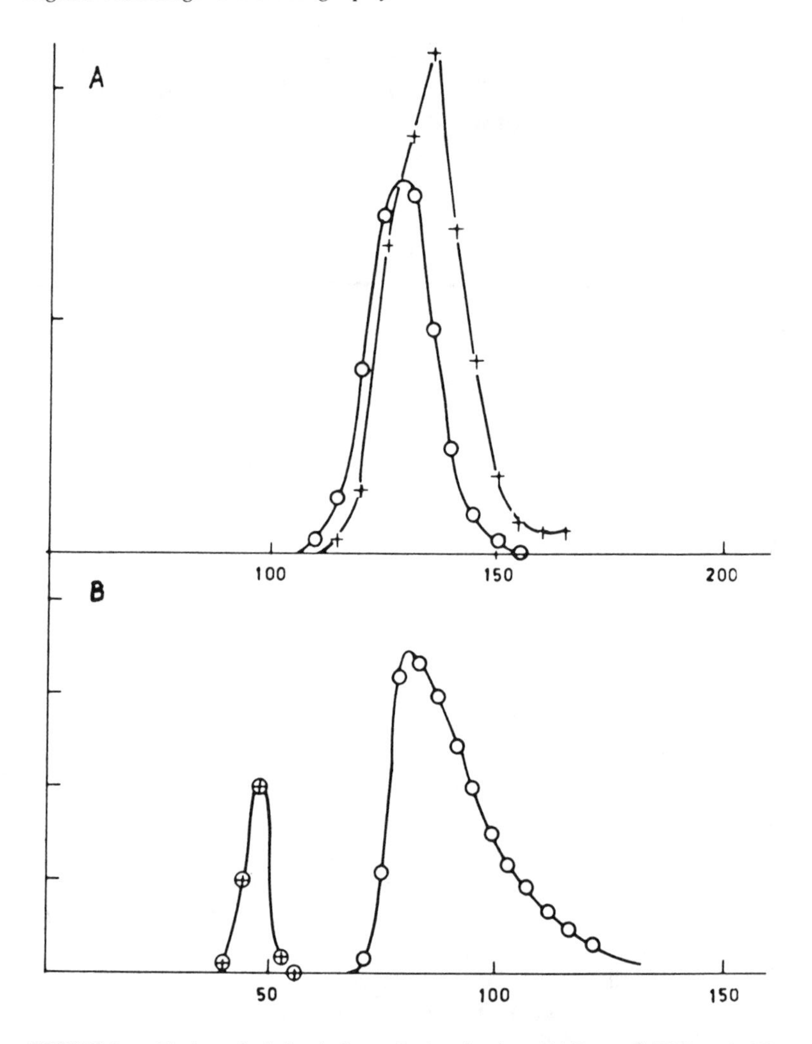

FIGURE 3. Elution of ethyleneimine and ethanolamine. (A) Dowex®-50W × 8, Ni form, 200-400 mesh; (B) Chelex®-100, Cu form, 200-400 mesh. In each case bulk column volume was 33 mℓ, eluent 1.0 *M* aqueous ammonia. Crosses, ethyleneimine; circles, ethanolamine. Ethanolamine determined by carbon-14; imine by color reaction with 1,2-naphthoquinone-4-sulfonic acid. (Courtesy of Shimomura, K., Ligand Exchange Chromatography, Ph.D. thesis, University of Colorado, Boulder, 1968. See also Reference 39.)

inversely on the first power. One can avoid interference by simply choosing the proper ammonia concentration.

Masters and Leyden[47] separated amino acids and amino sugars by LEC on copper-loaded silylated glass. They found the same amino sugar elution order as did Navratil et al. — glucosamine first, mannosamine last — but nearly all the amino acids were eluted after the amino sugars, not before. They used 0.1 *M* ammonia, whereas Navratil[45] used 1 *M* ammonia. Doury-Berthod et al.[48,49] showed that high ammonia concentration coupled with low metal concentration caused amino acids to be held weakly (see Figure 7 and the comments in References 48 and 49).

Amino sugars and their acetyl derivatives have been separated by thin layer chromatography. On copper-impregnated silica gel, with ammonia in 2-propanol as the mobile phase, glucosamine traveled faster than galactosamine, and the acetyl derivatives moved still faster.[30] Silica gel plates impregnated with cadmium, zinc, and manganese were used successfully for the separation of glucosamine, mannosamine, galactosamine, and *N*-acetylglucosamine.[50]

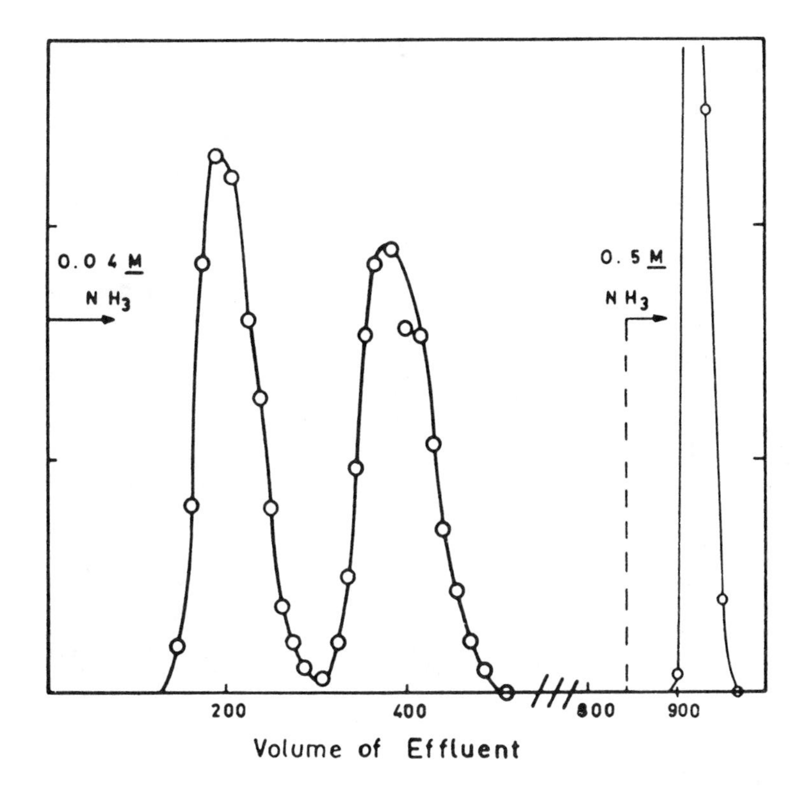

FIGURE 4. Elution of ethanolamines. Resin, Dowex®-50W × 8, 200-400 mesh, Ni form; column 30 cm × 0.43 cm^2. Elution order, triethanolamine, diethanolamine, monoethanolamine. Loading, 1.5 mg tri- , 2.5 mg di- , 5 mg monoethanolamine. Carbon-14 compounds used; ordinates are counting rates. (Courtesy of Shimomura, K., Ligand Exchange Chromatography, Ph.D. thesis, University of Colorado, Boulder, 1968.)

VII. ALKALOIDS

Alkaloids were separated on carboxylic resins carrying copper(II).[51] Again, the eluent was aqueous-alcoholic ammonia. The commonly available polyacrylic resin, Bio-Rex® 70, was not effective, because it held alkaloids too weakly. Several resins were tested, and a correlation was found between the ammonia:copper ratios in the resins and their ability to absorb alkaloids; this point was discussed above. Morphine, codeine, strychnine, papaverine, and narcotine could be separated, but the bands were broad, giving theoreticalplate heights of several millimeters. The difficulty seems to be that most alkaloids are secondary or tertiary amines, and in addition, the molecules are large and the basic nitrogen atoms are surrounded by other atoms. Steric hindrance to coordination is great, and as we have noted, ligand binding in metal-loaded exchangers is very much affected by steric hindrance.

VIII. MISCELLANEOUS NITROGEN-CONTAINING LIGANDS

Ribonucleic acid and DNA are fractionated on an aluminum-loaded gel-type sulfonated polystyrene resin.[52-55] A macroporous resin would probably have given better separations. Nucleosides and nucleic acid bases were absorbed on a copper-loaded chelating resin (Chelex® 100) and successively eluted with ammonia.[56-58]

Nucleotides were not absorbed and the bases were the most strongly absorbed. Thymine, cytosine, guanine, and adenine were eluted from a nickel-loaded chelating resin (Chelex®

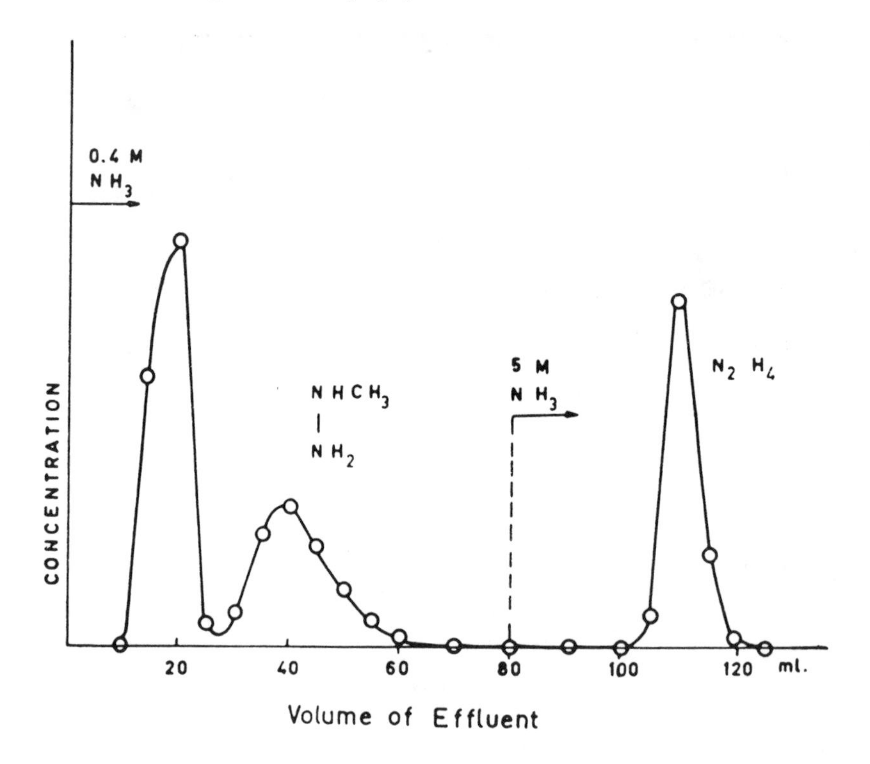

FIGURE 5. Elution of hydrazines. Resin, Dowex®-50W × 12, 200-400 mesh, Ni form; bulk column volume, 30 mℓ. Concentrations in arbitrary units (found by bromate titration). (From Shimonura, K., Dickson, L., and Walton, H. F., *Anal. Chim. Acta,* 37, 102, 1967. With permission.)

100) with 0.5 *M* ammonia in that order; if nickel ions were substituted for copper, the order of guanine and adenine was reversed.[13,35] Band widths are large with guanine and adenine, and LEC offers no advantage over conventional cation-exchange chromatography, except for the possibility of changing the elution order. However, Goldstein[56,57] recommended using a copper-loaded resin for the rapid separation of nucleotides, nucleosides, and nucleic acid bases. Nucleotides were not sorbed, the more weakly basic nucleosides were eluted with water, and the more strongly basic nucleosides eluted with 1 *M* ammonia. Uracil, guanine, adenine, and cytosine were finally eluted in that order with 2.5 *M* ammonia. Many purine derivatives and xanthines, including caffeine and theobromine, were cleanly separated on copper-loaded chelating resin by elution with 1 *M* ammonia; caffeine was the most strongly retained and could be easily measured in coffee and other beverages.[58] Analysis took about 60 min.

Faster and more efficient separation of nucleosides and nucleotides was reported by several investigators using ''outer-sphere complexation''.[59-61] Grushka and Chow[59] using a cobalt(II)-tris-ethylenediamine complex bonded to silica gel and a phosphate buffer of pH 6.4 as eluent, found that the nucleosides were not absorbed while the nucleotides were absorbed very strongly. To elute the nucleotides in a reasonable time, magnesium ions were added to the mobile phase; they formed complexes in solution and greatly hastened the elution. Adenosine mono-, di-, and tri-phosphate were separated, the triphosphate emerging last. This is the reverse of the order on the usual reverse-phase C_8 packings.

Concluding this ''miscellaneous'' section, we note the separation of 1-nitroso-2-naphthol and 2-nitroso-1-naphthol from each other and from aminophenols on an iron(III)-loaded sulfonated polystyrene resin. Aqueous-alcoholic ammonia was the eluent. Nitroso groups coordinated to the metal through the nitrogen atoms forming low-spin complexes, and the sorption-desorption process was accordingly slow.

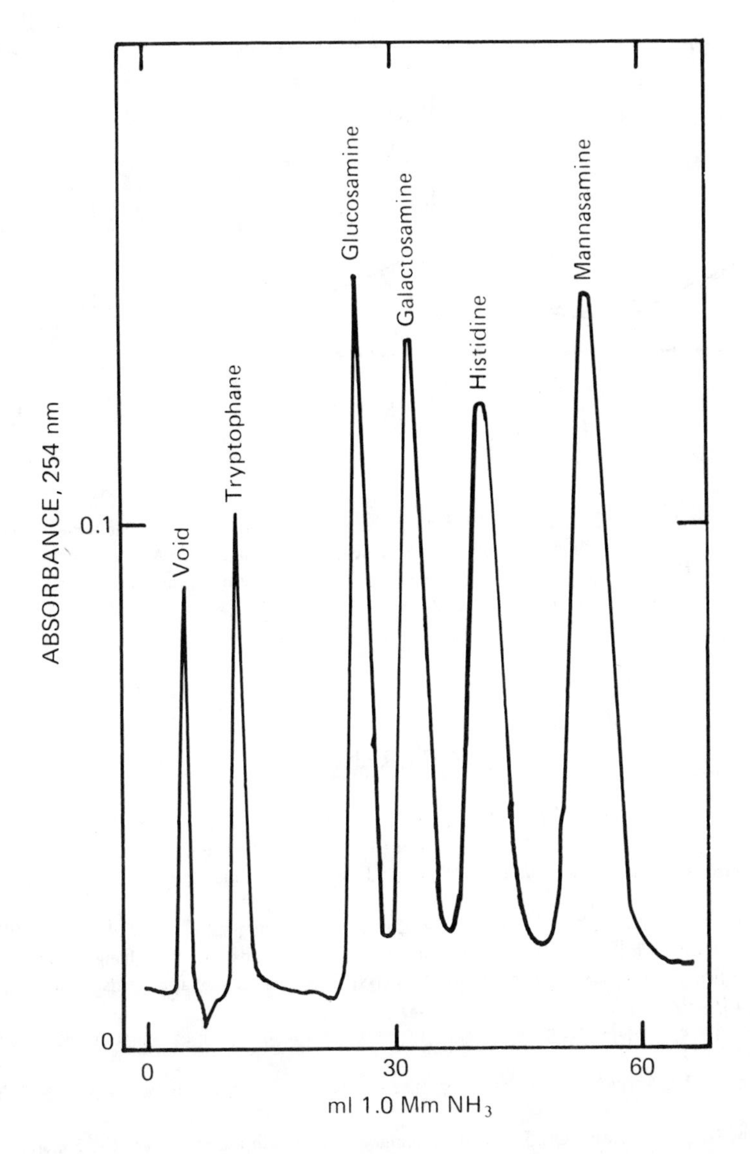

FIGURE 6. Chromatography of amino sugars and amino acids on carboxylic resin loaded with Cu(II). (From Navratil, J. D., Murgia, E., and Walton, H. F., *Anal. Chem.*, 47, 122, 1975. Copyright 1975 American Chemical Society. With permission.)

Finally, the use of a silver-loaded bonded exchanger, Zipax SCX, to retain and separate nitrogen-containing heterocyclic aromatic compounds should be noted.[61,63] The eluent was 1% CH_3CN in hexane. The mechanism undoubtedly involved silver-to-nitrogen coordination, as well as the π-electron interaction implied in "argentation chromatography" (see Chapter 7).

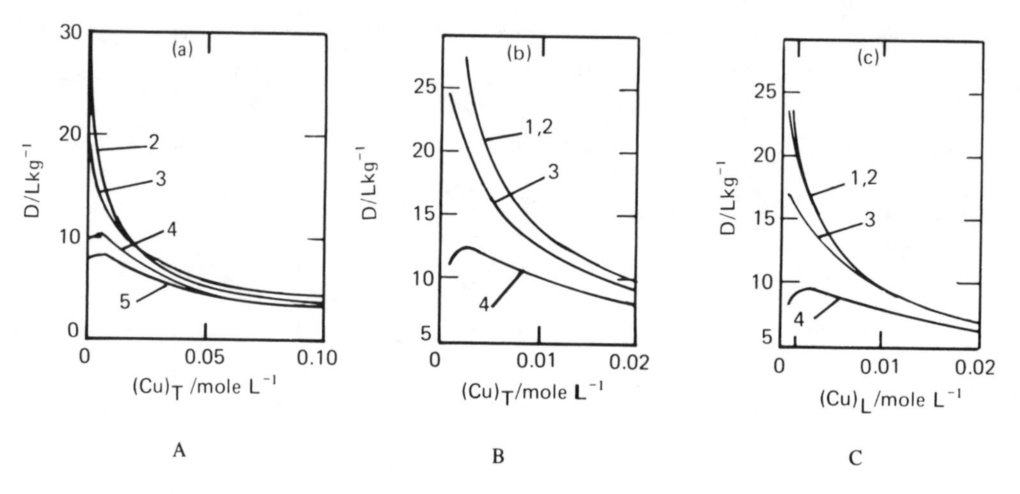

FIGURE 7. Distribution of amino acids between ammonia solutions and copper-loaded ion-exchange resins. (A) glycine on carboxylic resin; (B) and (C) alanine and leucine on phosphonic resin. Ammonia concentrations: 1, 0.2 *M*; 2, 0.3 *M*; 3, 0.5 *M*; 4, 1 *M*; 5, 1.5 *M*. Ordinates are distribution ratios in liters per kilogram. (From Doury-Berthod, M., Poitrenaud, C., and Trémillon, B., *J. Chromatogr.*, 131, 73, 1972. With permission.)

REFERENCES

1. **Walton, H. F. and Navratil, J. D.,** Ligand exchange chromatography, Li, N. N., Ed., *Recent Developments in Separation Science,* Vol. 6, CRC Press, Boca Raton, Fla., 1982, 65.
2. **Davankov, V. A. and Semechkin, A. V.,** Ligand exchange chromatography, *J. Chromatogr.*, 141, 313, 1977.
3. **Navratil, J. D. and Walton, H. F.,** Ligand exchange chromatography, *Am. Lab.*, 8, 69, 1976.
4. **Walton, H. F.,** Chromatography of non-ionic organic compounds on ion-exchange resins, in *Separation and Purification Methods 4,* Perry, E. S., Van Oss, C. J., and Grushka, E., Eds., Marcel Dekker, New York, 1975, 189.
5. **Walton, H. F.,** Liquid chromatography of organic compounds on ion exchange resins, *J. Chromatogr.*, 102, 57, 1974.
6. **Walton, H. F.,** *Advances in Ion Exchange and Solvent Extraction,* 4, Marcel Dekker, New York, 1973, chap. 2.
7. **Walton, H. F. and Latterell, J. J.,** Metal amine complexes in ion exchange. IV. Separation of amines by ligand exchange, *Analytical Chemistry 1962*, Elsevier, New York, 1963, 356.
8. **Tabak, S., Mauro, A. E., and Del'Acqua, A.,** Argentation thin-layer chromatography with silver oxide. II. Amines, unsaturated and aromatic carboxylic acids, *J. Chromatogr.*, 52, 500, 1970.
9. **Muzzarelli, R. A. A., Martelli, A. F., and Tubertini, O.,** Ligand exchange chromatography on thin layers and columns of natural and substituted celluloses, *Analyst,* 94, 616, 1969.
10. **Helfferich, F. G.,** Ligand exchange — a novel separation technique, *Nature (London),* 189, 1001, 1961.
11. **Helfferich, F. G.,** Ligand exchange. I. Equilibria. II. Separation of ligands having different coordinative valences, *J. Am. Chem. Soc.*, 84, 3237, 3242, 1962.
12. **Helfferich, F. G.,** Ligand exchange. I. Equilibria. II. Separation of ligands having different coordinative variances, *J. Am. Chem. Soc.*, 84, 3237, 3242, 1962.
13. **Shimonura, K., Dickson, L., and Walton, H. F.,** Separation of amines by ligand exchange. IV. Ligand exchange with chelating resins and cellulosic exchangers, *Anal. Chim. Acta,* 37, 102, 1967.
14. **Hill, A. G., Sedgley, R., and Walton, H. F.,** Separation of amines by ligand exchange. IV. A comparison of different cation exchangers, *Anal. Chim. Acta,* 33, 84, 1965.
15. **Barber, D. W., Phillips, C. S. G., Tusa, G. F., and Verdin, A.,** The chromatography of gases and vapors. VI. Use of the stearates of bivalent manganese, cobalt, nickel, copper, and zinc as column liquids in gas chromatography, *J. Chem. Soc.*, 18, 1959.
16. **Bak, C. M.,** Ligand exchange studies with an iminodiacetic acid ion exchange resin, *Daehan Hwahak Hwoejee,* 11, 56, 1967.

17. **Inczédy, J., Klatsmanyi-Gabor, P., and Erdey, L.,** The use of complex forming agents in ion exchange chromatography. VI, *Acta Chim. Acad. Sci. Hung.*, 69, 137, 265, 1971.
18. **deHernandez, C. M. and Walton, H. F.,** Liquid exchange chromatography of amphetamine drugs, *Anal. Chem.*, 44, 890, 1972.
19. **Funaska, W., Fujimura, K., and Kuriyama, S.,** Ligand-exchange chromatography. I. Separation of phenylenediamine isomers by ligand-exchange chromatography, *Bunseki Kagaku*, 18, 19, 1969.
20. **Funasaka, W., Fujimura, K., and Kuriyama, S.,** Ligand-exchange chromatography. II. Separation of aminobenzoic acid isomers by ligand-exchange chromatography, *Bunseki Kagaku*, 19, 104, 1970.
21. **Funasaka, W., Hanai, T., Fujimura, K., and Ando, T.,** Nonaqueous solvent chromatography. III. Complex chromatography between the metal ion of a cation-exchange resin and organic compounds in organic solvents, *J. Chromatogr.*, 78, 424, 1973.
22. **Vogt, C. R., Ryan, T. R., and Baxter, J. S.,** High-speed liquid chromatography on cadmium-modified silica gel, *J. Chromatogr.*, 136, 221, 1977.
23. **Chow, F. K. and Grushka, E.,** High performance liquid chromatography with metal-solute complexes, *Anal. Chem.*, 50, 1346, 1978.
24. **Chow, F. K. and Grushka, E.,** Separation of aromatic amine isomers by high pressure liquid chromatography with a copper (II)-bonded phase, *Anal. Chem.*, 49, 1756, 1977.
25. **Guyon, F., Tambute, A., Caude, M., and Rosset, R.,** Determination of *N*-methylpyridinium 2-aldoxime methylsulfate (contrathion) in rat plasma and urine by high-performance copper(II)-silica ligand exchange chromatography, *J. Chromatogr.*, 229, 475, 1982.
26. **Shimomura, K. and Walton, H. F.,** Thin-layer chromatography of amines by ligand exchange, *Sep. Sci.*, 3, 493, 1968.
27. **Yasuda, K.,** Thin-layer chromatography of chlorinated anilines on zinc salt-impregnated silica gel thin layers, *J. Chromatogr.*, 74, 142, 1972.
28. **Yasuda, K.,** Thin-Layer chromatography of aromatic amines on cadmium acetate-impregnated silica gel thin layers, *J. Chromatogr.*, 72, 413, 1972.
29. **Yasuda, K.,** Thin-layer chromatography of aromatic amines in cadmium sulphate-impregnated silica gel thin layers, *J. Chromatogr.*, 60, 144, 1971.
30. **Martz, M. D. and Krivis, A. F.,** Thin layer chromatography of hexosamines on copper impregnated sheets, *Anal. Chem.*, 43, 790, 1971.
31. **Kunzru, D. and Frei, R. W.,** Separation of aromatic amino isomers by high-pressure liquid chromatography on cadmium impregnated silica gel columns, *J. Chromatogr. Sci.*, 12, 191, 1974.
32. **Fujimura, K., Kitanaka, M., and Ando, T.,** Ligand-exchange gas chromatographic separation of aniline bases, *J. Chromatogr.*, 241, 295, 1982.
33. **Latterell, J. J. and Walton, H. F.,** Separation of amines by ligand exchange. II, *Anal. Chim. Acta*, 32, 101, 1965.
34. **Latterell, J. J.,** Separation of Amines by Ligand Exchange, Ph.D. thesis, University of Colorado, Boulder, 1964.
35. **Shimomura, K.,** Ligand Exchange Chromatography, Ph.D. thesis, University of Colorado, Boulder, 1968.
36. **Navratil, J. D. and Walton, H. F.,** Ligand exchange chromatography of diamines and polyamines, *Anal. Chem.*, 47, 2443, 1975.
37. **Gehrke, C. W., Kuo, K. C., Zumwalt, R. W., and Waalkes, T. P.,** Determination of polyamines in human urine by an automated ion-exchange method, *J. Chromatogr.*, 89, 231, 1974.
38. **Webster, P. V., Wilson, J. N., and Franks, M. C.,** Macroreticular ion-exchange resins: some analytical applications of petroleum products, *Anal. Chem. Acta*, 38, 193, 1967.
39. **Simomura, K., Hsu, T. J., and Walton, H. F.,** Ligand-exchange chromatography of aziridines and ethanolamines, *Anal. Chem.*, 45, 501, 1973.
40. **Pilgeram, L. O., Gal, L. M., Sassenrath, E. N., and Greenberg, O. N.,** Metabolic studies with ethanolamine-1,2-C^{14}, *J. Biol. Chem.*, 204, 367, 1953.
41. **Yoshino, Y., Kinoshita, H., and Sugiyama, H.,** Separation and determination of diethanolamine and monoethanolamine in triethanolamine by cation exchange chromatography, *Nippon Kagaky Zasshi*, 86, 405, 1965.
42. **Rebertus, R. L.,** Ion-exchange behavior of some substituted ammonium ions on zirconium phosphate, *Anal. Chem.*, 38, 1089, 1966.
43. **Tsuji, A. and Sekiguchi, K.,** The adsorption of isoniotinic acid hydrazide on cation-exchangers of various metal forms, *Nippon Kagaku Zasshi*, 81, 847, 1961.
44. **Tsuji, A. and Sekiguchi, K.,** Microdetermination of primary aromatic amines with ion-exchange resins, *Nippon Kagaku Zasshi*, 81, 847, 1961.
45. **Navratil, J. D.,** Ligand Exchange Chromatography of Non-Ionized Organic Compounds, Ph.D. thesis, University of Colorado, Boulder, 1975.

46. **Navratil, J. D., Murgia, E., and Walton, H. F.,** Ligand-exchange chromatography of diamines and polyamines, *Anal. Chem.*, 47, 2443, 1975.
47. **Masters, R. G. and Leyden, D. E.,** Ligand-exchange chromatography of amino sugars and amino acids on copper-loaded silylated control-led pore glass, *Anal. Chem. Acta,* 98, 9, 1978.
48. **Doury-Berthod, M., Poitrenaud, C., and Trémillon, B.,** Ligand exchange separations of amino acids. I. Distribution, equilibria of some amino acids between ammoniacal and copper(II) nitrate solutions and phosphonic, carboxylic, and iminodiacetic ion exchangers in the copper(II) form, *J. Chromatogr.*, 131, 73, 1977.
49. **Doury-Berthod, M., Poitrenaud, C., and Trémillion, B.,** Ligand-exchange separation of amino acids. II. Influence of the eluent composition and of the nature of the ion exchange, *J. Chromatogr.*, 179, 37, 1979.
50. **Reena,** Chromatographic behavior of amino sugars on metal salt impregnated thin layers, *Anal. Lett.*, 18, 753, 1985.
51. **Murgia, E. and Walton, H. F.,** Ligand-exchange chromatography of alkaloids, *J. Chromatogr.*, 104, 417, 1975.
52. **Kothari, R. M.,** Some aspects of fractionation of DNA on an IR-120 Al^{3+} column. VII. Effect of the tissue and source variation on the chromatographic profiles of DNA, *J. Chromatogr.*, 64, 85, 1972.
53. **Shankar, V. and Joshi, P. N.,** Fractionation of RNA on a metal ion equilibrated cation exchanger. I. Chromatographic profiles of RNA on an Amberlite® IR-120 (Al^{2+}) column, *J. Chromatogr.*, 90, 99, 1974.
54. **Shankar, V. and Joshi, P. N.,** Fractionation of RNA on a metal ion equilibrated cation exchanger. II. Chromatographic behaviour of RNA subjected to different treatments, on Amberlite® IR-120 Al^{3+} columns, *J. Chromatogr.*, 95, 65, 1974.
55. **Skorokhod, O. R. and Klizovich, L. I.,** Molecular and ligand sorption of pyridine on macroporous sulfonic cation-exchanger, *Kollid Zhur.*, 33, 268, 1971.
56. **Goldstein, G.,** Ligand-exchange chromatography of nucleotides, nucleosides, and nucleic acid bases, *Anal. Biochem.*, 20, 477, 1967.
57. **Burtis, C. A. and Goldstein, G.,** Terminal nucleoside assay of ribonucleic acid by ligand-exchange chromatography, *Anal. Biochem.*, 23, 502, 1968.
58. **Wolford, J. C., Dean, J. A., and Goldstein, G.,** Separation of oxypurines by ligand-exchange chromatography and determination of caffeine in beverages and pharmaceuticals, *J. Chromatogr.*, 62, 148, 1971.
59. **Chow, F. K. and Grushka, E.,** High performance liquid chromatography of nucleotides and nucleosides using outerspace and innerspace metal-solute complexes, *J. Chromatogr.*, 185, 361, 1979.
60. **Corradini, D., Sinibaldi, M., and Messina, A.,** Outer-sphere ligand exchange chromatography of nucleotides and related compounds on a modified polysaccharide gel, *J. Chromatogr.*, 235, 273, 1982.
61. **Sinibaldi, M., Carunchio, V., Messina, A., and Corradini, C.,** Ligand exchange chromatography on bonded silica gel modified with amino-complexes of cobalt(III), *Ann. Chem.*, 74, 175, 1984.
62. **Vivilecchia, R., Thiebaud, M., and Frei, R. W.,** Separation of polynuclear aza-heterocyclics by high-pressure liquid chromatography using a silver-impregnated adsorbent, *J. Chromatogr. Sci.*, 10, 411, 1972.
63. **Frei, R. W., Beall, K., and Cassidy, R. M.,** Determination of aromatic nitrogen heterocycles if air samples be high speed liquid chromatography, *Mikrochim. Acta,* 859, 1974.

Chapter 4

SEPARATION OF AMINO ACIDS, PEPTIDES, AND PROTEINS

V. A. Davankov

I. AMINO ACIDS AND PEPTIDES ON POLYMERIC SORBENTS

It is only natural that amino acids as typical complex-forming compounds have been frequently used to examine the separation ability of various ligand-exchanging chromatographic systems. Almost all types of natural and synthetic polymeric sorbents, as well as silica-based packing materials in combination with different transition metal ions, have been tested in chromatography of amino acids. The most important achievement in this area is separation of optical isomers of amino acids, which is dealt with in a special chapter. Separation of different amino acids from each other is a much less difficult task. Though it can be solved using the same chromatographic systems, sorbent-fixed ligands or complexing additives to the eluent do not need to be chiral. Conventional ion exchangers and other packing materials capable of retaining metal ions and their complexes are generally sufficient for separation of amino acids.

As early as 1966, Siegel and Degens[1] for the first time demonstrated the great potential of ligand-exchanging systems in selective sorption of amino acids. They used polystyrene-type resin, Chelex® 100, containing residues of iminodiacetate in the form of copper(II) complexes for binding selectively a variety of free amino acids from sea water. Because of the high ionic strength of sea water, sorption mechanisms other than complexation would fail to give the required selectivity. Sorbed amino acids were then displaced by ammonia solutions and subjected to a standard amino-acid analysis.

First attempts to solve another important practical problem, that of separating amino acids from peptides, using complexation reactions, can also be traced back to the middle of the 1960s. Cross-linked dextran gels, Sephadex® G-25, were charged with copper(II) ions under alkaline conditions.[2,3] By passing mixtures of amino acids with peptides through this sorbent, peptides were observed to elute first. Under basic conditions, peptides form stable copper complexes, stripping copper from Sephadex®; these complexes are then only slightly retained.[4] Peptides tend to coordinate copper ions through the end-positioned amino group and neighboring amide functions, with the latter losing herewith their amide protons. Copper-peptide complexes thus obtain negative charges when in alkaline media. Dextran gels should also be negatively charged at high pH values.

A similar fractionation mechanism should also be valid for mixtures of amino acids and peptides eluted with ammonia solutions through a copper-Chelex® 100 column.[5-9] Acidic and neutral peptides and also acidic amino acids, which form negatively charged copper complexes, elute first; neutral amino acids and basic peptides, forming neutral complexes, require 1.5 *M* ammonia eluents, whereas basic amino acids emerge with 6 *M* ammonia. Similarly, at pH 8.5 to 9.5, nickel,[10] zinc, or cobalt(II) forms[11] of Chelex® 100 are suitable for a selective retention and isolation of basic amino acids (lysine, ornithine, arginine, histidine). When introduced into a hydrophilic matrix like Sephadex® G-25,[12] nickel iminodiacetate again displays the highest affinity to basic amino acids[13] and basic peptides.[14] Herewith, the retention of neutral amino acids and oligopeptides increases significantly on rising the pH and ionic strength of the solution.[13,14] β-Amino acids form weak copper complexes and are scarcely retained on copper-Chelex® 100.[15] They can be easily separated from peptides and α-amino acids, with the latter group showing strongest retention.

If anion-exchanging resins such as DEAE or TEAE cellulose are used instead of negatively

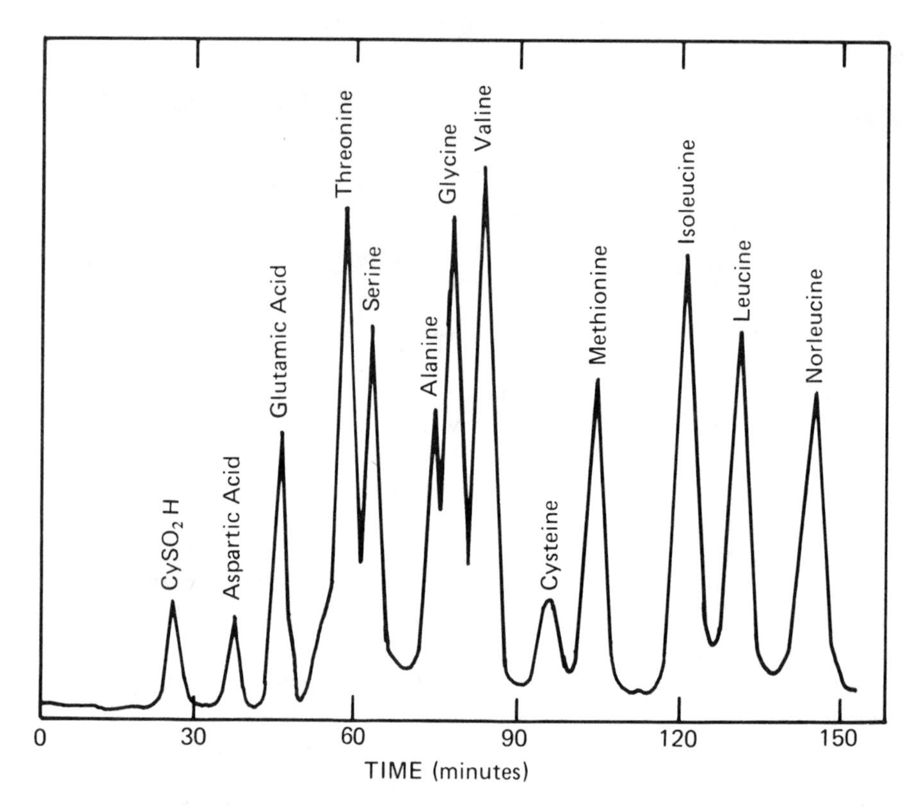

FIGURE 1. Chromatography of acidic and neutral amino acids on zinc(II)-loaded sulfonated polystyrene resin. Each peak represents 0.01 μmol of sample. Eluent, sodium acetate buffer, pH 4.10, $3.5 \cdot 10^{-4}$ *M* in zinc(II). (From Maeda, M., Tsuji, A., Ganno, S., and Onishi, Y., *J. Chromatogr.*, 77, 434, 1973. With permission.)

charged Chelex® 100 packings, copper complexes of acidic amino acids and peptides appear to be the stronger retained species.[16,17] This method of group separation of peptides from amino acids has since been applied to a number of biological fluids including urine,[18-20] serum,[21] cheese,[22] and wort.[23] By optimizing pH and ionic strength of alkaline sodium borate buffers, it has been possible to separate copper complexes of different peptides on a DEAE Sephadex® A-25 column.[24] Thus, peptides of a protein hydrolysate used in intravenous feeding (Aminosol) were examined, and oligoglycines, from glycine to pentaglycine, were completely separated in a step-wise elution process. Separation was ascribed to increasing net negative charge on the copper complexes of longer peptides.

Rather early efforts by Arikawa[25,26] resulted in the development of the Hitachi® Perkin-Elmer Model KLA-3B ligand exchange amino-acid analyzer. Using this instrument, a complete amino-acid analysis was performed in two steps. The acid and neutral amino acids (Figure 1) were eluted at 55°C from a sulfonated polystyrene cation exchanger with a buffer of pH 4.10 containing 4.10^{-4} *M* zinc acetate and $5.5.10^{-2}$ *M* sodium acetate. The elution of basic amino acids required 1.10^{-3} *M* zinc concentrations, 0.6 *M* sodium acetate, and pH 5.1. Improved buffer systems provided simultaneous analysis of all amino acids in a typical protein hydrolysate, as well as *S*-carboxymethylcysteine, *S*-(β-amino-ethyl) cysteine, glucosamine, galactosamine,[27] and γ-amino butyric acid.[28] Xylenol orange, pyrocatechol violet,[26] and pyridoxal[29] were proposed instead of ninhydrin as the color reagents for the more sensitive detection of amino acids in this method. However, the ligand exchange amino-acid analyzer did not receive further development and was unable to compete successfully with ion exchange amino-acid analyzers. One reason for the lack of acceptance was the difficulty of obtaining reproducible retention volumes.

It is worth mentioning here that in the presence of silver ions, sulfonated polystyrene resins acquire selective affinity toward sulfur-containing amino acids (methionine, cysteine) and methionine-containing peptides.[30]

The first chromatographic separations of amino acids on chelating resins were described by Hering and Heilmann[31] as early as 1966. The packing material was prepared by reacting cross-linked chloromethylated polystyrene with sarcosine (*N*-methylglycine). When saturated with copper or nickel ions, the sorbent could partially resolve simple amino-acid mixtures using pure water as the eluent. Though these first results seemed rather promising, no attempts were made to improve column efficiency. Similarly, several amino acids were partially separated on columns with copper-Chelex® 100[32] and copper-Chitosan,[33] as well as on thin layers of nickel-Chelex® 100,[34] but the results were far from being attractive for potential followers who were looking for new techniques to solve problems which other chromatographic methods fail to solve.

A thorough study into LEC of amino acids by Doury-Berthod et al., resulted in a theoretical description of distribution equilibria of amino acids, copper(II) ions, and ammonia molecules between the mobile and resin phases,[35] as well as in practical separation of complex mixtures of amino acids.[36] Three types of polymeric resins have been examined: (1) the acrylic acid polymer, Bio-Rex® 70, (2) phosphonated polystyrene, Bio-Rex® 63, and (3) Chelex® 100. In good agreement with the expressions derived theoretically, retention and separation selectivities of amino acids were found to rise as the concentration of copper ions and ammonia in the mobile phase decrease, unless, of course, the copper-ion concentration is so low that there is little copper in the resin. Graphs of distribution ratios of amino acids vs. copper-ion concentration in 1 *M* ammonia show maxima near [copper(II)] $= 2 \times 10^{-3}$ *M*. Amino acids were shown to form mixed-ligand copper complexes with ammonia molecules. Depending on the number of carboxylic and amino functions in the amino-acid molecule, the net charge of the mixed-ligand complexes varies from 0 to +2. Accordingly, acidic ligands are retained more weakly and basic ones more strongly than are neutral amino acids because all three resins investigated acquire a negative net charge in ammonia solutions. Within each group of amino acids, retention is governed by the stability of their copper ammonia complexes and by additional interactions with the resin matrix, with the latter being especially intensive in the case of hydrophobic amino acids and polystyrene-based resins. As shown in Figure 2, the column efficiency attained for the packing material of 50 to 65 μm in diameter (Bio-Rex® 70) was rather good. Polystyrene-based resins with phosphonic and iminodiacetic groups, Bio-Rex® 63 and Chelex® 100, respectively, showed lower efficiency. However, this could result from different cross-linking densities and swelling abilities of the resins concerned. The copper(II) form of Bio-Rex® 70 has been also found[37] to be suitable for the simultaneous analysis of amino-acid and amino-sugar mixtures (Figure 3).

Selective adsorption of cephalosporin from its mixture with deacetylcephalosporin and a series of amino acids and obtaining the antibiotic with purity of 99% seems to be of preparative value.[38] The ligand exchanger used was a polystyrene resin containing copper(II) complexes of L-lysine.

II. AMINO ACIDS AND PEPTIDES ON SILICA-BASED PACKINGS

Interest in LEC for the analysis of amino acids and peptides has revived with the introduction of highly efficient packings based on porous microparticulate silica. One of the simplest and most efficient ligand-exchanging systems suggested thus far is that prepared by Caude and Foucault.[39-44] An aqueous 10^{-2} to 2.10^{-2} *M* copper(II) sulfate solution in 1 *M* ammonia was allowed to percolate through commercial columns packed with 5-μm silica particles until equilibrium was reached and copper ions appeared in the effluent. Up to 4.8% (0.75 mmol/g) of copper was sorbed on a material having about 400 m^2/g inner surface

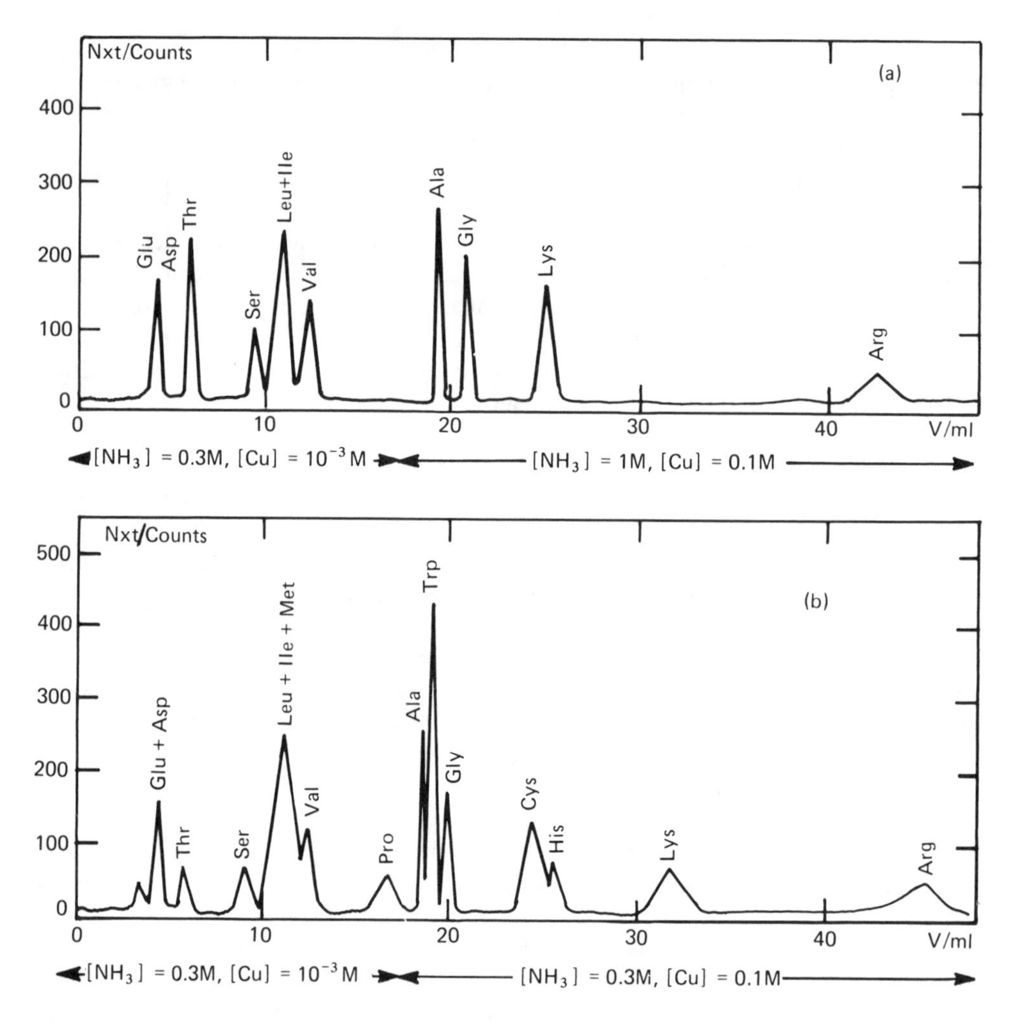

FIGURE 2. Separation of a mixture of ^{14}C-labeled amino acids on a carboxylic resin Bio-Rex® 70, 50 to 65 μm. Column, 45 × 0.4 cm. Continuous scintillation detection; amount injected, 4 nmol per amino acid corresponding to 0.02 to 0.07 μCi of activity; flow rate, 8.8 mℓ/hr; room temperature. (From Doury-Berthod, M., Poitrenaud, C., and Trémillon, B., *J. Chromatogr.*, 179, 37, 1979. With permission.)

area, most probably in the form of a superficial layer of copper silicate.* The latter seems to be a good support for LEC. As shown in Figures 4 and 5, mixtures of amino acids and peptides can be efficiently separated in an aqueous-organic ammonia solution. The ease of preparation of the metal-modified silica gel columns, their high efficiency and selectivity, together with the convenience of a sensitive photometric detection (5 to 50 ng for amino acids) of the solute-metal complexes eluted from the column, should make the method increasingly popular. The gradual dissolution of the silica in the ammonia-containing eluent can be effectively combated by the high proportion of the organic modifier in the eluent (30 to 80% acetonitrile) and insertion of a guard column filled with crude particles of copper(II)-modified silica gel between the pump and the injector.[46,47] Addition of trace amounts of copper sulfate to all eluents (1 ppm) stabilizes the quantity of active copper silicate in the analytical column and enhances reproducibility of retention data.

The coordination mechanism of retention of solutes in the system concerned is evident

* Good correlation has been found[45] between inner surface area of silica gels and the copper uptake, which can even be used for rapid quantitation of accessible silanol groups.

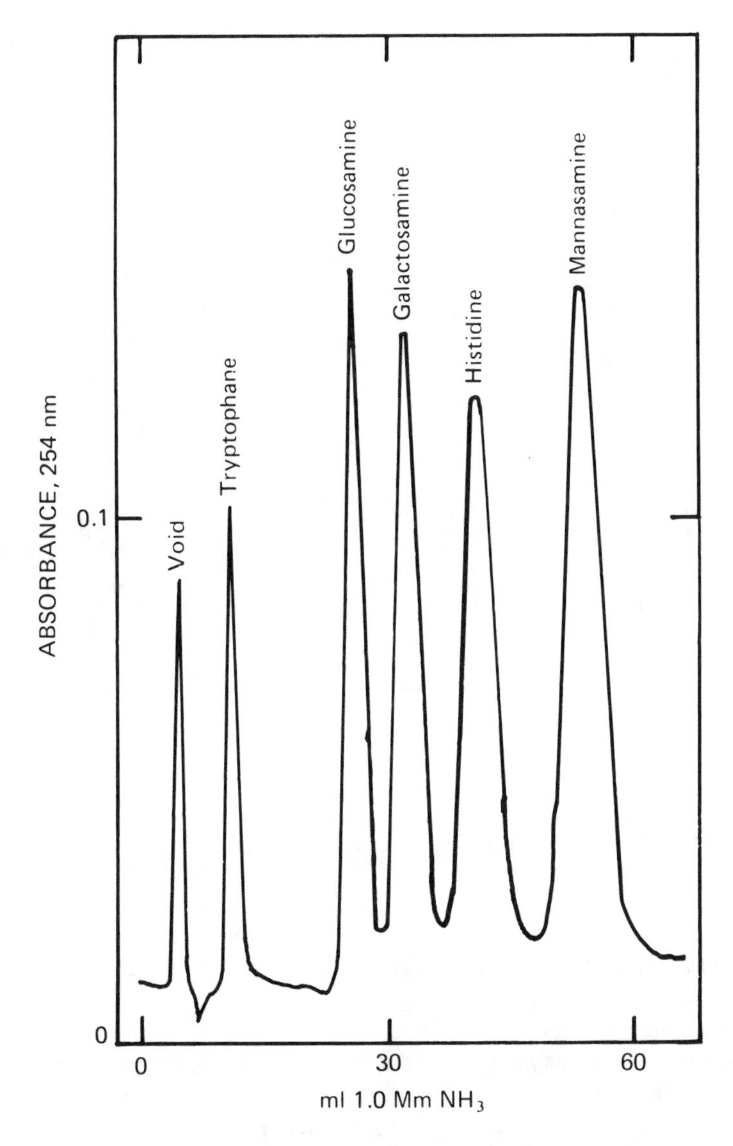

FIGURE 3. Chromatography of amino sugars and amino acids on a copper(II) loaded carboxylic resin Bio-Rex® 70. Column, 21 × 0.63 cm; flow rate 30 mℓ/hr; room temperature. (From Navratil, J. D., Murgia, E., and Walton, H. F., *Anal. Chem.*, 47, 122, 1975. With permission.)

from the fact that alanine, serine, and glycine are the most strongly retained species of all bifunctional amino acids due to low steric hindrance for coordination of these small molecules to the surface of copper silicate. Because of sterical reasons, higher peptides are more weakly retained. Aspartic and glutamic acids, being negatively charged in alkaline media, are repelled from silanol anions and are scarcely retained. On the contrary, basic amino acids and basic peptides display especially high affinity to copper-modified silica gel, both due to favorable electrostatic interactions and to coordination of several amino groups to copper. In this situation, gradient elution (Figure 6), e.g., from a water-acetonitrile mixture of 10:90 containing 0.1 M NH_3 to a water-acetonitrile mixture of 60:40 containing as much as 0.95 M NH_3, gives the opportunity of analyzing a whole palette of peptides and amino acids.[46] Figure 6 is especially remarkable in that it represents one of very few successful gradient elutions in LEC using direct detection of copper(II) complexes at 254 nm. Usually, gradient

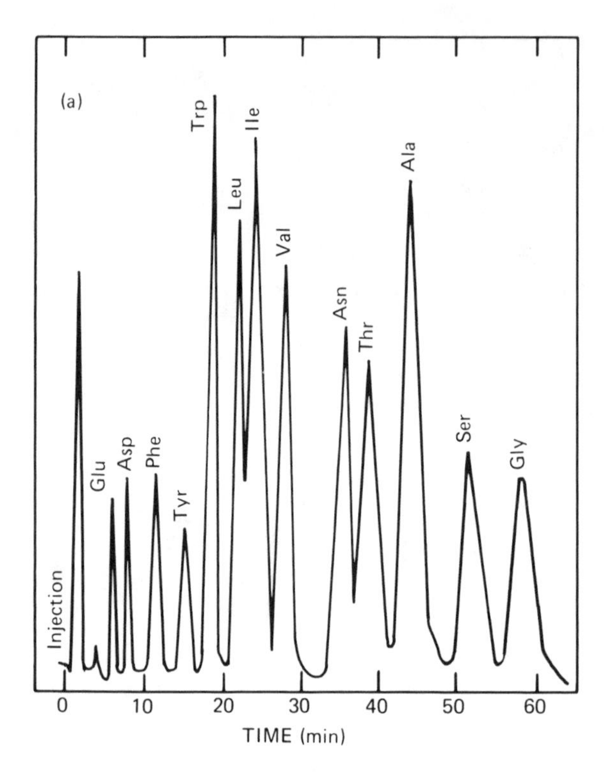

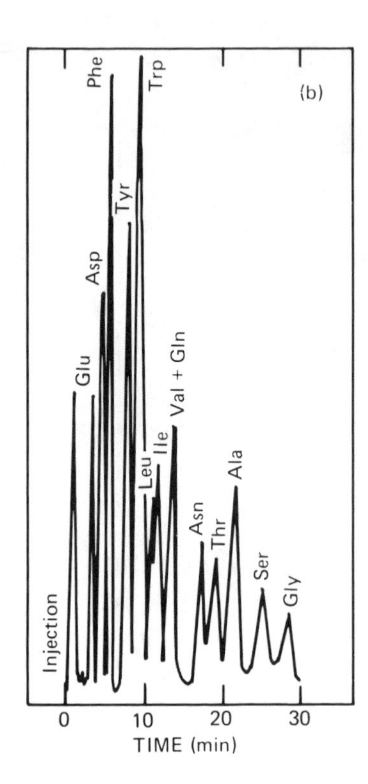

FIGURE 4. Separation of amino acids on copper(II) modified silica gel. Column, 15 × 0.48 cm. Packing, (a) Partisil® 5, (b) Spherosil® XOA 600. Mobile phase: (a) water acetonitrile (52:48)-0.15 *M* ammonia, (b) water acetonitrile (50:50)-0.17 *M* ammonia. Flow rate, (a) 1 mℓ/min, (b) 1.7 mℓ/min. Detection, ultraviolet at 210 nm. (From Foucault, A., Caude, M., and Oliveros, L., *J. Chromatogr.*, 185, 345, 1979. With permission.)

elution in LEC is complicated by the change in the equilibrium concentration of copper ions in the effluent, which causes a strong drift of the base line on the chromatogram.

The same authors[46] have presented an alternative method of fractionating mixtures of peptides and amino acids in a single chromatographic run under isocratic elution conditions. This was possible by substituting weakly basic *N,N,N′,N′*-tetramethylethylenediamine (TMED) for ammonia in the aqueous eluent. In an aqueous solution containing 5.10^{-3} *M* TMED and 2.10^{-5} *M* $CuSO_4$, the above mixtures were separated into four main groups: (1) tripeptides and higher polypeptides; (2) dipeptides, except for the basic ones; (3) free amino acids (in three major peaks) and some basic dipeptides like Lys–Asp; and (4) most of the basic dipeptides. The method has been suggested for fast screening of protein hydrolyzates and quantitation of dipeptides and amino-acid fractions.

Of chemically bonded chelating phases, those containing basic amino groups and prepared by reacting surface silanol groups with 3-aminopropyltriethoxysilane $(H_5C_2O)_3Si(CH_2)_3NH_2$, 3(2-aminoethylamino)propyltrimethoxysilane $(H_3CO)_3Si(CH_2)_3NH(CH_2)_2NH_2$, and 3[*N*-(2′-aminoethyl)-2-aminoethylamino]-propyltrimethoxysilane $(H_3CO)_3Si(CH_2)_3NH(CH_2)_2NH(CH_2)_2NH_2$ seem to be less suitable for LEC than are analogous polymer-based ion exchangers having similar fixed ligands of type of amino, ethylenediamine, and diethylenetriamine groups. Silica-based anion exchangers were found to be unstable in aqueous metal-ion-containing media of pH higher than four.[48,49] Nevertheless, on 600-NH amino columns (Alltech®) several peptides were separated in the presence of cadmium(II) ions in the eluent.[50] The use of zinc(II) makes it possible to separate peptides as a group from amino sugars, which are retained much weaker.[50] On the ethylenediamine-type bonded phase loaded with copper ions, some amino sugars and acids have been sep-

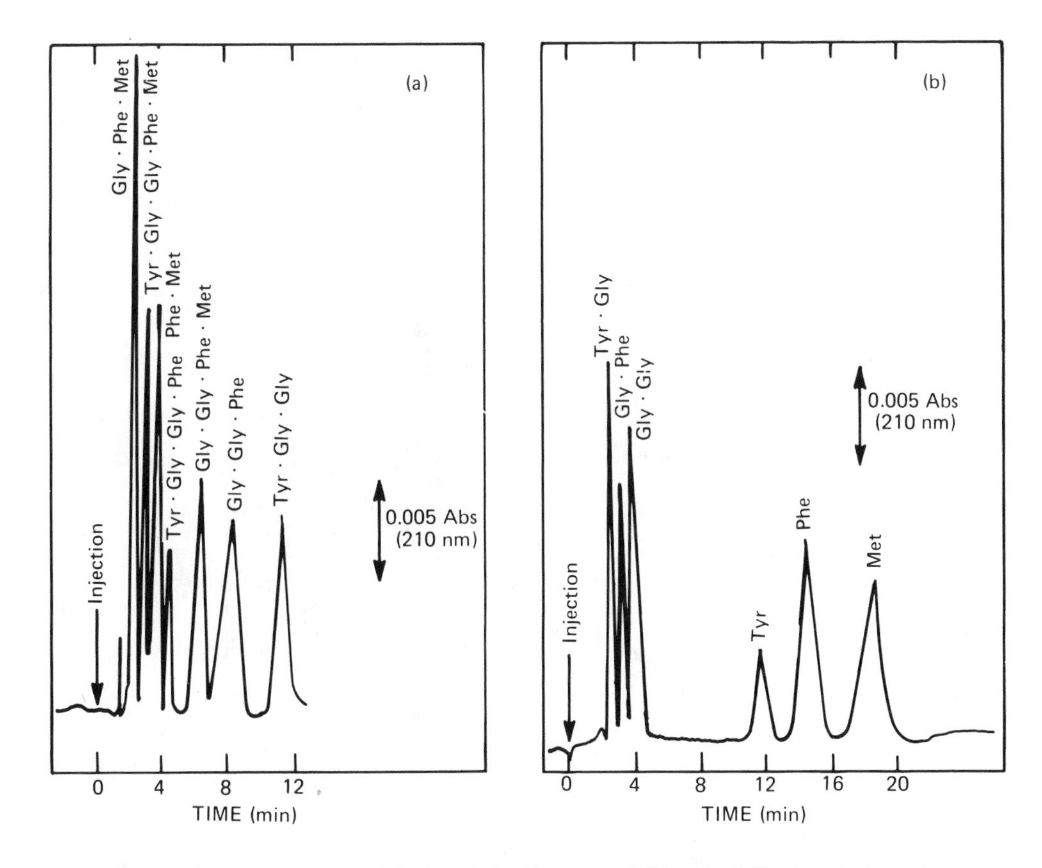

FIGURE 5. Separation of the possible degradation fragments of Met-enkephaline by two isocratic runs on copper(II)-modified silica gel Partisil®5. Column, 20 × 0.48 cm. Mobile phase, (a) water acetonitrile (13:87) 0.125 *M* ammonia, (b) water acetonitrile (70:30) 0.2 *M* ammonia. Flow rate, 1.5 mℓ/min. Detection, ultraviolet at 210 nm. (From Foucault, A., Caude, M., and Oliveros, L., *J. Chromatogr.*, 185, 345, 1979. With permission.)

arated,[48] and a mixture of seven dipeptides has been successfully resolved in an eluent containing 10^{-3} *M* cadmium sulfate, 5.10^{-2} *M* ammonium acetate, and 35% acetonitrile.[51] The high content of acetonitrile may enhance the hydrolytic stability of the amino phase. But much better stability, reproducibility, and — what is extremely important — column efficiency were observed on basic chelating phases prepared by dynamically coating conventional reversed-phase HPLC-columns.[51] Commercial LiChrosorb® C8 columns were operated in a water-acetonitrile (65:35) eluent containing zinc(II) ions and 4-dodecyldiethylenetriamine (C_{12}-dien), $CH_3(CH_2)_{11}N(CH_2CH_2NH_2)_2$. The hydrophobic chelating additive was strongly absorbed by the hydrocarbonaceous surface of the sorbent, thus producing a highly efficient ligand-exchanging packing. As shown in Figures 7 and 8, mixtures of several dansylamino acids and dipeptides were separated in a short period of time.

To conclude the chemically bonded chelating phases, mention should be made of the preparation of stable bonded silica supports[49] with ligands of the iminodiacetate and ethylenediamine-*N,N',N'*-triacetate types. The copper form of above IDA-silica was used to analyze the reaction mixture obtained on protecting the amino function of threonine with triphenylphosphonio-ethoxycarbonyl-chloride.[49] Finally, a complete analysis of all common α-amino acids in their native form under isocratic as well as under gradient elution conditions was reported,[52] using stable and efficient chiral bonded phases of the type

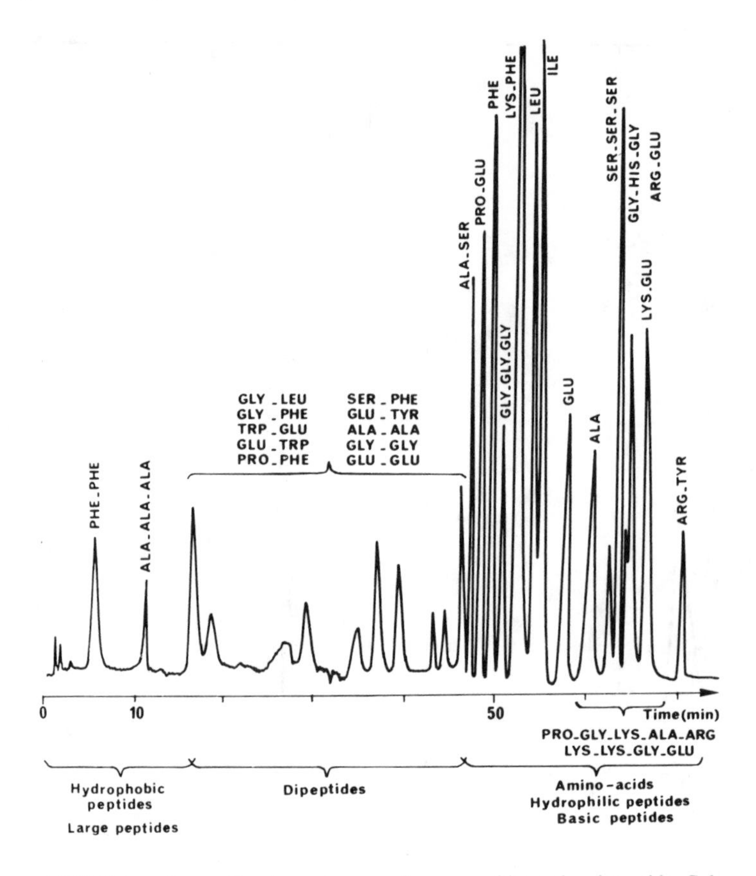

FIGURE 6. Composite chromatogram of some peptides and amino acids. Column, 15 × 0.58 cm. Packing, copper (II) modified LiChrosorb® Si 60, dp 7 μm. Mobile phase, gradient from A, water-acetonitrile 10/90 + 0.1 *M* ammonia + 1 ppm of Cu^{2+}, to B, water-acetonitrile 60/40 + 0.95 *M* ammonia + 1 ppm of Cu^{2+}. Flow rate, 2mℓ/min. Detection, 254 nm. Sample, 30 μℓ containing a few μg of the solutes. (From Foucault, A. and Rosset, R., *J. Chromatogr.*, 317, 41, 1984. With permission.)

$$\begin{array}{l} \qquad\qquad\quad CH_2—CH—R \\ \equiv Si—X—N \quad\; | \\ \qquad\qquad\quad \overset{*}{C}H—CH_2 \\ \qquad\qquad\quad\; | \\ \qquad\qquad\quad COOH \end{array}$$

where R = H or OH and X = $-CH_2-$, $-(CH_2)_3-$, $-(CH_2)_8-$, or $-CH_2-CH_2-C_6H_4-CH_2-$ in the presence of copper ions in the eluent. Synthesis, properties, and use of the above chiral phases for the resolution of amino-acid enantiomers are described in the detailed chapter on separation of enantiomers by LEC.

Closely related to LEC are separations of complexing solutes by adding transition metal ions to the eluent, with the stationary phase remaining indifferent to complexation interactions. Incorporation of the ligands to be separated into ligand-metal complexes can significantly change in the desired direction the retention and resolution of solutes. This may result from the change in electrostatic charges of solutes on complexing metal cations. Another important factor may be that ligands acquire new conformations when coordinated to a metal ion, and new functional groups become exposed to interaction with the stationary

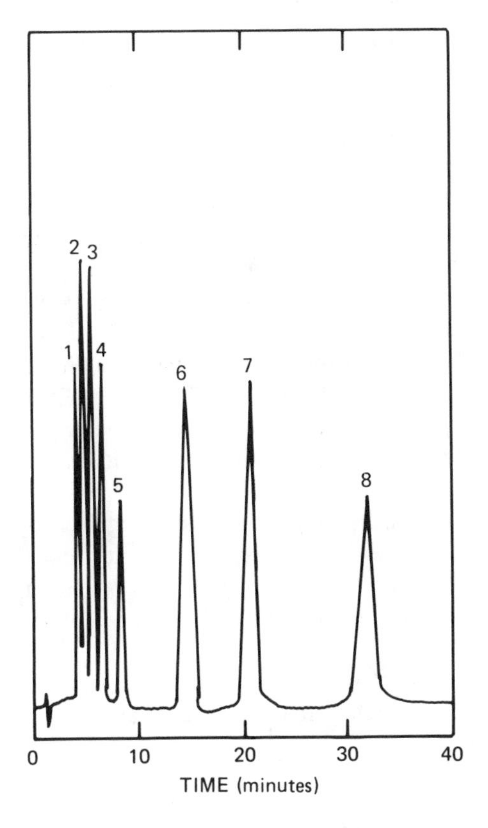

FIGURE 7. Separation of dansyl-amino acids by C_{12}-dien-Zn (II) chromatography on LiChrosorb® RP 8. Column, 25 × 0.46 cm. Mobile phase, 10^{-3} *M* $ZnSO_4$, 0.025% C_{12}-dien, 1% ammonium acetate, acetonitrile-water 35/65. Solutes: 1, glutamic acid; 2, γ-aminobutyric acid; 3, threonine; 4, serine; 5, α-aminobutyric acid; 6, norvaline; 7, leucine; 8, tryptophan. (From Cooke, N. H. C., Viavattene, R. L., Ekstein, R., Wong, W. S., Davies, G., and Kager, B. L., *J.. Chromatogr.*, 149, 391, 1978. With permission.)

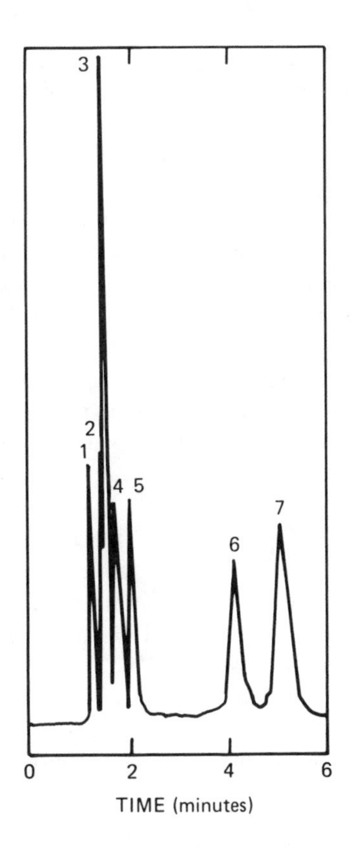

FIGURE 8. Separation of dipeptides by C_{12}-dien-zinc(II) chromatography on LiChrosorb® RP 8. Column and mobile phase as in Figure 7. Solutes: 1, Pro–Tyr; 2, Pro–Phe; 3, Pro–Trp; 4, Phe–Pro; 5, Tyr–Phe; 6, Trp–Phe; 7, Trp–Trp. (From Cooke, N. H. C., Viavattene, R. L., Ekstein, R., Wong, W. S., Davies, G., and Karger, B. L. *J. Chromatogr.*, 149, 391, 1978. With permission.)

phase. Thus, four possible diastereomers of pyroglutamyl-histidyl-3,3-dimethylprolineamide (L–L–D, L–D–L, L–L–L, and D–L–L) were found[53] to elute separately (in the above sequence) from a cyano-phase Spherisorb-CN after the aqueous-acetonitrile mobile phase (70:30) was provided with 10^{-5} *M* copper acetate. Similarly, in the presence of copper(II), L–Tyr–D–Ala–Gly–L–Phe–L–Met has been separated[54] from its D-Met[5] isomer. The cyano phase itself displays no affinity toward copper ions under experimental conditions (sorption is as low as 5 μmol/g), which means that metal-ion additives mainly modify the solute molecules.

The same should be valid for separation of amino acids and *N*-methyl amino acids on reversed-phase columns in the form of their copper complexes.[55,56] The main advantage of this approach is that copper amino-acid complexes strongly absorb ultraviolet radiation with λ_{max} around 230 nm. Thus, the solutes can easily be detected at relatively long wavelengths using simple photometric detectors. The linear range of detection was found to be more than four orders of magnitude with the detection limits as small as 10 ng/10 μℓ. Retention of amino acids increased with pH since more solute molecules entered into complexation.

III. METAL CHELATE AFFINITY CHROMATOGRAPHY OF PROTEINS

In LEC, the basic requirement to solutes under separation is the ability to form coordination bonds with transition metal ions. Many protein molecules incorporating residues of histidine, cysteine, tryptophan, and, to a lesser degree, tyrosine and lysine should meet this requirement. Indeed, Porath et al.[57] have shown that coordination interactions can be successfully applied to selective sorption and isolation of native enzymes and other proteins. This method utilizes a property of protein molecules which is different from those commonly used for fractionation purposes, such as molecular size, electrostatic charge, etc. Therefore, LEC is a powerful complementary technique which may be tried when size exclusion and ion-exchange chromatography have proved inadequate with regard to recovered activity, yield, and degree of purification.

LEC of proteins, or ''metal chelate affinity chromatography'', the term suggested by Porath et al. in 1975,[57] proved to possess the advantages of high sorption capacity, quantitative recovery of the protein with no detectable damage to its structure, and easy regeneration of the sorbent. Two short review articles appeared on the subject by Lönnerdal and Keen[58] and Davankov.[59]

The active sorption sites of sorbents in metal chelate affinity chromatography are usually represented by a transition metal ion iminodiacetate which is linked to the sorbent matrix with a sufficiently long spacer group, for example:[57,60]

$$\text{Sepharose}^{®}\text{–OCH}_2\text{CH(OH)CH}_2\text{–O(CH}_2)_4\text{O–CH}_2\text{CH(OH)CH}_2\text{–N}\begin{matrix}\diagup \text{CH}_2\text{–COO}\diagdown \\ \diagdown \text{CH}_2\text{–COO}\diagup\end{matrix}\text{Me}^{2+} \quad \text{(IDA-gel)}$$

$$\text{Sephadex}^{®}\text{–OCH}_2\text{CO–HN(CH}_2)_4\text{NH–COCH}_2\text{–N}\begin{matrix}\diagup \text{CH}_2\text{–COO}\diagdown \\ \diagdown \text{CH}_2\text{–COO}\diagup\end{matrix}\text{Me}^{2+}$$

It is obvious that the sorbent matrix should be accessible to large molecules of proteins in order to allow formation of mixed-ligand sorption complexes. Gels of high swelling capacity such as Sepharose® 6B, CM-Sephadex® C-50, Trisacryl GF 2000 (prepared by copolymerization of a new acrylic monomer, *N*-acryloyl-2-amino-2-hydroxymethyl-1,3-propanediol),[61] or other weakly cross-linked ion exchangers meet this requirement. Macroporous hydrophilic TSK-Gel® HW-55 was also shown[62] to be an excellent support for LEC of enzymes. There is no reason why bonded silica ligand exchangers should not be used for the same purpose. Indeed, first successful examples of LEC of proteins on silica gels with covalently bound IDA-chelates have been recently presented.[62,63] Of course, interaction types other than coordination of proteins to the matrix-fixed metal chelate should be largely excluded in order to prevent nonspecific sorption phenomena. Therefore, only sorbents with hydrophilic surfaces or hydrophilic cross-linked polymers should be used in LEC of proteins. Electrostatic interactions are usually minimized by using eluents of high ionic strength (1 *M* NaCl).

Binding of the protein to the matrix-fixed metal-iminodiacetate is certainly due to coordinate bond formation. Usually, it is accomplished in the pH range between 6 (acetate buffer) and 8 (phosphate buffer). The destruction of the sorption complex and elution of the bound protein can be performed by one of the three following protocols:

1. Lowering the pH of the eluent. This results in protonation of the protein donating group and its dissociation from the sorption complex. One has to bear in mind, however, that the matrix-fixed IDA groups can also be protonated, thus releasing the chelated metal ions. Bleaching of an IDA-Co^{2+} column (loss of Co^{2+}) can be visually observed already at pH 5. IDA-Zn^{2+} and, in particular, IDA-Cu^{2+} chelates are more stable.
2. Ligand exchange; coordinated protein molecules can be easily displaced by imidazole, which is a stronger ligand. Development of an IDA-Me^{2+} column with imidazole generally requires previous conversion of the starting packing into mixed ligand form, e.g., IDA-$Cu(Im)_2$, by equilibration with a 10 m*M* imidazole solution of pH 7.0 (1 *M* NaCl). In this case, dilute (1 m*M*) imidazole solutions would provide effective elution of proteins.
3. Chelate annihilation; sorption complexes can be destroyed at low pH by a mild chelating agent (histidine) or a strong chelating agent (EDTA) resulting in the release of bound proteins.

Any of the above types of elution procedures can be performed by a step-wise or gradient mode.

Metal chelate affinity chromatography has so far been mainly used as a preparative, not analytical, separation method. It is interesting to note that many enzymes isolated according to this technique belong to the class of metal proteins. Their molecules contain one or more strongly bonded metal ion(s) which is important for maintaining specific tertiary structure of protein molecules or is involved into interaction with specific substrates. It should be emphasized that these proteins have been recovered in their initial metal-saturated form. Thus, an iron-containing protein can be isolated on a copper(II)-chelate gel and subsequently eluted and recovered in its iron-saturated state. This implies that the metal-bonding sites in a metal-protein are not involved in a ligand-exchange process with the matrix-fixed metal chelate. Similarly, amino-acid residues that are responsible for retaining metal ions in the metal-protein molecule are not involved in interaction with the matrix-fixed metal ion. For example, lactoferrin has two binding sites for copper, yet this protein will, in its copper(II)-saturated state, bind to a copper-iminodiacetate gel as strongly as the apo-form of lactoferrin and iron-lactoferrin.[58] All three forms of lactoferrin behave identically; they elute at exactly the same position and maintain their native metal constituent. Another typical example is presented in Figure 9.

The above findings point out that only amino-acid residues that are exposed to the surface of the protein molecule take part in coordination to the matrix-fixed chelate. Just a few of several histidine or cysteine residues present in the protein molecule appear to be on its surface. This makes the method highly selective. Thus, in a series of related interferons, the affinity to different metal chelates was found to vary in a wide range, which can be of great analytical value in assessing the surface topography of protein molecules.[64]

In a highly elegant and convincing paper,[65] Sulkowski analyzes results of LEC on IDA-Me^{2+} gels of several proteins with known primary and secondary structures. This allows him to arrive at the following important conclusions:

1. A protein molecule which lacks histidine or tryptophan residues on its surface is not retained on IDA-Cu^{2+} gel. The contribution of cysteine remains to be demonstrated with a suitable protein model (metallothionein).
2. The presence of a single histidine residue on the surface of a protein is sufficient for the retention of the latter on IDA-Cu^{2+} at neutral pH.
3. The strength of binding on IDA-Cu^{2+} correlates positively with the multiplicity of available histidine residues.
4. The presence of a tryptophan residue on the surface of a protein can be recognized

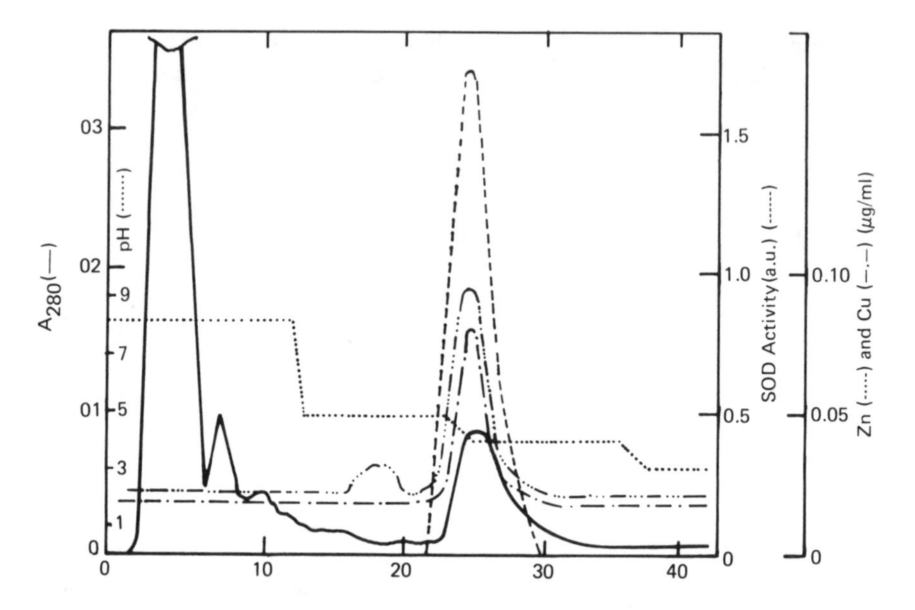

FIGURE 9. Metal chelate affinity chromatography of copper and zinc superoxide dismutase on copper(II) iminoidiacetate activated Sepharose® 4B. Mobile phase: 0.05 *M* tris-acetate, pH 8.2 in 0.5 *M* NaCl; gradient pH 5.0, 4.0, and 3.0 (stepwise) in the same buffer. Column, 20 × 1 cm. Flow rate, 1 mℓ/min. (From Lönnerdal, B. and Keen, C. L., *J. Appl. Biochem.*, 4, 203, 1982. With permission.)

by IDA-Cu^{2+}. However, it may take several tryptophan residues to result in binding of a protein.

5. Retention of a protein on IDA-Zn^{2+} and IDA-Co^{2+} requires the presence of two proximal histidine residues.

Table 1 contains a list of successful applications of LEC techniques for selective isolation of individual proteins. It can be noted that many metal ions have been tested, but until recently, there were just a few attempts to examine other ligand-exchanging sorbents than those initially suggested by Porath et al.[57]

Finally, mention should be made of a special technique named "labile ligand affinity chromatography". It was developed by Nexø[66] to overcome the difficulty of almost irreversible binding of some specific proteins to cobalamine that was taken as a matrix-fixed chelate. Instead, of a covalent immobilization of this chelate, hydroxocobalamine (or cyanocobalamine) was provisionally coordinated to a Sepharose® matrix containing primary amino groups. Then specific cobalamine-binding proteins were selectively sorbed at 4°C on the stationary phase obtained. These proteins were finally released again in the form of complexes with cobalamine by raising the temperature of the column to 37°C where dissociation of the cobalt-amino group coordination bonds occurred. In a similar manner, several transcobalamins,[66-72] haptocorrin,[73] and intrinsic factors,[72,74] were selectively sorbed and isolated.

In this connection, doubt may arise whether the method of "labile ligand affinity chromatography" should be considered together with the method of "metal chelate affinity chromatography". The two techniques have in common the fact that complex-forming metal ions are responsible for organizing interaction between the stationary phase and solutes under separation and the mixed-ligand sorption complexes materialize this interaction. The difference is that in metal chelate affinity chromatography the central metal ion remains attached to the matrix-fixed ligand before and after the interaction with the solute, whereas in labile

Table 1
METAL CHELATE AFFINITY CHROMATOGRAPHY OF PROTEINS

Sorbent[a]	Metal ion	Protein	Ref.
1	Zn, Cu	Human serum proteins	57
Tris-acryl-IDA	Cu	Human plasma globulins	61
1, 2	Ni, Fe	Human serum proteins	75
2	Tl	Serum proteins	76
1	Cu	Zn- and Cu-superoxide dismutase	58
1	Cu	Metallothioneins	58
TSK®-IDA	Cu	Exonuclease A 5	62
1	Co, Ni, Cu, Zn	Interferons HuIFN-α, HuIFN-β, HaIFN$_s$, MuIFN$_2$	64
1	Cu	Hamster interferon	80, 81
1	Zn	Human fibroblast interferon	82, 83
1	Cu	Human leucocyte interferon	84
1	Cu	Granule proteins of granulocytes	85
1	Zn	Plasminogen activator	86
1	Cu	Lactoferrin	58, 63, 87
1	Cu	Albumin from a crude Cohn extract	88
1	Fe	Enkephalin analogues	89
1	Cd	Cd-binding protein	90
1	Zn	Human clotting Factor XII	91
1	Zn	Human fibrinogen	92
1	Zn, Cu	Nucleosidediphosphatase of rat liver	93
1	Zn	Human inter-α-trypsin inhibitor	94
1	Cu	Thiol proteinase inhibitor	95
1	Zn	Human plasma α_2 macroglobulin and α_1 proteinase inhibitor	96, 97
1	Zn	Metalloproteinase inhibitor	98
1	Zn	Rabbit bone metalloproteinase	99
1	Zn	Phosphotyrosyl-protein phosphatase	100
1	Zn	Nitrate reductase	103
1	Cu	Glutathione reductase, lipoamide dehydrogenase	103
1	Zn	Human plasma α_2-SH glycoprotein	101
1	Zn	Epidermal proteinases	103
1	Ni, Zn	"Embryonin", bovine α_2-macroglobulin	104
1	Cu	RNA polymerase B stimulatory factor	105
1	Co	Brain neutral α-D-mannosidase	106
1	Ca	Seed lectin	107
1, 2	Zn, Cd	Human serum α_2 macroglobulin, hemopexin	108
3	Al	Histones	109
3	Hg	Calotropain FI	110

[a] 1 = agarose activated with 1,4-bis(2,3-epoxypropoxy)butane and coupled with iminodiacetate IDA-gel);[57] 2 = agarose activated with epichlorohydrin and coupled with ethylenediamine followed by alkylation with bromoacetate (TED-gel);[75] 3 = sulfonated polystyrene resin Amberlite® IR-120.

ligand affinity chromatography, one coordination bond is being formed during the sorption act, but another coordination bond will break during the desorption act so that the central metal ion (together with an additional ligand) changes its location from the stationary phase to the mobile ligand. Evidently, both variants are appropriate to be referred to as ligand exchange. From this point of view, a recent suggestion by Porath and Olin[75] to combine the terms "metal chelate affinity chromatography" and "LEC" under the new conception ''immobilized metal ion affinity chromatography'' does not seem to be appropriate since location of complex-forming metal ions in the stationary phase is not required for many LEC variants.

Apart from the above terminology questions, the paper[75] by Porath and Olin deserves special attention since it summarizes experiences of the first 10 years of protein chromatography using coordination interactions and gives important clues as to how further exploitation of the method might be accomplished. It also demonstrates especially elegantly the main advantage of fractionating proteins according to their coordinating properties, which is the unrivaled selectivity of this technique. By chromatography of total human serum proteins on two-bed tandem columns loaded with nickel(II) and iron(III) forms of iminodiacetate-Sepharose® 6B, individual proteins were shown to bind highly selectively to above metal forms, independent of the order in which the columns were connected. Moreover, entirely different proteins were found to adsorb on nickel(II) and iron(III) forms of a similar Sepharose® 6B matrix, but containing tris(carboxymethyl)ethylenediamine-type ligands (TED-gel):

$$\text{Sepharose}^{®}\text{–}CH_2CH(OH)CH_2\text{–}N(CH_2COOH)\text{–}CH_2CH_2\text{–}N(CH_2COOH)_2$$

Being potentially pentadentate ligands, in contrast to terdentate iminodiacetate groups, the former give different binding possibilities for the protein-donating functions as compared to those given by IDA-chelates of octahedral metal ions. Thus, immunoglobulins are adsorbed on the nickel(II)-IDA bed, whereas hemopexin is adsorbed entirely on the nickel(II)-TED column.[75] Of the group IIIA metal ions (Al^{3+}, Ga^{3+}, In^{3+}, and Tl^{3+}) examined[76] in combination with TED gel, alone, Tl^{3+} displayed coordination ability toward proteins. Sepharose® 6B-TED-Tl^{3+} was found to behave similarly in many respects to the corresponding Ni^{2+} column, but not identically. Dependence of protein binding to the above metal chelates on type and concentration of inorganic salts in the mobile phase, addition of organic solvents, detergents, or urea to the eluent, and finally, on pH has also been examined[75] pointing out great possiblities for selective elution of individual components from the sorbed palette of proteins. Thus, new types of matrix-fixed ligands and new elution techniques may be most promising directions of further development.

More recently,[77-79] new types of Sepharose® 6B based chelating sorbents have been suggested, possessing α-amino hydroxamic acid functions:

$$\text{Sepharose}^{®}\text{:–}CH_2CH(OH)CH_2\text{–}NH\text{–}CH_2\text{–}CONHOH$$

$$\text{Sepharose}^{®}\text{:–}CH_2CH(OH)CH_2\text{–}N(CH_2CONHOH)_2$$

$$\text{Sepharose}^{®}\text{:–}CH_2CH(OH)CH_2\text{–}N(CH_2CONHOH)\text{–}CH_2CH_2\text{–}N(CH_2CONHOH)_2$$

Hydroxamic acids are known to form stable chelates* with Fe^{3+} ions:

$$-HN-CH_2-C\genfrac{}{}{0pt}{}{\nearrow O\cdots Fe^{3+}}{\searrow NH-O\;\;\;|}$$

Mono-, *bis*-, and *tris*-hydroxamic acid Sepharose® sorbents in their Fe^{3+} form were found to behave mainly as weak cation exchangers. However, they fractionate serum proteins in a different manner as compared to metal-free ion exchangers. Noteworthy is the high affinity of Fe^{3+}-hydroxamate functions to aspartic and glutamic acids, and especially to cysteine.[75,76]

Another prospective field of development of metal chelate affinity chromatography should be analysis of protein mixtures using high performance LEC.

REFERENCES

1. **Siegel, A. and Degens, E. T.,** Concentration of dissolved amino acids from saline waters by ligand exchange chromatography, *Science,* 151, 1098, 1966.
2. **Grasbeck, R. and Karlsoon, R.,** Continuous microdetermination of protein with a Sephadex copper[64] detector column, *Acta Chem. Scand.*, 17, 1, 1963.
3. **Fazakerley, S. and Best, D. R.,** Separation of amino acids, as copper chelates, from amino acid, protein, and peptide mixtures, *Anal. Biochem.*, 12, 290, 1965.
4. **Rothenbühler, E., Waibel, R., and Solms, J.,** An improved method for the separation of peptides and α-amino acids on copper Sephadex, *Anal. Biochem.*, 97, 367, 1979.
5. **Buist, N. R. M. and O'Brien, D.,** The separation of peptides from amino acids in urine by ligand exchange chromatography, *J. Chromatogr.*, 29, 398, 1967.
6. **Bellinger, J. F. and Buist, N. R. M.,** The separation of peptides from amino acids by ligand exchange chromatography, *J. Chromatogr.*, 87, 513, 1973.
7. **Boisseau, J. and Jouan, P.,** Separation des oligopeptides et des acides amines par chromatographie sur resin echangeuse d'ions, Chelex-X-100, *J. Chromatogr.*, 54, 231, 1971.
8. **Boisseau, J. and Jouan, P.,** Separation des oligopeptides et des acides amines par chromatographie par echange de coordinates, *Bull. Soc. Chim. Fr.*, 153, 1973.
9. **Maurer, R.,** Separation of enkephalin degradation products by ligand exchange chromatography, *J. Biochem. Biophys. Meth.*, 2, 183, 1980.
10. **Hemmasi, B.,** Ligand exchange chromatography of amino acids on nickel-Chelex 100, *J. Chromatogr.*, 104, 367, 1975.
11. **Hemmasi, B. and Bayer, E.,** Ligand exchange chromatography of amino acids on copper-, cobalt-, and zinc-Chelex 100, *J. Chromatogr.*, 109, 43, 1975.
12. **Hemdan, E. S. and Porath, J.,** Development of immobilized metal affinity chromatography. I. Comparison of two iminodiacetate gels, *J. Chromatogr.*, 323, 247, 1985.
13. **Hemdan, E. S. and Porath, J.,** Development of immobilized metal affinity chromatography. II. Interaction of amino acids with immobilized nickel iminodiacetate, *J. Chromatogr.*, 323, 255, 1985.
14. **Hemdan, E. S. and Porath, J.,** Development of immobilized metal affinity chromatography. III. Interaction of oligopeptides with immobilized nickel iminodiacetate, *J. Chromatogr.*, 323, 265, 1985.
15. **Antonelli, M. L., Bucci, R., and Carunchio, V.,** An application of the ligand exchange chromatography to the analysis of some protein components, *J. Liq. Chromatogr.*, 3, 885, 1980.
16. **Tommel, D. K. J., Vliegenthart, J. F. G., Penders, T. J., and Arens, J. F.,** A method for the separation of peptides and α-amino acids, *Biochem. J.*, 99, 48P, 1966.
17. **Tommel, D. K. J., Vliegenthart, J. F. G., Penders, T. J., and Arens, J. F.,** A method for the separation of peptides and α-amino acids, *Biochem. J.*, 107, 335, 1968.
18. **Niederwieser, A. and Curtius, H. C.,** Separation of peptides and amino acids in urine, *J. Chromatogr.*, 51, 491, 1970.

* Some authors[74,75] suggest additional coordination of the α-amino group to the same metal ion, which is impossible for sterical reasons since the sp^2 hybridization state of the carbonyl carbon atom requires flat arrangement of its three substituents.

19. **Lutz, W., Markiewicz, K., and Klyszejko-Stefanowicz, L.,** Oligopeptides excreted in the urine of healthy humans and of patients with nephrotic syndrome, *Clin. Chim. Acta,* 39, 425, 1972.
20. **Sampson, B., Barlow, B., and Wilkinson, A. W.,** Metabolism of 1-^{14}C-labeled glycyl dipeptides in mice, *Biochem. Soc. Trans.,* 3, 684, 1975.
21. **Lutz, W., Markiewicz, K., and Klyszejko-Stefanowicz, L.,** Oligopeptides in blood plasma of healthy human and of patients with nephrotic syndrome, *Clin. Chim. Acta,* 39, 319, 1972.
22. **Polzhoffer, K. P. and Ney, K. H.,** Isolierung von Peptiden aus Cheddar-Käse. Trennung von Aminosäure/Peptide-Copper(II)-Komplexen an Sephadex QAE, *Tetrahedron,* 28, 1721, 1972.
23. **Clapperton, J. F.,** Simple peptides of wort and beer, *J. Inst. Brew. London,* 77, 177, 1971.
24. **Sampson, B. and Barlow, G. B.,** Separation of peptides and amino acids by ion exchange chromatography of their copper complexes, *J. Chromatogr.,* 183, 9, 1980.
25. **Arikawa, Y. and Tochida, K.,** Multipurpose liquid chromatography, *Hitachi Rev.,* 16, 236, 1967; *Chem. Abstr.,* 69, 16008u, 1968.
26. **Arikawa, Y.,** Automatic liquid chromatography and chromatograph, British Patent 1,173,996, 1967; *Chem Abstr.,* 72, 45513X, 1970.
27. **Wagner, F. W. and Shepherd, S. L.,** Ligand exchange amino acid analysis — resolution of some amino sugars and cysteine derivatives, *Anal. Biochem.,* 41, 314, 1971.
28. **Wagner, F. W. and Liliedahl, R. L.,** A rapid method for the quantitative analysis of α-aminobutyric acid in hypothalamus homogenates by ligand exchange chromatography, *J. Chromatogr.,* 71, 567, 1972.
29. **Maeda, M., Tsuji, A., Ganno, S., and Onishi, Y.,** Fluorophotometric assay of amino acids by using automated ligand exchange chromatography and pyridoxal zinc(II) reagent, *J. Chromatogr.,* 77, 434, 1973.
30. **Shaw, D. S. and West, C. E.,** The isolation of methionine and ethionine by silver ligand chromatography and application to methionine containing peptides, *J. Chromatogr.,* 200, 185, 1980.
31. **Hering, R. and Heilmann, K.,** Die trennung von Aminosäuren an Schwermetallformen des Sarkosin-Harzes durch Wasserelution, *J. Prakt. Chem.,* 4. Reihe, 32, 59, 1966.
32. **Sinjavski, V. G. and Dzyubenko, A. V.,** Separation of neutral amino acids on a chelating cation exchanger with iminodiacetate functional groups, in Methods of Obtaining and Analysis of Biochemical Preparations, *Abstracts of Papers, Riga, VNIITEKhim,* 21, 1977.
33. **Muzzarelli, R. A. A., Tanfani, F., Muzzarelli, M. G., Scarpini, G., and Rocchetti, R.,** Ligand exchange chromatography of amino acids on copper loaded chitozan, *Sep. Sci. Technol.,* 13, 869, 1978.
34. **Antonelli, M. L., Marino, A., Massina, A., and Petronio, B. M.,** A proposal for the application of ligand exchange chromatography on thin layers, *Chromatographia,* 13, 167, 1980.
35. **Doury-Berthod, M., Poitrenaud, C., and Trémillon, B.,** Ligand exchange separations of amino acids. I. Distribution equilibria of some amino acids between ammoniacal copper(II) nitrate solutions and phosphonic, carboxylic, and iminodiacetic ion exchangers in the copper(II) form, *J. Chromatogr.,* 131, 73, 1977.
36. **Doury-Berthod, M., Poitrenaud, C., and Trémillon, B.,** Ligand exchange separation of amino acids. II. Influence of the eluent composition and of the nature of the ion exchanger, *J. Chromatogr.,* 179, 37, 1979.
37. **Navratil, J. D., Murgia, E., and Walton, H. F.,** Ligand exchange chromatography of amino sugars, *Anal. Chem.,* 47, 122, 1975.
38. **Sacco, D. and Dellacherie, E.,** Ligand exchange chromatography of cephalosporin C on polystyrene resins containing copper complexes of lysine derivatives, *J. Liquid Chromatogr.,* 6, 2543, 1983.
39. **Caude, M. and Foucault, A.,** Ligand exchange chromatography of amino acids on copper(II) modified silica gel with ultraviolet spectrophotometric detection at 210 nanometers, *Anal. Chem.,* 51, 459, 1979.
40. **Schmidt, E., Foucault, A., Caude, M., and Rosset, R.,** Chromatographie d'echange de ligandes sur silice chargee en cuivre(II) — application a la separation d'acides amines, *Analusis,* 7, 366, 1979.
41. **Foucault, A., Caude, M., and Oliveros, L.,** Ligand exchange chromatography of the enantiomeric amino acids on copper-loaded chiral bonded silica gel and of amino acids on copper(II) modified silica gel, *J. Chromatogr.,* 185, 345, 1979.
42. **Guyon, F., Foucault, A., and Caude, M.,** Ligand exchange chromatography of small peptides on copper(II) modified silica gel — application to the study of the enzymatic degradation of methionine-enkephalin, *J. Chromatogr.,* 186, 677, 1979.
43. **Guyon, A., Roques, B. P., Guyon, F., Foucault, A., Perdrisot, R., Swerts, J.-P., and Schwartz, J.-C.,** Enkephalin degradation in mouse brain studied by a new H.P.L.C. method: further evidence for the involvement of carboxypeptidase, *Life Sci.,* 25, 1605, 1979.
44. **Caude, M. and Foucault, A.,** Ligand exchange chromatography of amino acids, *Spectra Phys. Chromatogr. Rev.,* 6(2), 4, 1980.
45. **Jennings, E. C., Jr.,** Surface activity of silica as measured by a copper complex, 8th Int. Symp. Column Liquid Chromatography, May 20 to 25, 1984, New York, 2p-30.

46. **Foucault, A. and Rosset, R.**, Ligand exchange chromatography on copper(II) modified silica gel — improvements and use for screening of protein hydrolyzate and quantitation of dipeptides and amino acid fractions, *J. Chromatogr.*, 317, 41, 1984.
47. **Guyon, F., Chardonett, L., Caude, M., and Rosset, R.**, Study of silica gel and copper silicate gel solubilities in high performance liquid chromatography, *Chromatographia*, 20, 30, 1985.
48. **Masters, R. G. and Leyden, D. E.**, Ligand exchange chromatography of amino sugars and amino acids on copper loaded silylated controlled-pore glass, *Anal. Chim. Acta*, 98, 9, 1978.
49. **Gimpel, M. and Unger, K.**, Hydrolytically stable chemically bonded silica supports with metal complexating ligands — synthesis, characterization and use in high-performance ligand exchange chromatography (HPLEC), *Chromatographia*, 16, 117, 1982.
50. **Dua, V. K. and Bush, C. A.**, High-performance liquid chromatographic separation of amino sugars and peptides with metal ion modified mobile phases, *J. Chromatogr.*, 244, 128, 1982.
51. **Cooke, N. H. C., Viavattene, R. L., Ekstein, R., Wong, W. S., Davies, G., and Karger, B. L.**, Use of metal ions for selective separations in high-performance liquid chromatography, *J. Chromatogr.*, 149, 391, 1978.
52. **Roumeliotis, P.**, Complete analysis of native α-amino acid mixtures by means of ligand exchange column liquid chromatography, in 7th Int. Symp. on Column Liquid Chromatography, Baden-Baden, West Germany, 1983, 45.
53. **Sugden, K., Hunter, C., and Lloyd-Jones, J. G.**, Separation of the diasteromers of pyroglutamyl-histidyl-3,3-dimethylprolineamide by ligand exchange chromatography, *J. Chromatogr.*, 204, 195, 1981.
54. **Hunter, C., Sugden, K., and Lloyd-Jones, J. C.**, HPLC of peptides and peptide diastereomers on ODS- and cyanopropyl-silica gel column packing materials, *J. Liquid Chromatogr.*, 3, 1335, 1980.
55. **Grushka, E., Levin, S., and Gilon, C.**, Separation of amino acids on reversed phase columns as their copper(II) complexes, *J. Chromatogr.*, 235, 401, 1982.
56. **Grushka, E., Atamna, J., Gilon, C., and Chorly, M.**, Liquid chromatographic separation and detection of *N*-methyl amino acids using mobile phases containing copper(II) ions, *J. Chromatogr.*, 281, 125, 1983.
57. **Porath, J., Carlsson, J., Olsson, I., and Belfrage, G.**, Metal chelate affinity chromatography — a new approach to protein fractionation, *Nature (London)*, 258, 598, 1975.
58. **Lönnerdal, B. and Keen, C. L.**, Metal chelate affinity chromatography of proteins, *J. Appl. Biochem.*, 4, 203, 1982.
59. **Davankov, V. A.**, Review of ligand exchange chromatography, in *Handbook of HPLC for the Separation of Amino Acids, Peptides, and Proteins*, Vol. 1, Hancock, W. S., Ed., CRC Press, Boca Raton, Fla., 1984, 393.
60. **Gozdzicka-Josefiak, A. and Augustyniak, J.**, Preparation of chelating exchangers with a polysaccharide network and low cross-linkage, *J. Chromatogr.*, 131, 91, 1977.
61. **Moroux, Y., Boschetti, E., and Egly, J. M.**, Preparation of metal-chelating trisacryl gels for chromatographic applications, *Sci. Tools*, 32(1), 1, 1985.
62. **Varlamov, V. P., Lopatin, S. A., and Rogozhin, S. V.**, Ligand exchange chromatography of enzymes. I. Synthesis of chelating sorbents and purification of exonuclease A5 from actinomyces, *Bioorgan. Khim.*, 10, 927, 1984.
63. **Fanou-Ayi, L. and Vijayalakshmi, M.**, Metal-chelate affinity chromatography as a separation tool, *Ann. N.Y. Acad. Sci.*, 413, 300, 1983.
64. **Sulkowski, E., Vastola, K., Oleszek, D., and Von Muenchhausen, W.**, Surface topography of interferons — a probe by metal chelate chromatography, in *Affinity Chromatography and Related Techniques*, Gribnau, T. C. J., Visser, J., and Nivard, R. J. F., Eds., Elsevier, Amsterdam, 1982, 313.
65. **Sulkowski, E.**, Purification of Proteins by IMAC, *Trends Biotechnol.*, 3, 1, 1985.
66. **Nexø, E.**, A new principle in biospecific affinity chromatography used for purification of cobalamin-binding proteins, *Biochim. Biophys. Acta*, 379, 189, 1975.
67. **Nexø, E.**, Purification of rabbit transcobalamin II by labile ligand affinity chromatography, *Can. J. Physiol. Pharmacol.*, 55, 923, 1977.
68. **Nexø, E., Olesen, H., Bucher, D., and Thomsen, J.**, Purification and characterization of rabbit transcobalamin. II, *Biochim. Biophys. Acta*, 494, 395, 1977.
69. **Nexø, E.**, Trancobalamin I and other human R-binders — purification, structural, spectral, and physiological studies, *Scand. J. Haemotol.*, 20, 221, 1978.
70. **Lindemans, J., Van Kapel, J., and Abels, J.**, Purification of human transcobalamin II cyanocobalamin by affinity chromatography using thermolabile immobilization of cyanocobalamin, *Biochim. Biophys. Acta*, 579, 40, 1979.
71. **Van Kapel, J., Loef, B. G., Lindemans, J., and Abels, J.**, An improved method for large scale purification of human holo-transcobalamin II, *Biochim. Biophys. Acta*, 676, 307, 1981.
72. **Jacobsen, D. W., Montejano, Y. D., and Huennekens, F. M.**, Rapid purification of cobalamin-binding proteins using immobilized aminopropylcobalamin, *Anal. Biochem.*, 113, 164, 1981.

73. **Nexø, E. and Olesen, H.,** Purification and characterization of rabbit haptocorrin, *Biochim. Biophys. Acta,* 667, 370, 1981.
74. **Bucher, D., Thomsen, J., and Nexø, E.,** Amino terminal sequence of hog intrinsic factor, *Comp. Biochem. Physiol.,* 62B, 175, 1979.
75. **Porath, J. and Olin, B.,** Immobilized metal ion affinity adsorption and immobilized metal ion affinity chromatography of biomaterials. Serum protein affinities for gel-immobilized iron and nickel ions, *Biochemistry,* 22, 1621, 1983.
76. **Porath, J., Olin, B., and Granstrand, B.,** Immobilized-metal affinity chromatography of serum proteins on gel-immobilized group III A metal ions, *Arch. Biochem. Biophys.,* 225, 543, 1983.
77. **Ramadan, N. and Porath, J.,** α-Aminoacyl hydroxamate adsorbents — a new type of immobilized chelator, *J. Chromatogr.,* 321, 81, 1985.
78. **Ramadan, N. and Porath, J.,** Fe^{3+}-hydroxamate as immobilized metal affinity-adsorbent for protein chromatography, *J. Chromatogr.,* 321, 93, 1985.
79. **Ramadan, N. and Porath, J.,** Separation of serum proteins on a Fe^{3+}-monohydroxamate adsorbent, *J. Chromatogr.,* 321, 105, 1985.
80. **Bollin, E. and Sulkowski, E.,** Metal chelate affinity chromatography of hamster interferon, *Arch. Virol.,* 58, 149, 1978.
81. **Bollin, E., Jr.,** Production, partial purification and characterization of syrian hamster interferon, *Methods Enzymol.,* 78, 178, 1981.
82. **Edy, V. G., Billiau, A., and De Sommer, P.,** Purification of human fibroblast interferon by zinc chelate affinity chromatography, *J. Biol. Chem.,* 252, 5934, 1977.
83. **Heine, K. J. W. and Billiau, A.,** Purification of human fibroblast interferon by adsorption to controlled-pore glass and zinc-chelate chromatography, *Methods Enzymol.,* 78, 448, 1981.
84. **Berg, K. and Heron, J.,** SDS-polyacrylamide gel electrophoresis of purified human leucocyte interferon and the antiviral and anticellular activities of the different interferon species, *J. Gen. Virol.,* 50, 441, 1980.
85. **Torres, A. R., Peterson, E. A., Evans, W. H., Mage, M. G., and Wilson, S. M.,** Fractionation of granule proteins of granulocytes by copper chelate chromatography, *Biochim. Biophys. Acta,* 576, 385, 1979.
86. **Rijken, D. C., Wijngaards, G., Zaal-De Jong, M., and Welbergen, J.,** Purification and partial characterization of plasminogen activator from human uterine tissue, *Biochim. Biophys. Acta,* 580, 140, 1979.
87. **Lonnerdal, B., Carlsson, J., and Porath, J.,** Isolation of lactoferrin from human milk by metal-chelate affinity chromatography, *FEBS Lett.,* 75, 89, 1977.
88. **Hansson, H. and Kagedal, L.,** Adsorption and desorption of proteins in metal chelate affinity chromatography. Purification of albumin, *J. Chromatogr.,* 215, 333, 1981.
89. **Porath, J.,** Explorations into the field of charge-transfer adsorption, *J. Chromatogr.,* 159, 13, 1978.
90. **Khazeli, M. B. and Mitra, R. S.,** Cadmium-binding component in Escherichia coli during accommodation to low levels of this ion, *Appl. Environ. Microbiol.,* 41, 46, 1981.
91. **Weerasinghe, K., Scully, M. F., and Kakkar, V. V.,** Purification of human clotting factor XII by Zn^{2+} chelate chromatography, *Biochem. Soc. Trans.,* 9, 336, 1981.
92. **Scully, M. F. and Kakkar, V. V.,** Zn^{2+} chelate chromatography of human fibrinogen, *Biochem. Soc. Trans.,* 9, 335, 1981.
93. **Ohkubo, J., Kondo, T., and Taniguchi, N.,** Purification of nucleoside diphosphates of rat liver by metal-chelate affinity chromatography, *Biochim. Biophys. Acta,* 616, 89, 1980.
94. **Salier, J. P., Martin, J. P., Lambin, P., McPhee, H., and Hochstrasser, K.,** Purification of the human serum inter-α-trypsin inhibitor by zinc chelate and hydrophobic interaction chromatography, *Anal. Biochem.,* 109, 273, 1980.
95. **Ryley, H. C.,** Isolation and partial characterization of a thiol proteinase inhibitor from human plasma, *Biochem. Biophys. Res. Commun.,* 89, 871, 1979.
96. **Kurecki, T., Kress, L. F., and Laskowski, M.,** Purification of human plasma α2 macroglobulin and α1 proteinase inhibitor using zinc chelate chromatography, *Anal. Biochem.,* 99, 415, 1979.
97. **Wunderwald, P., Schrenk, W. J., Port, H., and Kresze, G.,** Removal of endoproteinases from biological fluids by "sandwich affinity chromatography" with α2-macroglobulin bound to zinc chelate-sepharose, *J. Appl. Biochem.,* 5, 31, 1983.
98. **Bunning, R. A. D., Murphy, G., Kumar, S., Phillips, P., and Reynolds, J. J.,** Metalloproteinase inhibitors from bovine cartilage and body fluids, *Eur. J. Biochem.,* 139, 75, 1984.
99. **Gallowy, W. A., Murphy, G., Sandy, J. D., Gavrilovic, J., Cawston, T. E., and Reynolds, J. J.,** Purification and characterization of a rabbit bone metalloproteinase that degrades proteoglycan and other connective-tissue components, *Biochem. J.,* 209, 741, 1983.
100. **Horlein, D., Gallis, B., Brautigan, O. L., and Bornstein, P.,** Partial purification and characterization of phosphotyrosyl-protein phosphatase from Ehrlich ascites tumor cells, *Biochemstry,* 21, 5577, 1982.
101. **Lebreton, J. P.,** Purification of the human plasma $alpha_2$-SH glycoprotein by zinc chelate affinity chromatography, *FEBS Lett.,* 80, 351, 1977.

102. **Horie, N., Fukuyama, K., Ito, Y., and Epstein, W. L.,** Detection and characterization of epidermal proteinases by polyacrylamide gel electrophoresis, *Comp. Biochem. Physiol.*, 77B, 349, 1984.
103. **Smarrelli, J. and Campbell, W. H.,** Heavy metal inactivation and chelator stimulation of higher plant nitrate reductase, *Biochim. Biophys. Acta*, 742, 435, 1983.
104. **Feldman, S. R., Gonias, S. L., Neg, K. A., Pratt, S. W., and Pizzo, S. V.,** Identification of "embryonin" as bovine a_2-macroglobulin, *J. Biol. Chem.*, 259, 4458, 1984.
105. **Kikuchi, H. and Watanabe, M.,** Significance of use of amino acids and histamine for the elution of nonhistone proteins in copper-chelate chromatography, *Anal. Biochem.*, 115, 109, 1981.
106. **Mathur, R. and Balasubramanian, A. S.,** Cobalt-ion chelate affinity chromatography for the purification of brain neutral α-D-mannosidase and its separation from acid α-D-mannosidase, *Biochem. J.*, 222, 261, 1984.
107. **Borrebaeck, C. A. K., Lonnerdal, B., and Etzler, M. E.,** Metal chelate affinity chromatography of the dolichos biflorus seed lectin and its subunits, *FEBS Lett.*, 130, 194, 1981.
108. **Andersson, L.,** Fractionation of human serum proteins by immobilized metal affinity chromatography, *J. Chromatogr.*, 315, 167, 1984.
109. **Diwan, A. M. and Joshi, P. N.,** Fractionation of histones on a metal ion equilibrated cation exchanger I. Chromatographic profiles on an Amberlite® IR-120 (Al^{3+}) column, *J. Chromatogr.*, 173, 373, 1979.
110. **Abraham, K. I. and Joshi, P. N.,** Isolation of calotropain FI, *J. Chromatogr.*, 168, 284, 1979.

Chapter 5

SEPARATION OF ENANTIOMERS

V. A. Davankov

I. INTRODUCTION AND BACKGROUND: THREE-POINT INTERACTION

A triumph of LEC has been the separation of optical isomers of unmodified α-amino acids, which is one of the very important, but difficult to solve problems.

Optical isomers, or enantiomers, are absolutely identical to each other with respect to all their physical and chemical properties, so that the difference in their spatial structures can manifest itself only through their interaction with a chiral structure or field (optically active reagent, chiral surface, circular polarized radiation). Therefore, just as in any other process of chiral recognition and discrimination of two enantiomeric structures, the chromatographic resolution of two enantiomeric molecules nessarily requires a special chiral resolving agent to be involved into the interaction with these enantiomers.

A purely geometric consideration unambiguously shows that two enantiomers can be recognized by the chiral resolving agent only if at least three active positions of the agent (A, B, and C) simultaneously interact with appropriate positions (A′, B′, and C′) of one enantiomer. In this case, the corresponding positions of the second enantiomer would appear in a wrong sequence (A″, C″, and B″) with respect to the chiral reagent as shown in Figure 1. This would result in a definite difference in the total interaction energy of the chiral reagent with the aforesaid enantiomers, i.e., in the enantioselectivity of the reagent, enabling it to discriminate between the enantiomers.

A two-point attachment of the chiral agent to the enantiomeric structure is, as a matter of principle, insufficient for chiral recognition (Figure 1).

Enantioselectivity of the chiral resolving agent can be of a kinetic or a thermodynamic nature. Since the pioneering discoveries of Louis Pasteur, both these types of enantioselective processes have beeen successfully used for the practical resolution of racemic compounds into constituent optically active components. Without being converted into a mixture of diastereomeric compounds, the two enantiomers can be separated by a kinetically enantio-selective enzyme-catalyzed reaction or a thermodynamically enantioselective crystallization process. In these processes, the chiral resolving agents are the active site of the enzymatic catalyst and the elementary cells of the growing chiral crystal, respectively.

In order to achieve a significant discrimination of the enantiomers in the above "single-step" interaction processes with the resolving agents, the enantioselectivity of the latter must be very high. To obtain one of the enantiomers with the optical purity of 99% (i.e., to achieve a discrimination by a factor, α, of 100), the difference in the free energy of interactions, $\delta\Delta G$, within the two diastereomeric isomer-reagent adducts should amount to 11.3 kJ/mol at 298 K. There are but a limited number of crystallizing systems that display enantioselectivity of this order of magnitude. Much more often, a chemist has to repeat a crystallization dozens of times in order to obtain, at the expense of the yield, a more or less optically pure product. On the other hand, enzymatic reactions are restricted to naturally occurring classes of organic compounds.

However, the chromatography method of Tswett, one of the most powerful separation techniques, succeeds in completely separating compounds which differ in the free energy of interaction with the stationary phase as little as $\delta\Delta G = 40$ J/mol (selectivity $\alpha = 1.01$). Even preparative-scale separations are easy to perform in systems displaying $\delta\Delta G$ values of about 1 kJ/mol ($\alpha = 1.5$). The success of a chromatographic resolution is due to the fact

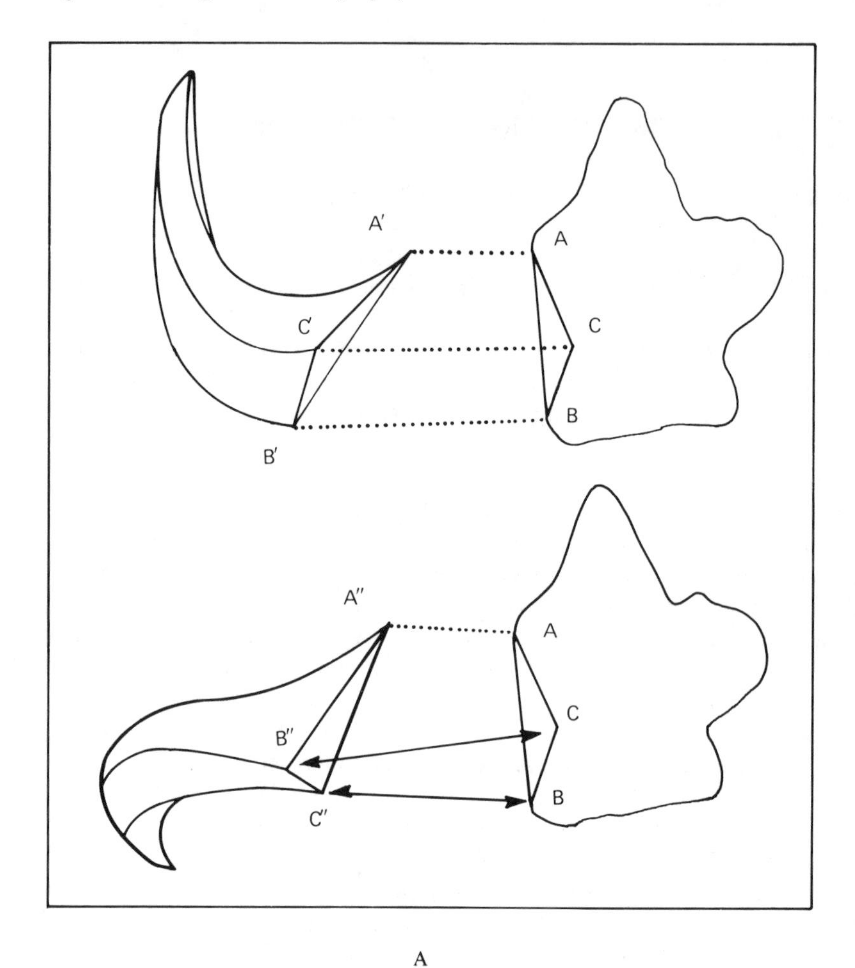

A

FIGURE 1. (A) A chiral recognition requires a three-point interaction of the chiral resolving agent with one of the enantiomers; (B) two-point interaction is inefficient (From Davankov, V. A. and Kurganov, A. A., *Chromatographia*, 17, 686, 1983. With permission.)

that the interactions of the enantiomers to be separated with the resolving agent are repeated many times as the enantiomers advance along the chiral stationary phase, with each single step of interaction helping to increase the distance between the two enantiomeric molecules.

Chromatography makes it possible to exploit effectively a great number of weak intermolecular interactions having low enantioselectivity and to apply one chiral resolving agent to a whole series of classes of racemic compounds.

Further, chromatography possesses another principle advantage against all the classical "single-step" procedures of resolving the racemates; using chromatographic techniques, it is possible to obtain quantitatively and in the optically pure state both enantiomers, even in the case where the chiral resolving agent has an optical purity lower than 100%. This statement, made by Davankov[1-3] on a theoretical consideration of the problem, is familiar today to every chromatographer, but was quite embarrassing for many researchers involved in stereochemistry who knew that the insufficient optical purity of the resolving agent necessarily led to an equivalent loss in optical purity or the yield of the resolved product.[4] Ten years later, Beitler and Feibush[5] derived and proved experimentally the following relation between the enantiomers separation selectivity α and α^x on chiral stationary phases of optical purity P and 100%, respectively:

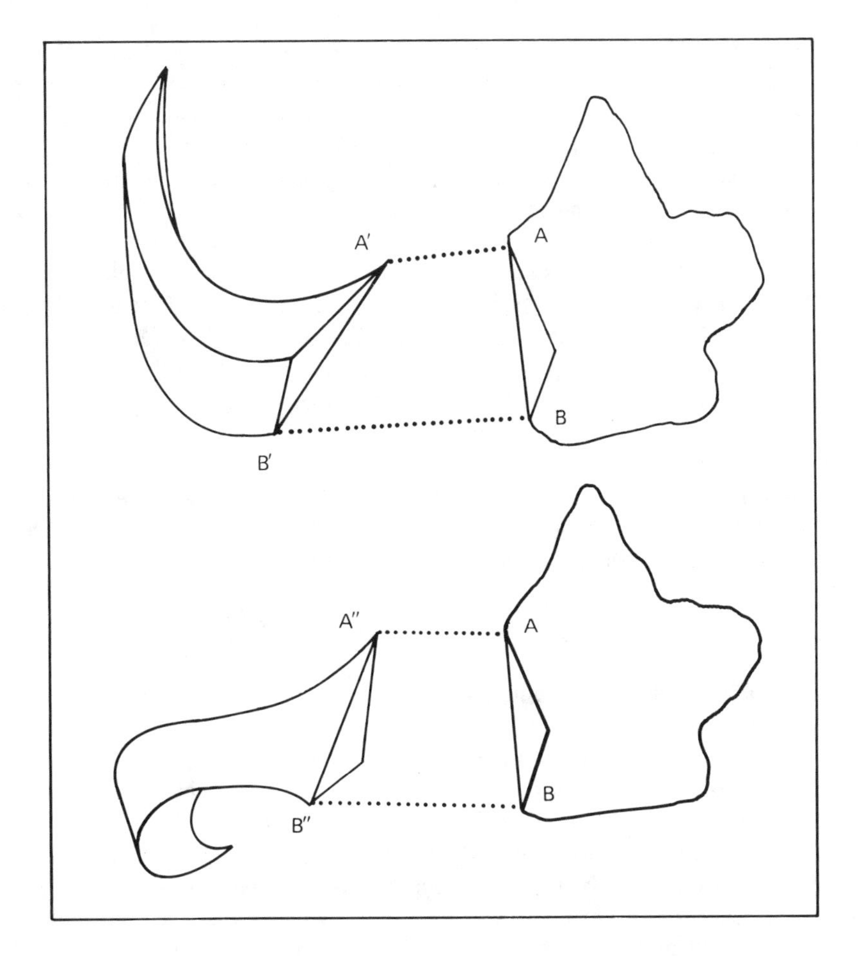

FIGURE 1B.

$$\alpha^{x} = \frac{(\alpha + 1)\,P + (\alpha - 1)\,100}{(\alpha + 1)\,P - (\alpha - 1)\,100}$$

Naturally, it is better to use chiral resolving agents of high optical purity, but the only theoretical requirement for a complete chromatographic resolution is that $\alpha^{x} > 1$ and $P > 0$. The practical selectivity level is, however, $\alpha \geqslant 1.01$ if the plate requirement is to remain below 10^5.

As the analytical method of choice for determining the enantiomeric composition of a chiral compound, chromatography is beyond any competition, for it is highly sensitive, reliable and rapid, and fairly independent of impurities that might accompany the compound of interest.

As the preparative method for obtaining optically active compounds, chromatography is distinguished by very high yields of enantiomers resolved and minimal losses of the valuable chiral resolving agent, i.e., chiral stationary phase.

Several review articles discuss numerous attempts — some failures, other partially successful — to achieve racemate resolutions on an analytical and preparative scale using chromatographic technique.[6-21] The last seven reviews are mainly devoted to LEC of optical isomers which has focused the greatest part of the effort in the last decade and developed especially fruitfully, contributing much to the theory and practice of LEC.

Returning to the "three-point interaction" model, we have to emphasize that it does not specify the nature of interactions between the corresponding active sites of the chiral resolving agent and enantiomers. Any kind of attractive or repulsive interactions within the diastereomeric adducts should prove productive in respect of the chiral recognition. In LEC, the resolving agent is a chiral ligand coordinated to a transition metal ion, and at least one of its interaction links to enantiomers is a coordination bond. Thus, the interaction leads to formation of mixed-ligand complexes.

Containing the chiral resolving agent of a definite sterical configuration and one or another of the two enantiomeric molecules, the two possible mixed-ligand complexes are diastereomeric. These diastereomeric adducts can possess different stabilities if the resolving agent, due to some other additional interactions, recognizes the configuration of the second ligand. A difference in the binding energy of two enantiomeric molecules to the resolving ligand is a necessary prerequisite for the resolution of the former. Thus, LEC of optical isomers is based on thermodynamic enantioselectivity in the formation of kinetically labile mixed-ligand complexes.

Depending on whether the chiral resolving ligand is bonded to the sorbent matrix or is added to the eluent, two main modifications of the LEC technique of enantiomers are described in the literature. One uses chiral sorbents and the other chiral eluents.

II. CHIRAL STATIONARY PHASES

A. Chiral Polymeric Ligand Exchangers

1. Synthesis of Chiral Resins

The first chiral ligand exchangers suitable for testing the idea of LEC of enantiomers were obtained by Davankov and Rogozhin starting from a cross-linked chloromethylated polystyrene and natural α-amino acids. At first sight, it looked feasible to preserve the coordination ability of both the α-amino and carboxy functions of the starting chiral component of the resin by simply alkylating the amino group with polymeric chloromethyl groups. But, much effort was needed to develop a method which would lead to a resin with a high exchange capacity and a strictly defined structure of chiral chelating ligands. Any possibility of racemization of asymmetric α-carbon atoms should be avoided.

First of all, to introduce chloromethyl groups entirely into the parapositions of the phenyl rings in a cross-linked polystyrene, the temperature of the chloromethylation procedure should be kept below 25°C.[22] It was found that a quantitative chloromethylation with monochlorodimethyl ether in dichloroethane at room temperature requires 2 to 10 days in the case of copolymers containing 0.3 to 5.0% divinylbenzene and $SnCl_4$ as the catalyst.[23]

As chloromethylpolystyrene was found to alkylate amino acids too slowly under mild conditions, appreciably more reactive[24] iodomethylated and bromomethylated copolymers were synthesized for the first time.[23,25] Later, a very simple method for obtaining *p*-bromo- and *p*-iodo-methylated polystyrene was found:[26,27] boiling the chloromethylated product with lithium bromide in a mixture of acetone with dioxane and sodium iodide in acetone. Conversion in the latter case amounts to 75 to 96%, depending on the degree of cross-linking of copolymers and the chlorine content. Finally, another significant improvement in the resin synthesis has been introduced: it was shown that sodium iodide was an excellent catalyst in reactions of chloromethylated styrene polymers with nucleophilic compounds in dioxane-methanol mixtures of 6:1 at 50°C,[24,28] so that the special conversion of chloromethylated copolymers into iodomethylated was not required.

Nevertheless, a direct interaction of the activated chloromethylpolystyrene with amino acids was found to suffer from an undesirable partial alkylation of the carboxy function of amino acids. This happened because in polar organic solvents or water, amino acids tend to exist in the form of a zwitter-ion with the amino group passivated via protonation and the carboxy group activated via deprotonation. Of the many natural amino acids investigated, only proline reacts almost entirely through its amino group. L-proline-containing resin (having an exchange capacity of 2.0 to 2.8 mmol/g dry material ligands) was thus the first chiral sorbent[29] to be applied for LEC of optical isomers. Structure of its repeating unit is shown below:

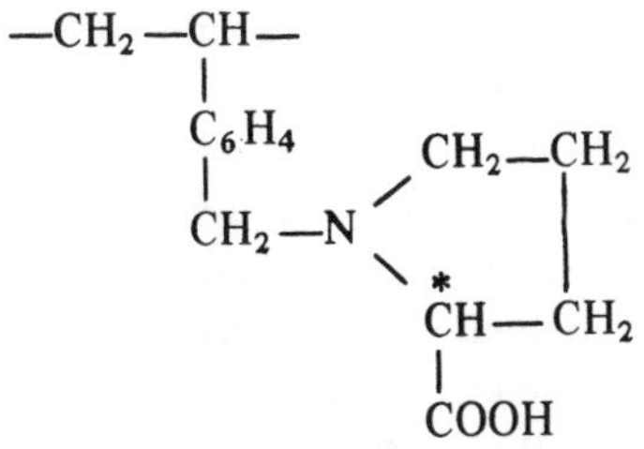

$$
\begin{array}{l}
-CH_2-CH- \\
\quad\quad\;\; | \\
\quad\quad C_6H_4 \quad\quad CH_2-CH_2 \\
\quad\quad\;\; | \quad\quad\quad\; / \quad\quad\quad | \\
\quad\quad CH_2-N \quad\quad\quad\quad\; | \\
\quad\quad\quad\quad\quad\;\; \backslash \;\; * \quad\quad\; | \\
\quad\quad\quad\quad\quad\quad CH-CH_2 \\
\quad\quad\quad\quad\quad\quad\; | \\
\quad\quad\quad\quad\quad\quad COOH
\end{array}
$$

Other amino acids can react easily with halogenomethylated styrene copolymers in the form of esters or amides.[28] These amino-acid derivatives are usually available as stable hydrochlorides. It was found that in the presence of $NaHCO_3$, hydrochlorides of amino esters can be used effectively in reactions with chloromethylpolystyrene.[30] Thus, it has become possible to obtain chiral resins of high exchange capacity starting from *p*-chloromethylated polystyrene and amino ester hydrochlorides taken in the molar ratio of 1.0:1.1 under very mild conditions, excluding any observable racemization.[2]

The situation was not equally favorable with respect to the removing of the carboxyl protecting groups in the resin — the ester hydrolysis still required prolonged boiling with alkaline solutions. Another significant problem was the very low swelling ability of resins obtained in the zwitter-ionic or metal-chelated form. Contrary to this, in acid or alkaline media, the weakly cross-linked resins swelled too much, so that changing the pH of the eluent often ruined the homogenity of the chromatographic columns packing. The use of a "macronet isoporous" polystyrene matrix helped to solve these two problems.

Davankov et al.[31] suggested cross-linking the linear polystyrene chains in the dissolved state with different bifunctional compounds. Since by this approach the cross-linking agents are distributed evenly along the whole of the polymer solution at the onset of the reaction, the cross-bridges also should be distributed evenly along the three-dimensional network produced. Such forming structures are related to the isoporous type. Reacting 4,4′-*bis*-chloromethyldiphenyl (CMDP) or *p*-xylilene dichloride (XDC) with polystyrene according to the Friedel-Crafts reaction, cross-bridges between two chains are formed which are two phenyl rings longer than the starter cross-linking agent:

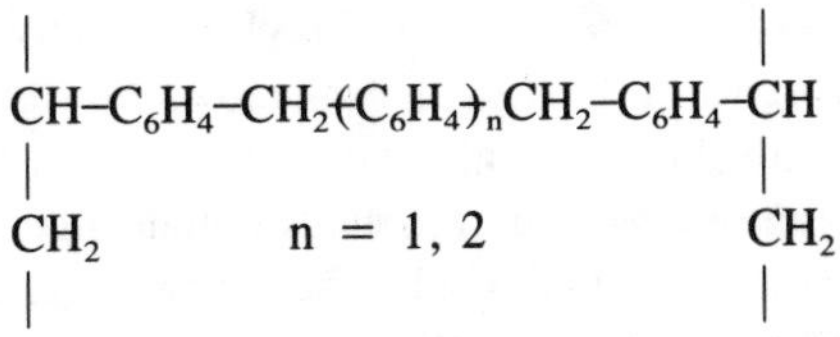

$$
\begin{array}{ccc}
| & & | \\
CH-C_6H_4-CH_2 & \!\!\!(C_6H_4)_n\!\!\! & CH_2-C_6H_4-CH \\
| & & | \\
CH_2 & n = 1, 2 & CH_2 \\
| & & |
\end{array}
$$

One can also use a strictly defined amount of monochlorodimethyl ether (MCDE) as the linking agent in a similar manner. In all cases the dimensions of the cross-bridges significantly exceed those of divinylbenzene (DVB) which is the standard cross-linking agent for styrene monomers. Therefore, the gels are appropriately referred to as macronet structures.

To obtain a macronet isoporous matrix in the form of regular beads, one has to use granular copolymers of styrene with small amounts of DVB (0.3 to 2.0%) in the maximally swollen state. It is convenient to accomplish the cross-linking process in boiling dichloroethane and to carry out the subsequent chloromethylation in the same vessel, at room temperature, after adding the necessary amount of MCDE and stannic chloride.

Of all the peculiar properties of the macronet isoporous styrene copolymers, on which more detailed information can be found in review articles by Davankov et al.,[32-34] the most important for the problem of LEC is the enhanced swelling ability of these three-dimensional networks.[35-39] There is a marked tendency characteristic of highly cross-linked (hyper cross-linked) polystyrene gels to maintain a spacious structure and a large volume in which their rigid matrix was formed and fixed. Unlike conventional styrene DVB copolymers, hyper cross-linked polystyrene is capable of swelling not only in solvating media like toluene, but also in liquids which do not dissolve linear polystyrene like methanol, hexane, water, etc.

This tendency appreciably improves the swelling properties of chiral resins prepared from macronet isoporous styrene copolymers. Their enhanced permeability is evident already from the fact that the hydrolysis of the ester groups in the products of amino ester reacting with chloromethylated macronet isoporous polystyrene is now possible even at room temperature. This makes the chiral resins containing a whole series of natural amino acid residues more available. Besides, due to a comparatively high degree of cross-linking, the resins obtained are mechanically stable and exhibit only a minimal change in their volume on changing pH of the eluent (Table 1), which makes their use in chromatographic columns very convenient. By the way, data presented in Table 1 are remarkable in that they contradict the commonly accepted conviction that increasing the degree of cross-linking of a copolymer must diminish its swelling ability. Apparently, additional cross-linking of highly swollen copolymers, as is the case with the preparation of macronet isoporous polystyrene matrices, brings about an opposite effect.[39]

Table 2 presents a list of chiral ligand exchangers synthesized and studied in LEC by Davankov[40] and his group. Here, in the majority of cases, amino acid methyl esters were introduced into the sodium-iodide-catalyzed reaction with chloromethylated macronet isoporous styrene copolymers. Valine and isoleucine (2 and 3) were taken in the form of *tert*-butyl esters, with the hydrolysis being carried out by boiling in a dioxane 20% aqueous HCl mixture.[41]

Some difficulties can arise with three-functional amino acids. For example, dimethylglutamate will suffer under cyclization reaction during an attempt to hydrolyze the ester groups in the polymer with alkali. The residues of pyroglutamic acid (11) formed herewith are unidentate ligands and are unable to chelate metal ions; unfortunately, they cannot be converted back into glutamate residues.[42] A similar cyclization reaction was observed with 2,4-diaminobutyric acid methyl ester leading to a mixture of two ligand structures (12). However, acid hydrolysis of the cyclic amide bonds can be achieved in this instance to give linear tridentate polymer-fixed ligands (13).[43]

Both the carboxy and α-amino groups of lysine can be protected effectively by chelating copper(II) ions, which gives the opportunity to alkylate selectively the ε-amino groups.[44] Since water is the only solvent for $Cu(Lys)_2$, instead of chloromethylated copolymer, the product of its interaction with dimethyl sulfide should be used to alkylate the complex.[45] Copper complexes of 2,4-diaminobutyric acid were found to be inert in this reaction, and those of ornithine react very slowly. Obviously, the γ-amino groups are blocked by copper ions in the 2,4-diaminobutyric acid, thus acting as a terdentate ligand, while the δ-amino groups in ornithine are only partially chelated.[44]

Histidine and methionine methyl esters react with chloromethylated polystyrene under mild conditions of catalytic alkylation only with their α-amino groups, which are more nucleophilic than imidazole and sulfide (15 and 16).[46,47] Contrary to this, one can make use of the high nucleophilicity of the cysteine HS-group to prepare the resin (20)

Table 1
IMPROVING THE SWELLING ABILITY OF CHIRAL LIGAND EXCHANGERS BY AN ADDITIONAL CROSS-LINKING OF THEIR POLYSTYRENE-TYPE MATRICES

Fixed ligand	Cross-linking of the matrix	Exchange capacity (mmol/g)	Solvent uptake (g/g)				
			Water	0.5 *N* HCl	0.5 *N* KOH	Dioxane	Methanol
–NH–CH–COOH with side chain CH_2–COOH on the CH	0.8% DVB	2.4	0.16	0.40	0.41	0.42	0.31
	0.8% DVB 5% CMDP	2.1	1.39	1.55	1.66	1.10	1.05
–N< in a five-membered ring: N–C(=O)–CH_2–CH_2–CH(COOH)–N	0.8% DVB	1.6	0.19	0.27	0.46	0.70	0.40
	0.8% DVB + 5% CMDP	2.0	0.95	1.32	1.45	1.00	0.83

From Davankov, V. A., Tsyurupa, M. P., and Rogozhin, S. V., *Angew. Makromol. Chem.*, 53, 19, 1976. With permission.

Table 2
CHIRAL CHELATING RESINS BASED ON CHLOROMETHYLATED MACRONET ISOPOROUS POLYSTYRENE AND α-AMINO ACIDS

Number	Fixed ligand	Capacity (mmol/g)	Swelling ability (wt %)			
			Water	0.5 *N* HCl	0.5 *N* KOH	Ethanol
1	Alanine[30] $-NHCH(COOH)CH_3$	2.2	89	102	122	108
2	Valine[41] $-NHCH(COOH)CH(CH_3)_2$	2.3	84	118	156	117
3	Isoleucine[41] $-NHCH(COOH)CH(CH_3)C_2H_5$	2.5	54	82	132	101
4	Proline[40] $-N<$ ($CH_2—CH_2$ / $CH—CH_2$, ring; CH–COOH)	2.3	88	135	130	98
5	Serine[30] $-NHCH(COOH)CH_2OH$	3.0	93	108	127	91
6	Threonine[30] $-NHCH(COOH)CH(OH)CH_3$	2.9	94	109	149	113
7	Tyrosine[30] $-NHCH(COOH)CH_2C_6H_4OH$	2.2	81	88	109	76
8	Hydroxyproline[30] $-N<$ ($CH_2—CH–OH$ / $CH—CH_2$, ring; CH–COOH)	3.0	136	165	211	76

9	allo-hydroxyproline[40] 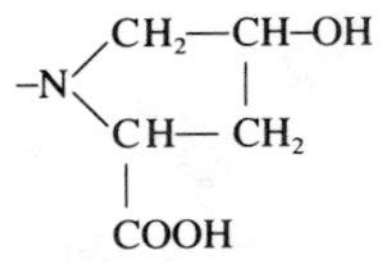 $-N(CH_2-CH(OH)-CH_2-CH(COOH))$ (ring)	3.2	149	170	210	84
10	Aspartic acid[42] $-NHCH(COOH)CH_2COOH$	2.5	139	155	166	94
11	Pyroglutamic acid[42] 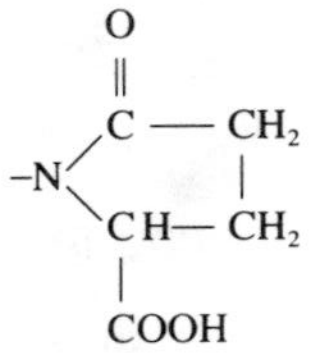 $-N(C(=O)-CH_2-CH_2-CH(COOH))$ (ring)	2.2	95	151	132	80
12	Diaminobutyric acid (cycl.)[43] 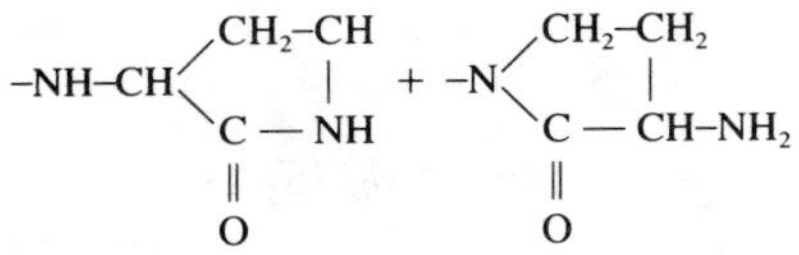 $-NH-CH(CH_2-CH-NH-C(=O))$ (ring) + $-N(CH_2-CH_2-CH(NH_2)-C(=O))$ (ring)	2.3	83	97	83	73
13	Diaminobutyric acid[43] $-NHCH(COOH)CH_2CH_2NH_2$ + $-NHCH_2CH_2CH(NH_2)COOH$	2.2	143	112	156	76
14	Lysine[44] $-NH(CH_2)_4CH(NH_2)COOH$	1.7	120	130	127	102

Table 2 (continued)
CHIRAL CHELATING RESINS BASED ON CHLOROMETHYLATED MACRONET ISOPOROUS POLYSTYRENE AND α-AMINO ACIDS

Number	Fixed ligand	Capacity (mmol/g)	Swelling ability (wt %)			
			Water	0.5 *N* HCl	0.5 *N* KOH	Ethanol
15	Histidine[46] –NHCH(COOH)CH_2–C(=CH–NH–CH=N–) (imidazolyl)	2.0	110	131	122	92
16	Methionine[47] –NHCH(COOH)$CH_2CH_2SCH_3$	2.1	100	105	129	82
17	Methionine-(*dl*)-sulfoxide[56] –NHCH(COOH)$CH_2CH_2S(\rightarrow O)CH_3$	2.1	131	133	160	92
18	Methionine-(*d*)-sulfoxide[57] –NHCH(COOH)$CH_2CH_2S(\rightarrow O)CH_3$	2.2	85	—	—	—
19	Methionine-(*l*)-sulfoxide[57] –NHCH(COOH)$CH_2CH_2S(\rightarrow O)CH_3$	2.2	80	—	—	—
20	Cysteine[45] –$SCH_2CH(NH_2)COOH$	2.3	98	90	114	34
21	Cysteic acid[49] –NHCH(COOH)CH_2SO_3H	1.2	112	108	127	105

22	*O*-hydroxyproline[40] –O–CH —— CH_2 │ │ CH_2 CH–COOH ╲ ╱ NH	1.0	17	20	19	53
23	Azetidine carboxylic acid[58] –N — CH–COOH │ │ CH_2–CH_2	2.4[a]	140	—	—	—
24	Phenylalanine[50] –NHCH(COOH)$CH_2C_6H_5$	2.1[a]	51	53	59	32
25	Ar-phenylalanine[50] –C_6H_4-CH_2CH(NH_2)COOH	1.1[a]	95[b]	100	110	55
26	*N*-carboxymethylvaline[59] –N(CH_2COOH)CH(COOH)CH$(CH_3)_2$	1.5	100	83	120	112
27	*N*-carboxymethylaspartic acid[59] –N(CH_2COOH)CH(COOH)CH_2COOH	1.2	44	28	62	68
28	*S*-(2-aminoethyl)cysteine[51] –NHCH(COOH)$CH_2SCH_2CH_2NH_2$	1.8[a]	240	240	270	230
29	*S*-carboxymethylcysteine[51] –NHCH(COOH)CH_2SCH_2COOH	1.6[a]	82	94	86	58
30	*S,S'*-ethylene-*bis*-cysteine[51] COOH HOOC │ │ –NHCH $CHNH_2$ │ │ $CH_2SCH_2CH_2SCH_2$	1.1[a]	55	65	75	67

Table 2 (continued)
CHIRAL CHELATING RESINS BASED ON CHLOROMETHYLATED MACRONET ISOPOROUS POLYSTYRENE AND α-AMINO ACIDS

			Swelling ability (wt %)			
Number	Fixed ligand	Capacity (mmol/g)	Water	0.5 *N* HCl	0.5 *N* KOH	Ethanol
31	*N,N'*-ethylene-*bis*-methionine[51] $-N(CH(CH_2CH_2SCH_3)COOH)-CH_2CH_2-NH(CH(CH_2CH_2SCH_3)COOH)$	0.9[a]	53	58	59	46
32	α-Amino benzylphosphonic acid[52] $-NHCH(C_6H_5)P(O)(OH)_2$	2.0[c]	57	56	129	50
33	α-Amino benzylphosphonic acid monoethyl ester[52] $-NHCH(C_6H_5)P(O)(OH)OC_2H_5$	2.2[c]	67	54	101	65
34	α-Amino-α-methylbenzyl-phosphonic acid[53] $-NHC(CH_3)(C_6H_5)P(O)(OH)_2$	0.9	45	48	45	46
35	α-Amino-α-methylpropyl-phosphonic acid[53] $-NCH(CH_3)(C_2H_5)P(O)(OH)_2$	0.6	42	41	45	47
36	α-amino-α-methylpentyl-phosphonic acid[53] $-NHC(CH_3)(C_4H_9)P(O)(OH)_2$	0.8	64	60	52	65
37	α-amino-α-methylpentyl-phosphonic acid monoethyl ester[53] $-NHC(CH_3)(C_4H_9)P(O)(OH)(OC_2H_5)$	1.3	45	48	52	61

[a] Degree of cross-linking 10% (MCDE).
[b] The resin was additionally sulfonated (1.3 mmol/g).
[c] Degree of cross-linking 7.5% (0.5% DVB + 7% XDC); degree of cross-linking of other resins 6% (0.8% DVB + 5% CMDP).

$$\begin{array}{l} | \\ CH{-}C_6H_4{-}CH_2Cl \\ | \\ CH_2 \\ | \end{array} + HSCH_2\overset{*}{C}HCOOH\underset{|}{} \ (NH_2\cdot HCl) \xrightarrow{NaI} \begin{array}{l} | \\ CH{-}C_6H_4{-}CH_2{-}S{-}CH_2\overset{*}{C}HCOOH \\ | \qquad\qquad\qquad\qquad\qquad\qquad | \\ CH_2 \qquad\qquad\qquad\qquad\qquad NH_2 \\ | \end{array}$$

in a simpler way[45] than Roberts and Haigh's synthesis.[48]

Cysteic acid (21) was brought into reaction with a chloromethylated copolymer in the presence of 2 mol ethyldiisopropylamine necessary for neutralizing two acid functions. As expected, a noticeable portion of the amino acid was linked to the polymer matrix via ester groups.[49]

An attempt was made to direct the alkylation process toward the hydroxyproline HO-group (the amino and carboxy functions being protected by formylation and esterification, respectively),[40] and toward the phenylalanine aromatic nucleus (22 and 25).[50]

The synthesis of chiral resins with polyfunctional ligands, especially sulfur-containing polyfunctional ligands like 28 through 31, is quite laborious because several selective protecting groups are required to prevent undesirable routes of reaction with a chloromethylated copolymer.[51]

Similar persistence proved the synthesis of chiral resins containing optically active α-amino phosphonic acid residues (32 through 37),[52,53] since the enantiomeric resolution of the racemic amino phosphonic acids was rather tedious.[54,55]

Altogether, about 40 chiral chelating polystyrene-type resins based on α-amino carboxylic and α-amino phosphonic acids have been prepared and investigated so far by Davankov and his group. Though the properties of many of these resins are still far from being optimal, an extensive search was considered necessary for selecting structures which really deserved a detailed study. We will see later that L-proline, first bonded onto a polystyrene matrix, happened to give one of the most efficient chiral ligand exchangers of the above amino acid series.

Several other research groups also made use of natural amino acids and cross-linked styrene copolymers for the synthesis of chiral ligand exchangers. The copolymers were, however, of conventional styrene-DVB type and not of the macronet isoporous type. This probably accounts for lower exchange capacity of resins obtained in these synthesis as compared to those of sorbents in Table 2.

As early as 1971, Angelici et al.[60,61] prepared a resin containing 0.5 mmol/g residues of *N*-carboxymethyl-L-valine in a styrene 0.8% DVB copolymer (compared to 1.5 mmol/g in a macronet isoporous matrix of a degree of cross-linking as high as 6% [Table 2] 26). Chloromethyl groups of the initial copolymer were found to require activation, either via conversion into iodomethyl groups or dimethylsulfonium groups, prior to the alkylation of the amino component:

−CH−CH₂− (C₆H₄−CH₂Cl) —NaI, acetone→ −CH−CH₂− (C₆H₄−CH₂I) —1. CM—L—Val—OEt₂; 2. NaOH→ −CH−CH₂− (C₆H₄−CH₂−N−*CH(CH(CH₃)₂)−COOH; N−CH₂−COOH)

−CH−CH₂− (C₆H₄−CH₂Cl) —S(CH₃)₂, i−ProH+H₂O→ −CH−CH₂− (C₆H₄−CH₂S⁺(CH₃)₂Cl⁻) —(CM−L−Val)Na2, i−ProH+H₂O→ −CH−CH₂− (C₆H₄−CH₂−N−*CH(CH(CH₃)₂)−COOH; N−CH₂−COOH)

The second route was found to give better results. However, we would like to draw attention to the fact that because of the absence of protecting groups at the carboxyl functions of carboxymethyl-L-valine, a portion of chiral ligands in the resin obtained most probably was linked via ester bonds.

French workers[62,63] arrived at similar conclusions on searching for optimal conditions of binding various amino acids (phenylalanine, proline, alanine, threonine, glutamic acid, and hydroxyproline) onto copolymers of styrene with 1 to 8% DVB. The best achievement was the introduction of 1.5 mmol L-proline per gram copolymer with 1% DVB (4 in Table 2 — 2.3 mmol/g). Again, iodomethylated copolymers showed lower attachment yields than did their sulfonium analogs when reacted with free, unesterified amino acids.

Another French group was able to bind 0.6 mmol L-histidine per gram conventional copolymer with 2% DVB (15 in Table 2 — 2.0 mmol/g). Somewhat higher histidine content (0.8 mmol/g) in the resin was achieved along with copolymerization of *N*-*p*-vinylbenzyl-L-histidine methyl ester with hydroxyethyl methacrylate and ethylene dimethacrylate.[64]

Tsuchida et al.[65,66] have chosen another route to prepare a polystyrene resin with fixed L-leucine ligands:

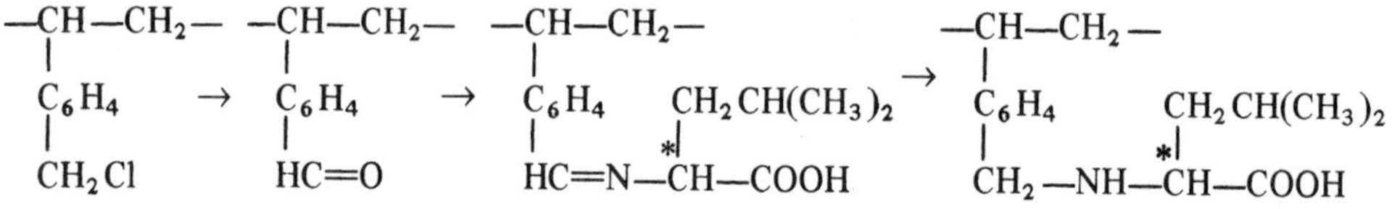

$$\begin{array}{c} -CH-CH_2- \\ | \\ C_6H_4 \\ | \\ CH_2Cl \end{array} \rightarrow \begin{array}{c} -CH-CH_2- \\ | \\ C_6H_4 \\ | \\ HC{=}O \end{array} \rightarrow \begin{array}{l} -CH-CH_2- \\ \;| \\ C_6H_4 \qquad CH_2CH(CH_3)_2 \\ \;| \qquad\qquad\quad *| \\ HC{=}N-CH-COOH \end{array} \rightarrow \begin{array}{l} -CH-CH_2- \\ \;| \\ C_6H_4 \qquad\quad CH_2CH(CH_3)_2 \\ \;| \qquad\qquad\qquad *| \\ CH_2-NH-CH-COOH \end{array}$$

The exchange capacity of the polymer amounted to 1.48 mmol/g, but all tranformations were made on linear polystyrene, and the final sorbent, being a powder, was less suitable for column packing.

Thus, the main problems with chiral sorbents containing amino acids in a cross-linked polystyrene matrix is their low exchange capacity, and consequently, low swelling capacity and slow mass transfer. Thus far, only the macronet isoporous polystyrene matrices afford a possibility to overcome these obstacles.

Naturally, one can enhance the water uptake of the chiral polystyrene resin by a subsequent partial sulfonation as was done with the sorbent (25 in Table 2) containing L-phenylalanine linked to the matrix through its aromatic nucleus.[50] Partially sulfonated resins are automatically obtained by the approach of Vesa,[67] who reacted a chlorosulfonated polystyrene copolymer with L-phenylalanine to give the following structure:

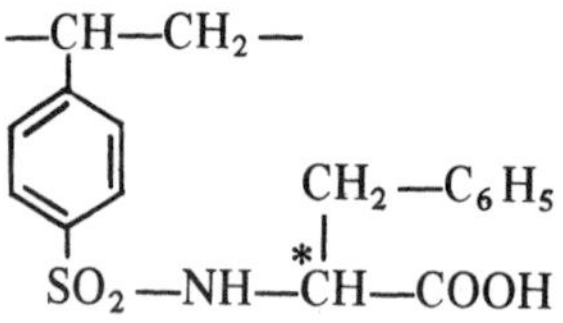

$$\begin{array}{l} -CH-CH_2- \\ \;| \\ \;C_6H_4 \qquad\qquad CH_2-C_6H_5 \\ \;| \qquad\qquad\qquad *| \\ SO_2-NH-CH-COOH \end{array} \qquad 2.42\ \text{mmol/g}$$

Several other chiral amino acids were also successfully bonded onto chlorosulfonated polystyrene.[62,68] The ability of the sulfonamide group to coordinate metal ions is weaker than that of the free amino group of the above sorbents. This does not imply, however, that sorbents based on chlorosulfonated styrene copolymers (that are easily available) should not be examined intensively in LEC of optical isomers.

Finally, the last polystyrene-based series of chiral ligand exchangers should be concerned with containing optically active amine or polyamine as the matrix-fixed ligands.

Patent applications[69,70] were made known claiming optical resolution of racemic amino acids with the aid of chiral tetramine metal complexes. In particular, nickel(II) chelate of a polystyrene-type resin containing 2.5 mmol/g of fixed ligands of the structure:

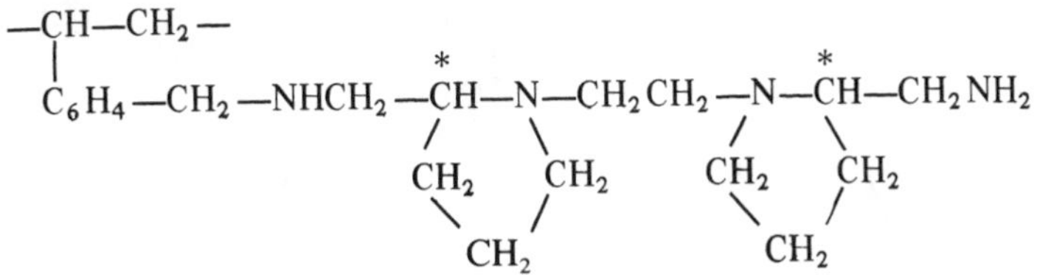

```
—CH—CH2—
 |
 |                        *                    *
C6H4—CH2—NHCH2—CH—N—CH2CH2—N—CH—CH2NH2
                 /    \          /    \
              CH2     CH2     CH2     CH2
                 \    /          \    /
                  CH2              CH2
```

were used for LEC of DL-alanine to give a total resolution of 81 to 86%.

Kurganov et al.[71] described synthesis of four chiral anion exchangers based on a macronet isoporous styrene copolymer and derivatives of (*S*)-1-phenylethylamine or (R)-1,2-propanediamine. The repeating units of the sorbents are represented by the following structures:

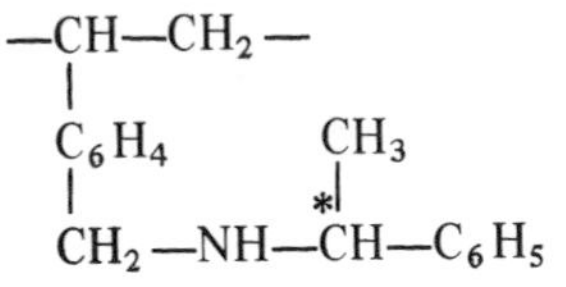

```
—CH—CH2—
 |
C6H4        CH3
 |          *|
CH2—NH—CH—C6H5

             2.88 mmol/g
```

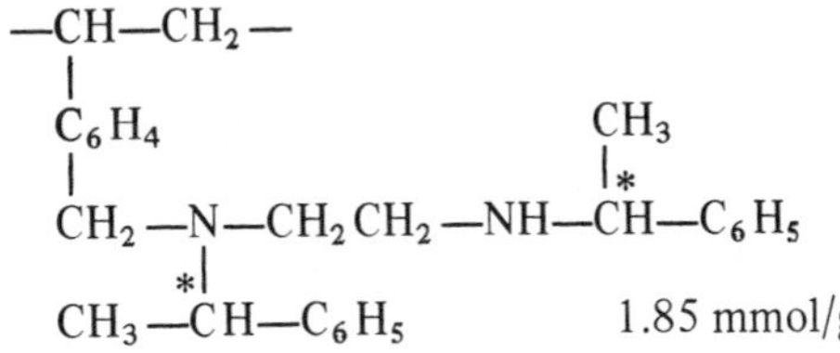

```
—CH—CH2—
 |
C6H4                        CH3
 |                           |*
CH2—N—CH2CH2—NH—CH—C6H5
     *|
CH3—CH—C6H5             1.85 mmol/g
```

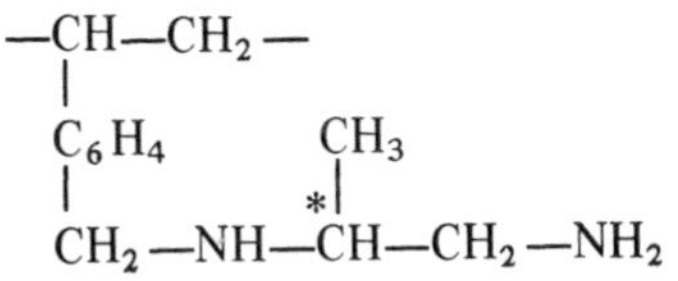

```
—CH—CH2—
 |
C6H4        CH3
 |          *|
CH2—NH—CH—CH2—NH2

             2.45 mmol/g
```

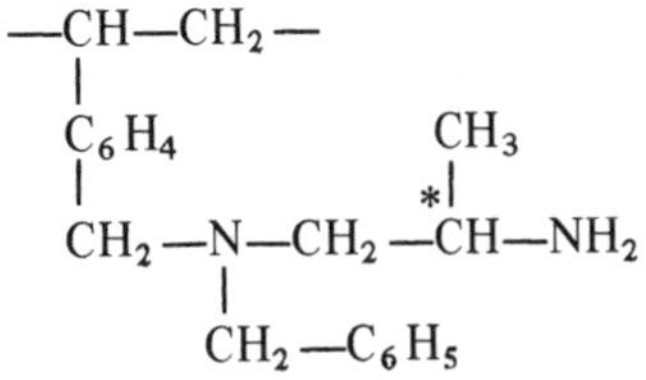

```
—CH—CH2—
 |
C6H4              CH3
 |                *|
CH2—N—CH2—CH—NH2
     |
    CH2—C6H5            2.71 mmol/g
```

Unexpectedly, the first two resins showed poor affinity to metal ions, probably because of steric overcrowding and low basicity of benzylated nitrogens. On the contrary, the last two sorbents due to chelation effects retained copper(II) ions satisfactorily and proved rather efficient in the chromatographic resolution of racemic amino acids.[71] Though propylenediamine is a bifunctional compound, the resins based on it had a well-defined structure of active sites. This was achieved through selective protection of the N^1 or N^2 atoms of propanediamine with benzaldehyde which forms Schiff bases with primary amino groups, thus preventing their alkylation. The authors[72] were lucky to find that the reaction of 1,2-propanediamine with benzaldehyde can be carried out selectively, with the N^1/N^2 ratio of mono-*N*-benzylidene-1,2-propanediamine isomers being 9:1. On the basis of these findings, the following synthetic routes were realized:

$$\underset{}{H_2NCH(CH_3)CH_2NH_2} \longrightarrow H_2NCH(CH_3)CH_2N{=}HCC_6H_5 \longrightarrow H_2NCH(CH_3)CH_2NH(CH_2C_6H_5) \longrightarrow C_6H_5CH{=}NCH(CH_3)CH_2NH(CH_2C_6H_5)$$

$$H_2NCH(CH_3)CH_2N{=}HCC_6H_5 \longrightarrow {-}CH(C_6H_4CH_2NHCH(CH_3)CH_2N{=}HCC_6H_5){-}CH_2{-} \longrightarrow {-}CH(C_6H_4CH_2{-}NHCH(CH_3)CH_2NH_2){-}CH_2{-}$$

$$C_6H_5CH{=}NCH(CH_3)CH_2NH(CH_2C_6H_5) \longrightarrow {-}CH(C_6H_4CH_2N(CH_2C_6H_5)CH_2CH(CH_3)N{=}HCC_6H_5){-}CH_2{-} \longrightarrow {-}CH(C_6H_4CH_2{-}N(CH_2C_6H_5)CH_2CH(CH_3)NH_2){-}CH_2{-}$$

The resins obtained readily absorb copper(II) ions from solutions in an acetate buffer. The highest sorption value corresponds to the formation of complexes with the composition of 1 mol of diamine per 1 mol of copper(II) in the resin phase and is easily achieved at relatively low concentrations of copper(II) in solution. However, the sorption capacity of the resins unexpectedly drops sharply in the presence of NH_3 (Figure 2), and at higher NH_3 concentrations, copper ions are not sorbed at all. In contrast, in the case of the previously described aminocarboxylate-type sorbents, the presence of ammonia in high concentrations exerts no effect on the sorption (60 to 70% of the theoretical capacity calculated for 2:1 complexes even in 2 to 3 *M* ammonia). This sharp difference in the behavior toward copper(II) is presumably connected with the unfavorable charge of the anion-exchanger phase during chelation: the diamine polymer complex with copper has a positive charge which leads to the exclusion of the positively charged Cu^{2+} and $Cu(NH_3)_4{}^{2+}$ ions from the sorbent phase. This phenomenon is opposite to the enhanced complexing properties of the negatively charged (in an alkaline media) amino acid-type resins. In the case of acetate buffer, most of the copper exists in the form of electrically neutral acetate complexes which easily enter the anion-exchanger phase.

In a Swiss patent,[73] a chiral ligand exchanger is described, prepared by reacting optically active *N*-acetyl-1,2-propanediamine with a chloromethylated styrene copolymer, followed by a hydrolysis of the acetamide bonds with 6 *N* H_2SO_4, and finally alkylated with sodium chloroacetate. The resin contained two to three acetate residues per propanediamine molecule, and taken in the copper(II) form, partially resolved some amino acids.

Of course, cross-linked polystyrene, though extremely popular in synthesis of various ion exchangers and other sorbents, is not the only possible matrix for a chiral ligand exchanger. Being aware of difficulties arising from the hydrophobic nature of conventional styrene-DBV-type matrices, Lefebvre et al.[74] suggested an alternative, highly hydrophilic matrix which is a cross-linked polyacrylamide. With the aid of formaldehyde, they grafted L-proline onto small beads of an acrylamide-methylenebisacrylamide copolymer to obtain a resin of the following structure:

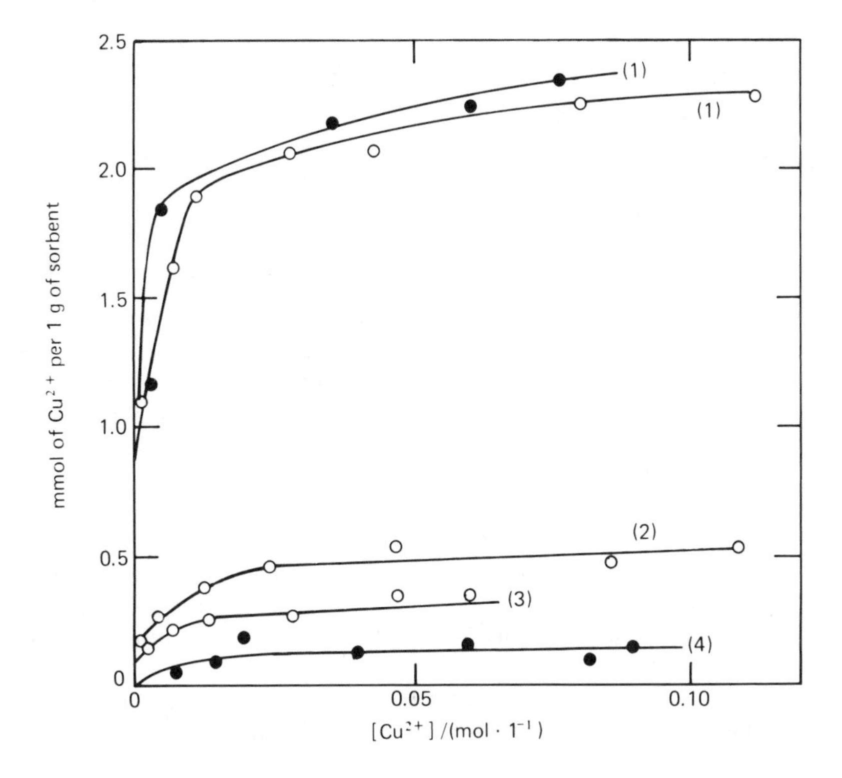

FIGURE 2. Isotherms of Cu^{2+} sorption to chiral anion exchangers incorporating residues of 1,2-propanediamine (○) and N^1-benzyl-1,2-propanediamine (●). (1) 1 *M* sodium acetate, pH 6; (2) 0.2 *M* NH_4OH; (3) 0.4 *M* NH_4OH; (4) 0.5 M NH_4OH. (From Kurganov, A. A., Zhuchokova, L. Ya., and Davankov, V. A., *Makromol. Chem.*, 180, 2101, 1976. With permission.)

```
–CH–CH2–
 |
 C = O
 |           CH2–CH2
 |          /       |
 NH–CH2–N           |
            \   *   |
             CH–CH2
             |
             COOH
```

Due to the low molecular weight of the repeating unit of polyacrylamide, the amount of the grafted ligands can be relatively high, varying between zero and 3 mmol/g. Besides proline, several other chiral amino acids have been grafted onto polyacrylamide: alanine, valine, threonine, phenylglycine, histidine, azetidine carboxylic acid, hydroxyproline, pipecolic acid, and phenylalanine.[75] Herewith, a whole palette of hydrophilic chiral ligand exchangers has become available. The majority of them, just like the majority of the polystyrene-type chiral sorbents, still remains only superficially investigated.

Finally, two other types of hydrophilic polymers have been used for binding chiral amino acids. Russian workers[76] made use of macroporous poly(2,3-epoxypropyl methacrylate), G-Gel-60 and Separon H1000, to bind hydroxyproline and proline, respectively. Japanese investigators[77] took advantage of highly porous polyethylene glycol gel, TSK-Gel® 2000 PW, the primary alcohol groups of which were treated with epichlorohydrine, followed by coupling with L-tryptophan, L-phenylalanine, or L-histidine. In all these cases, epoxypropyl groups of the matrix alkylate the amino groups of the ligands to be fixed according to the scheme:

$$\bar{P}\text{–O–CH}_2\text{–}\underset{\diagdown\,O\,\diagup}{\text{CH – CH}_2} + \text{H}_2\text{N}\overset{*}{\text{C}}\text{H(R)COOH} \rightarrow \bar{P}\text{–O–CH}_2\text{–}\underset{\text{OH}}{\text{CH}}\text{–CH}_2\text{–NH–}\underset{\text{COOH}}{\overset{*}{\text{C}}\text{H}}\text{–R}$$

where $\bar{P}$ stands for polymer.

No doubt, several other polymers, chiral ligands, and linkage types will follow the above described structures. Today, after the principle of LEC of optical isomers was invented and its usefulness shown, it is appropriate to make improvements on the resin structures. Generally for a resin to be convenient in practical use, it has to meet several requirements:[15]

1. The exchange capacity of the resin should be as high as possible (this is not necessary for packings used in analytical columns).
2. The structure of the resin elementary unit should be unique and exactly known.
3. The racemization of the optically active starting component should be excluded during the synthesis of the resin.
4. The resin should be mechanically, chemically, and configurationally (with respect to asymmetric centers) stable.
5. The swelling ability of the resin has to be enough to ensure a rapid establishment of the interphase equilibria and constant enough to prevent sealing off or channeling of the column on changing the eluent.
6. The final resin should not be very expensive.

As far as the configurational stability of resins described is concerned, data are available only on structures with a methylene bridge between polystyrene and α-amino groups of an aminocarboxylic ligand. It was known that metal ions catalyze racemization of optically active amino acids, and indeed, the chiral complex $Cu(\text{L-Ala})_2$ was found to lose half of its optical activity in 2 *N* NH_4OH in 24 hr. Fortunately, *N*-benzyl derivatives of L-alanine and other amino acids proved to be stable under the same set of conditions,[78] which coincides with the fact that only a minimal loss in the resolving power of resins containing alanine and threonine could be detected after 1 year of using them, and no change whatsoever was observed in respect to resins with L-proline and L-hydroxyproline after 7 years of usage. Chemical and configurational stability of polystyrene sulfonamide-type resins and those based on polyacrylamide still needs to be examined.

The final comment in this section addresses the ability of chiral resins to retain transition metal ions. The relative stability of metal complexes with carboxylic and amino donors decreases in the following series of M(II) ions: Cu > Ni > Zn > Cd. Except for unidentate fixed ligands like pyroglutamic acid and 1-phenylethylamine, the majority of other ligands investigated bind metal ions with sufficient strength, especially in the case of high concentrations of fixed ligands in the resin phase. Trifunctional and polyfunctional ligands exhibit an enhanced metal binding capacity (lysine, methionine, and methioninesulfoxide behave like bidentate ligands). There is a surprisingly high affinity of cysteine and cysteic acid for nickel(II) and aminophosphonic acids for iron(III).

2. *Chromatographic Resolution of Racemic Amino Acids*

LEC on chiral sorbents was suggested in 1968 by Rogozhin and Davankov as a general approach to resolve racemates of compunds capable of forming complexes with transition metal ions.[79] The first publications[80-83] described already a complete semipreparative resolution of 0.5 g DL-proline on a polystyrene-type ligand exchanger containing residues of L-proline as the matrix-fixed chiral ligands. The procedure was described above, in Chapter 1. Today, no one would start investigating a new chromatographic technique with column loadings that high. But at that time, sophisticated chromatographic equipment was hardly available so that a common methodology was to use big columns and loadings large enough to permit a quantitative analysis of a big number of the effluent fractions. By measuring their optical activity and knowing the amount of the racemate applied and the theoretical optical activity of its pure enantiomers, it was easy to calculate the extent of resolution.[84] i.e., the total optical yield. This is why the results of testing of the first chiral resins synthesized (see Table 2) in LEC of typical racemic amino acids are represented (see Table 3) in the form of ''degree of resolution''. Now, this parameter allows one to have an idea of the prospects of applying one or another resin to preparative-scale resolutions.

As was shown,[85] there are several ways of achieving solute desorption in LEC: (1) elution by replacing ligands (ammonia, pyridine, ethylenediamine); (2) elution with complexing metal ions; (3) lowering the pH of the eluent; and (4) increasing the temperature of the column. It was convenient to start with pure water and then gradually raise the ammonia concentration in the eluent until both enantiomers of the initial racemate were displaced from the column. The packing materials were used in the copper(II) and nickel(II) forms and sometimes in the zinc(II) and cadmium(II) forms. The retention of the majority of solutes was found to decrease in the same order of metal ions, coinciding with the decreasing stabilities of amino acid complexes of these metal ions.

From Table 3 one can easily recognize that in most resolutions of bifunctional amino acids and mandelic acid (2-hydroxy-2-phenyl-acetic acid), the best results are obtained with copper(II) ions. With tridentate fixed and mobile ligands, octahedral nickel(II) ions often display greater enantioselectivity compared to copper. Apparently, the separation of optical isomers becomes most efficient when the number of coordinate bonds formed by the stationary and mobile ligands is equal to the coordination number of the central metal ion. In this instance the mobile ligand can coordinate only in one way, with the sorption complex having a single possible structure.

Chiral resins containing $\overline{R}$Ala, $\overline{R}$Pro, $\overline{R}$aHyp,$\overline{R}$Tyr, $\overline{R}$Met, $\overline{R}$MetSO, $\overline{R}$His, and $\overline{R}$Asp as the polystyrene-fixed ligands of the L configuration usually form more stable copper(II) complexes with the D-enantiomers of bidentate mobile amino acid ligand than with L-enantiomers. Contrary to this, mixed-ligands sorption complexes with trifunctional mobile ligands (threonine, aspartic acid, ornithine), as well as bulky fixed ligands ($\overline{R}$Val, $\overline{R}$Ile) often exhibit higher stability when both constituent ligands belong to the same configurational series (see underlined numbers in Table 3). In respect to the *N*-carboxylmethyl-L-valine resin, the data of Table 3 are consistent with earlier observations of Snyder et al.[61] in that the D-valine and D-proline are eluted from the column ahead of the L-enantiomers; however, higher selectivity was found for proline than for valine. (Aside from these two solutes, these authors[61] reported the same elution order of enantiomers and falling resolutions for isoleucine, norvaline, aminobutyric acid, and alanine).

Although it is hardly possible to find strict regularities ruling the resolution efficiency, still one can point out rising enantioselectivity with the growing size of α-radicals in the fixed and mobile ligands. Amino acids having additional functional groups, but unable to get involved in interaction with the central metal ion, as is the case with lysine and glutamic acid, are resolved to a lesser extent than their analogues with nonpolar aliphatic side chains.

α-Amino phosphonic acids as the polystyrene-fixed chiral ligands apparently yield in their resolving power to those of amino carboxylic-type ligands. Of the sorbents listed in Table 3, most promising are the structures containing proline, hydroxyproline, allo-hydroxyproline, methionine-sulfoxide, isoleucine, and histidine. It does not imply, however, that other resin matrices or other types of bonded ligands should not produce different regularities.

Table 3
DEGREE OF RESOLUTION (%) OF 0.1 TO 0.2 g RACEMIC AMINO ACIDS[a] AND MANDELIC ACID ON 30-mℓ COLUMNS PACKED WITH METAL CHELATES OF CHIRAL RESINS[b] (SEE TABLE 2) BY ELUTION WITH WATER OR AMMONIA SOLUTIONS[c]

No.[d]	Resin	Me^{2+}	Ala	Val	Iva	Ile	Pro	Ser	Thr	Asp	Orn	His	aThr	Met	Mandelic acid
1	RAla[90]	Cu	13	32	41	20	64	37	20	8	0	—	—	—	13
		Ni	—	6	—	—	3	—	8	13	7	—	—	—	1
		Zn	—	—	—	—	3	—	5	7	3	—	—	—	4
2	RVal[90]	Cu	19	31	37	37	17	15	20	15	6	—	—	—	16
		Ni	—	5	—	—	11	—	24	10	1	—	—	—	6
		Zn	—	—	—	—	2	—	4	5	0	—	—	—	8
3	RIle[90]	Cu	26	23	68	23	21	10	21	9	4	—	—	—	8
		Ni	—	6	—	—	9	—	25	25	4	—	—	—	7
		Zn	—	—	—	—	8	—	10	0	0	—	—	—	5
4	RPro[84]	Cu	—	55	87	43	100	—	—	—	3	—	—	38	28
		Ni	—	—	20	10	10	—	0	—	—	—	—	—	—
5	RSer[91]	Cu	—	23	17	—	12	—	22	13	0	—	—	—	18
		Ni	—	13	—	—	22	—	12	1	0	—	—	—	1
		Zn	—	—	—	—	3	—	4	0	0	—	—	—	3
6	RThr[91]	Cu	—	23	40	49	19	8	21	14	0	—	—	—	28
		Ni	—	7	—	—	32	—	17	12	6	—	—	—	4
		Zn	—	—	—	—	19	—	11	4	4	—	—	—	4
7	RTyr[91]	Cu	—	21	54	21	21	30	14	13	18	—	—	—	17
		Ni	—	9	—	—	13	—	10	13	8	—	—	—	4
		Zn	—	—	—	—	1	—	18	0	0	—	—	—	6
8	RHyp[84]	Cu	35	90	36	100	93	79	100	—	10	—	—	46	33
		Ni	—	—	22	24	16	—	14	—	—	—	—	—	—
9	RaHyp[40]	Cu	0	—	—	—	100	—	—	0	—	100	—	—	45
10	RAsp[42]	Cu	8	—	40	27	90	—	16	12	0	—	—	—	21
		Ni	6	—	8	18	6	—	2	12	0	—	—	—	1
11	RPyGlu[42]	Cu	—	2	—	—	4	—	4	0	—	—	5	—	3
		Ni	—	0	—	—	2	—	2	0	—	—	—	—	0
12	RDAB-cycl.[43]	Cu	—	12	—	—	18	2	3	3	3	—	—	—	18
		Ni	—	8	—	—	16	—	11	13	0	—	—	—	3
13	RDAB[43]	Cu	—	12	—	—	40	—	1	15	4	—	—	—	1
		Ni	—	5	—	—	12	—	1	17	0	—	—	—	0
14	R-ξ-NH_2Lys[40]	Cu	5	8	5	31	3	—	7	0	2	—	10	—	1
		Ni	8	13	—	84	9	—	10	0	5	—	22	—	1
15	RHis[92]	Cu	—	7	9	3	94	—	10	35	18	29	14	0	8
		Ni	—	4	4	15	47	—	14	36	8	54	12	0	1
		Cd	—	—	—	—	1	—	5	0	5	—	—	—	—
16	RMet[92]	Cu	7	30	27	9	48	2	11	—	—	—	—	—	17
		Ni	—	—	—	—	7	—	—	16	—	—	21	—	—
17	RMetSO-(*dl*)[92]	Cu	8	60	15	60	60	3	14	—	0	—	—	—	16
		Ni	—	—	—	—	8	—	—	60	—	—	24	0	—
20	R-*S*-Cys[40]	Cu	—	11	—	—	65	—	7	0	0	—	—	0	3
		Ni	—	7	—	—	14	—	4	0	0	—	—	0	1
21	RCysSO_3H[49]	Cu	2	15	18	28	33	—	9	0	1	—	5	—	2
		Ni	2	5	—	34	14	—	2	0	11	—	16	0	2

Table 3 (continued)
DEGREE OF RESOLUTION (%) OF 0.1 TO 0.2 g RACEMIC AMINO ACIDS[a] AND MANDELIC ACID ON 30-mℓ COLUMNS PACKED WITH METAL CHELATES OF CHIRAL RESINS[b] (SEE TABLE 2) BY ELUTION WITH WATER OR AMMONIA SOLUTIONS[c]

No.[d]	Resin	Me^{2+}	Ala	Val	Iva	Ile	Pro	Ser	Thr	Asp	Orn	His	aThr	Met	Mandelic acid
26	RCMVal[59]	Cu	—	$\underline{12}$	—	—	$\underline{15}$	—	$\underline{10}$	$\underline{6}$	$\underline{4}$	—	—	—	0
		Ni	—	$\underline{4}$	—	—	$\underline{5}$	—	$\underline{1}$	$\underline{14}$	12	—	—	—	$\underline{3}$
27	RCMAsp[59]	Cu	—	—	—	—	$\underline{17}$	—	—	—	—	—	—	—	—
	Aminophosphonic Acids														
32	Benzyl[52]	Cu	—	—	—	—	3	—	—	—	—	—	—	—	—
33	Benzyl OC_2H_5[52]	Cu	—	—	—	—	3	—	—	—	—	—	—	—	—
34	Methyl, benzyl[53]	Cu	—	—	—	—	9	—	—	—	—	—	—	—	—
35	Methyl, propyl[53]	Cu	—	—	—	—	2	—	—	—	—	—	—	—	—
36	Methyl, pentyl[53]	Cu	—	9	—	—	34	—	25	1	0	—	—	—	1
37	Methyl, pentyl OC_2H_5[53]	Cu	—	—	—	—	12	—	—	—	—	—	—	—	—

[a] Numbers underlined indicate systems displaying an "inverse" (D before L) elution order of enantiomers.
[b] Particle size, 100 to 200 μm.
[c] Flow rate, 10 to 15 mℓ/hr.
[d] Referring to Table 2.

It is only natural that amino acids, being strong complexing agents, partially strip metal ions from the metal-saturated column packing. Since for the polarimetric measurements metal-free effluents were wanted, additional small columns were introduced charged with the same chiral resin, or better, with Dowex® A-1 in the metal-free form.[79,80] Later, it was recognized that the (partial) elution of amino acids in the form of metal chelates provides the opportunity of easy detection in the effluent using simple ultraviolet detectors operating at 206[58,86,87] or 254 nm.[71,88,89] This made it possible to use much smaller quantities of both the racemate and the chiral resin and to evaluate enantioselectivity as the ratio of (corrected) elution volumes (i.e., capacity factors), $\alpha = k'_D/k'_L$, of the two enantiomers chromatographed separately or as a racemic mixture.

In a more convenient manner, Table 4 sums up the results of evaluating the enantioselectivity of racemic amino acid chromatography on a further series of the polystyrene-type chiral resins. Here, $\alpha > 1$ when D-enantiomers are retained longer than their L-antipodes on packings incorporating natural L-amino-acid residues as the fixed ligands. Conversely, $\alpha < 1$ when $k'_D < k'_L$. One can see again that copper(II) ions should be preferred to nickel(II) ions when chromatography of bifunctional mobile ligands proceed on resins with bidentate fixed ligands of the L-phenylalanine type bound to polystyrene via α-amino or phenyl groups. Alone, tridentate histidine exhibits better resolutions with nickel(II) ions. The situation becomes quite different when the fixed ligands are potentially terdentate (*S*-aminoethyl-cysteine and *S*-carboxymethyl-cysteine) or hexadentate (*S,S'*-ethylene-*bis*-methionine and *N,N'*-ethylene-*bis*-methionine). In this instance, metal ions with the coordination number of six take priority over copper(II) ions.

The very high resolving power of chiral sorbents operating according to the ligand exchange mechanisms is evident from the fact that enantioselectivity observed in many systems reaches and even exceeds that limit ($\alpha \geqslant 1.5$ or $\alpha \leqslant 0.67$) which is usually used as a criterion for separating two components on a preparative scale. This is especially true with the five most

Table 4
VALUES OF SELECTIVITY, $\alpha = k'_D/k'_L$, OF RESOLUTION OF RACEMIC AMINO ACIDS ON THE POLYSTYRENE-TYPE CHIRAL LIGAND EXCHANGERS (SEE TABLE 2) BY ELUTION WITH WATER OR AMMONIA SOLUTIONS[50,51,57,71]

No.[a]	Resin	Me^{2+}	Ala	Val	Leu	Pro	Phe	His	Thr	Ser	Met
18	RMetSO*d*	Cu	1.07	1.25	1.30	1.35	1.48	0.65	1.24	—	1.09
		Ni	1.00	1.00	1.11	1.38	1.21	0.45	1.14	—	—
19	RMetSO*l*	Cu	1.19	1.29	1.21	1.78	1.40	0.55	1.27	—	1.28
		Ni	1.00	1.03	1.07	1.07	1.19	0.81	1.20	—	—
24	RPhe	Cu	1.11	1.19	1.32	1.72	2.00	0.67	1.35	0.74	0.88
		Ni	—	1.06	1.11	1.23	1.54	1.64	—	—	—
25	RarPhe	Cu	1.16	1.40	0.67	0.76	0.70	0.79	0.77	1.46	0.59
		Ni	—	1.12	0.90	0.83	0.80	0.73	—	—	—
28	RAECys	Cu	1.19	0.74	1.31	1.31	0.74	0.52	1.49	—	0.86
		Ni	1.11	1.14	1.43	1.18	0.63	1.82	1.04	—	0.57
29	RCMCys	Cu	0.96	1.16	0.90	1.18	1.58	1.75	1.31	—	0.77
		Ni	0.90	1.14	1.18	1.41	0.87	0.57	0.88	—	0.79
30	REBCys	Cu	1.14	1.35	1.54	1.33	1.69	2.02	0.77	—	1.33
		Ni	0.93	1.16	1.26	0.83	1.56	0.50	1.19	—	1.58
31	REBMet	Cu	0.98	1.19	0.87	1.11	1.62	1.69	0.33	—	1.09
		Ni	0.79	0.63	1.41	1.35	0.63	2.15	0.82	—	0.54
	Rpn[b]	Cu	1.05	1.22	—	1.28	—	0.87	—	—	—

[a] Referring to Table 2.
[b] Resin fixed ligand, (R)-1,2-propanediamine; eluent, 1 *M* sodium acetate of pH 6.

efficient chiral resins of the whole above polystyrene-type sorbents described thus far. They contain cyclic amino acids L-proline, L-hydroxyproline, (its diastereomer) L-allo-hydroxyproline, and L-azetidine carboxylic acid, as well as (R)-N^1-benzyl-1,2-propanediamine as the chiral resolving ligands. These resins also possess the widest application range when combined with copper(II) as complexing metal ions. Table 5 sums up enantioselectivity values attained, showing that all common α-amino acid racemates can be successfully resolved using one or another of these five chiral resins.

Figure 3 represents resolutions of 13 typical amino acids on a macronet isoporous polystyrene resin containing 3.86 mmol/g of L-hydroxyproline residues. Here, 14-cm-long columns were used, which are rather short for the 50-μm resin particles of an irregular shape. From these results[88] it has become clear that LEC can be very useful in rapidly determining the enantiomeric composition of amino acids. It would have a significant advantage over the well known gas-chromatographic method in that the preparation of volatile derivatives is fully superfluous and that no purification of amino acids from mineral salts and other accompanying compounds is necessary.

The last statement needs some additional comments. It will be shown below that copper(II) ions, as well as ions of any other metal, would always partition between the resin phase and eluent depending on their composition. Complexing by-products and neutral salts in the eluent, as well as its pH, would influence the partition coefficient. Therefore, if the sample

Table 5
SELECTIVITY VALUES, $\alpha = k'_D/k'_L$, OF RESOLUTION OF RACEMIC AMINO ACIDS ON THE COPPER(II) FORMS OF POLYSTYRENE-TYPE RESINS CONTAINING RESIDUES OF L-PROLINE,[58] L-HYDROXYPROLINE,[86] L-ALLO-HYDROXYPROLINE,[87] L-AZETIDINE CARBOXYLIC ACID,[58] AND N^1-BENZYL-(R)-PROPANEDIAMINE-1,2[93]

Racemic amino acid	$\overline{R}$Pro	$\overline{R}$Hyp	$\overline{R}$aHyp	$\overline{R}$AzCA	$\overline{R}$Bznp
Alanine	1.08	1.04	1.04	1.06	0.70
Aminobutyric acid	1.17	1.22	1.18	1.29	0.49
Norvaline	1.34	1.65	1.42	1.24	0.48
Norleucine	1.54	2.20	1.46	1.40	0.49
Valine	1.29	1.61	1.58	1.76	0.31
Isovaline	—	1.25	—	—	—
Leucine	1.27	1.70	1.54	1.24	0.49
Isoleucine	1.50	1.89	1.74	1.68	0.62
Serine	1.09	1.29	1.24	2.15	0.62
Threonine	1.38	1.52	1.48	0.78	0.44
allo-Threonine	1.55	1.45	—	—	—
Homoserine	—	1.25	—	—	—
Methionine	1.04	1.22	1.52	1.29	0.60
Asparagine	1.18	1.17	1.20	1.44	0.70
Glutamine	1.20	1.50	1.40	1.25	—
Phenylglycine	1.67	2.22	1.78	1.38	0.49
Phenylalanine	1.61	2.89	3.10	1.86	0.52
α-Phenyl-α-alanine	—	1.07	—	—	0.38
Tyrosine	2.46	2.23	2.36	1.78	—
Phenylserine	—	1.82	—	—	—
Proline	4.05	3.95	1.84	2.48	0.47
Hydroxyproline	3.85	3.17	1.63	2.25	—
allo-Hydroxyproline	0.43	0.61	1.48	1.46	—
Azetidine carboxylic acid	—	2.25	—	—	—
Ornithine	1.0	1.0	1.20	1.0	1.0
Lysine	1.10	1.22	1.33	1.06	1.0
Histidine	0.37	0.36	1.32	0.56	0.85
Tryptophan	1.40	1.77	1.10	1.13	—
Aspartic acid	0.91	1.0	0.81	0.88	1.0
Glutamic acid	0.62	0.82	0.69	0.77	—

under analysis contains large amounts of contaminating compounds, the base line in the chromatogram is no longer as stable as during the calibration experiments. Mineral salts, through increasing the ionic strength of the eluent, cause additional desorption of copper ions which would be registered by an UV detector as a separate peak followed by a drop of the baseline corresponding to the reestablishment of the distribution equilibria. By changing concentration of copper ions in the initial eluent, it is possible to influence the position of the "negative" peak, yet these artifacts can interfere with enantiomeric peaks of interest in the chromatogram. The best method of suppressing the fluctuations of the base line when analyzing protein hydrolysates or other contaminated samples[94] is to apply eluents sufficiently strong in ionic strength and buffer capacity. These ingredients are needed, therefore, for improving the detection, but not the resolution of the enantiomers. Paper[94] is also interesting in that very short columns of 1 to 2 cm in length were used to shorten down to a few minutes the analysis time of the enantiomeric composition of tryptophan, valine, proline, phenylalanine, and tyrosine (as well as their esters), and that conditions have been found to analyze selectively aromatic amino acids in their mixtures with other amino acids and peptides.

Aside from the above analytical applications, the same L-hydroxyproline incorporating

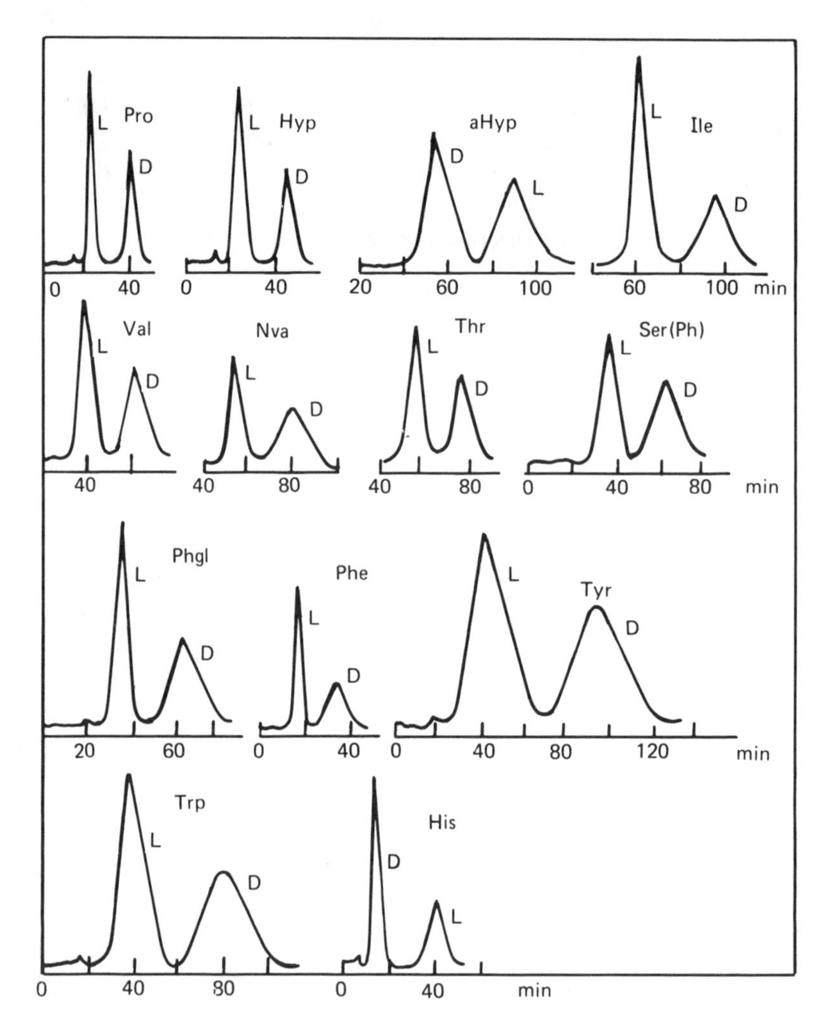

FIGURE 3. Chromatography of about 1 mg racemic acids on a 7.8 × 140-mm column containing 6.3 mℓ of the copper loaded polystyrene-type resin with L-hydroxyproline ligands (capacity, 3.86 mmol/g; degree of cross-linking, 6%; particles of irregular shape, about 50 μm). Detector, Uvicord I without light filter. Proline: 65% copper, 1.0 *M* NH_4OH, 20 mℓ/hr. Hydroxyproline and allo-hydroxyproline: 65% copper, 0.5 *M* NH_4OH, 20 mℓ/hr. Isoleucine: 65% copper, 0.1 *M* NH_4OH, 13 mℓ/hr. Valine: 65% copper, 0.1 *M* NH_4OH, 13 mℓ/hr. Norvaline: 65% copper, 0.05 *M* NH_4OH, 16 mℓ/hr. Threonine: 80% copper, 0.05 *M* NH_4OH, 20 mℓ/hr. Phenylserine: 45% copper, 0.05 *M* NH_4OH, 14 mℓ/hr. Phenylglycine: 65% copper, 0.1 *M* NH_4OH, 13 mℓ/hr. Phenylalanine: 45% copper, 0.1 *M* NH_4OH, 20 mℓ/hr. Tyrosine: 65% copper, 0.1 *M* NH_4OH, 16 mℓ/hr. Tryptophan: 30% copper, 0.4 *M* NH_4OH, 20 mℓ/hr. Histidine: 30% copper, 0.5 *M* NH_4OH, 25 mℓ/hr. (From Davankov, V. A., Zolotarev, Yu, A., and Tevlin, A. V., *Bioorg. Khim.*, 4, 1164, 1978. With permission.)

polystyrene-type resin can be used successfully for preparative scale separations. A column of 1 ℓ in volume (about 300 g of resin which swelling capacity amounts to 250%) completely resolves in one chromatographic run up to 20 g DL-proline (α = 3.95), 5 g DL-threonine (α = 1.52), or 5 g DL-mandelic acid (α = 1.65) into optically pure enantiomers.[89] Micro-preparative resolutions of DL-leucine,[95] as well as tritium containing DL-[^{3}H] valine[96,97] and DL-[^{3}H]histidine,[97] are described in detail. LEC is the method of choice for the commercial-scale production of the optically pure tritium-labeled amino acids of very high radioactivity, since no crystallization procedures can be applied to these products undergoing a rapid decomposition in a condensed state or concentrated solution. Some earlier publications[1,63,79,83] describing preparative resolution of DL-proline on polystyrene-type resins with fixed L-proline ligands should also be mentioned here.

Examples of baseline resolutions are given[89] for racemates representing other than α-amino acids, classes of organic compounds, β-phenyl-β-alanine, mandelic acid, 2-amino-propanol-1, and N^1-benzyl-1,2-propanediamine which molecular structures are given below:

$$C_6H_5\overset{*}{C}H(NH_2)CH_2COOH$$

$$C_6H_5\overset{*}{C}H(OH)COOH$$

$$CH_3\overset{*}{C}H(NH_2)CH_2OH$$

$$CH_2(NHCH_2C_6H_5)\overset{*}{C}H(NH_2)CH_3$$

The L-proline-containing resin resolves racemates of ephedrine, valine amide, and a stable radical[79,84,89] having the following formula

$$C_6H_5\overset{*}{C}H(OH)\overset{*}{C}H(NHCH_3)CH_3, \qquad (CH_3)_2CH\overset{*}{C}H(NH_2)CONH_2$$

```
             *
         CH2 —CH—NH2
  H3C     |     |    CH3
     \    |     |   /
       C        C
     /   \     /  \
  H3C      \  /     CH3
            N
            |
            O•
```

respectively. Thus, amino acids are by no means the only object of resolutions using LEC, but rather a convenient model for developing the general technique.

When trying to understand the process of chiral recognition of amino-acid enantiomers in the polystyrene-type chiral resins, one has to bear in mind that the formation of ternary sorption complexes proceeds in the resin phase swollen with water. On several occasions it has been noticed that the higher the exchange capacity and the swelling ability of the polystyrene-type resins, the higher their resolving power and the chromatography efficiency.[63,88] This leads to a suggestion that neither the macromolecular matrix structure nor the neighboring chain segments or repeating units do significantly affect the stereochemical situation within the individual diastereomeric ternary sorption complexes. If this is true, it should be possible to simulate the structure of a sorption complex in the resin phase by a low-molecular-weight model, comprised of an amino acid, copper ion, and *N*-benzyl-L-proline, with the latter representing the structure of the L-proline-containing polystyrene resin.

Copper(II) complexes of *N*-benzyl-proline,[98-103] among those of other *N*-alkylated amino acids, have been intensively examined by Davankov and Kurganov. We will leave considering the stereochemical aspects of these studies for a later point, and deal now only with some thermodynamic results of these and related experiments. It has been found that of the two diastereomeric *bis*-complexes, CuLL, and CuDL, where D and L stand for the *N*-benzyl-proline ligands of the opposite configurations, the latter complex is by 4.8 ± 0.8 kJ/mol more stable. (This was the first example[98] where thermodynamic enantioselectivity was detected in a *bis*-complex of copper with bidentate amino acid ligands). Due to the very unusual fact that the entropic contribution in this particular complex throughout the studied temperature range happened to predominate over the enthalpic contribution to the net enantioselectivity, the excess stability of the CuDL complex increased somewhat with rise in temperature.[103] This growing enantioselectivity at a rising temperature agrees with the results of LEC: racemic *N*-benzyl-proline resolves on the L-proline-containing resin with higher α-values when chromatographed at elevated temperatures, with the D-isomer being always the stronger retained component.

Similarly, in several carefully investigated systems, a good qualitative consistency has been found between the elution order of two enantiomers and the difference in stabilities of the two corresponding diastereomeric complexes in solution. Thus, D-valine,[104] D-proline,[104-106] and L-histidine[105-107] are the stronger coordinated species in both the L-proline-containing resin and in the ternary model-complexes with *N*-benzyl-L-proline. Likewise, *N*-benzyl-valine of the L configuration has the higher affinity toward the copper complexes of both the polymeric[15] and soluble[108] *N*-benzyl-L-valine ligands. The same preferred coordination of the L-isomers of valine, leucine, phenylalanine, and alanine was observed to occur in ternary complexes with *N*-carboxymethyl-L-valine and in the polystyrene resin containing these ligands.[61,109,110] There is even good quantitative agreement between the (by 3.6 kJ/mol) higher sorption energy of D-proline on the copper-loaded L-proline containing resin[58] on the one hand, and the enantioselectivity value (of 3.8 ± 1.0 kJ/mol) estimated for corresponding ternary model complexes with *N*-benzyl-L-roline and D- or L-proline on the other hand.[104]

However, Tsuchida et al.[65] were confused by the opposite signs of enantioselectivity in ternary copper complexes containing polymeric or soluble ligands of the *N*-benzyl-L-leucine type: the L-leucine-containing resin retained more strongly the D-isomers of valine, leucine, and alanine, whereas *N*-benzyl-L-leucine displayed higher affinity to the L-isomers. Therefore, the authors were forced to assume that the fixed ligands in their resin happened to possess a different structure than had been expected, namely, that of *N,N*-dibenzyl-L-leucine.

Another contradiction that arose with the polystyrene-type chiral resins is that French authors[105,106] report the absence of any noticeable enantioselectivity in the chromatography of DL-phenylalanine on the copper(II) form of the L-proline resin, as well as equal stability constants of two corresponding ternary complexes of phenylalanine enantiomers with *N*-benzyl-L-phenylalanine, whereas a Russian group[58] has found the discriminating ability of the resin to be sufficiently high for the same racemate ($\alpha = 1.63$). This discrepancy may be attributed to the difference in the performance of the two resins used: the first contained 1% DVB, 1.53 mmol L-proline per gram dry resin, and swelled with water to about 60%;[63,106] the second had a macronet isoporous type matrix with cross-linking degree of 11%, exchange capacity of 2.78 mmol/g, and swelling ability of 170%.[58] The enantioselectivity value of $\alpha = 1.63$ would correspond to the difference in stability constants of two diastereomeric ternary complexes of about 0.20 logarithmic units, which hardly exceeds the error limits of potentiometric determination of the enantioselectivity effects in labile complexes.

Spassky et al.[64] reexamined another polystyrene-type resin containing L-histidine as the fixed ligands. Being loaded with nickel(II) ions, the resin displayed higher affinity to the D-isomer of histidine, in full agreement with the earlier reported results.[92] This particular case is interesting in that both the fixed and mobile ligands are terdentate in accordance with the coordination number of six of the octahedral nickel(II) ion.

When testing the chiral resin prepared by reacting a chlorosulfonated polystyrene with L-phenylalanine, Vesa[67] reported that it partially resolved (in the copper form) racemates of proline, valine, serine, threonine, and to a lesser degree, some other amino acids. Recently,[111] enantioselectivity has been found in ternary copper(II) complexes containing valine, phenylalanine, or proline, besides the *N*-benzenesulfonyl-L-phenylalanine. In consistence with the chromatographic data reported, ternary complexes with L-Val, D-Pro, and L-Phe are preferred diastereomers.

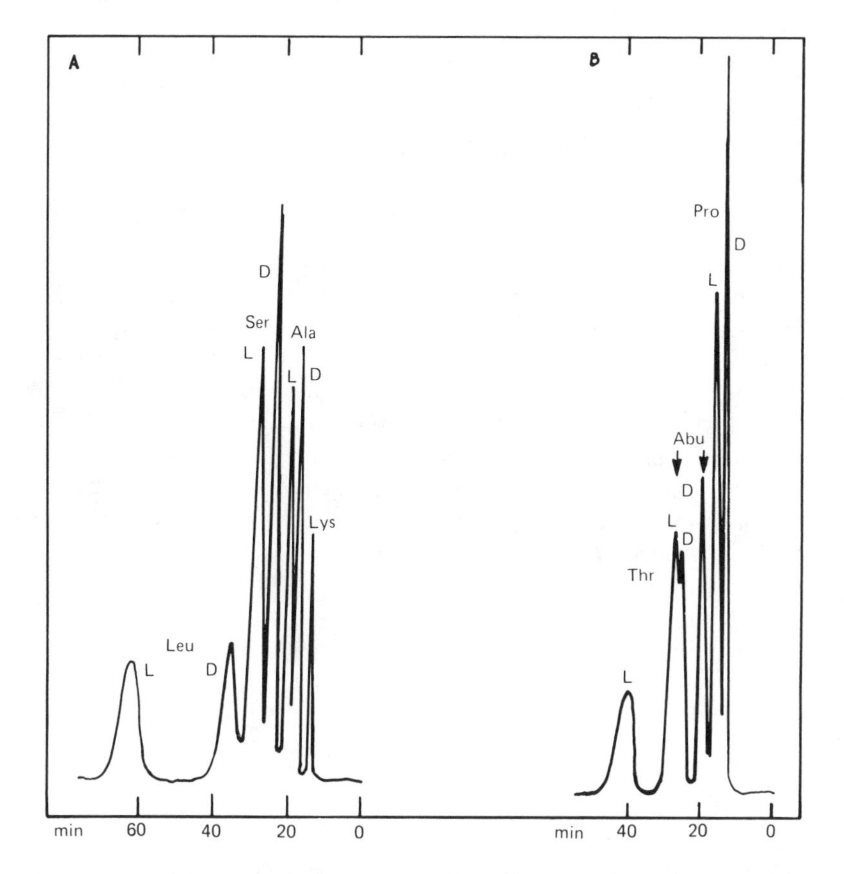

FIGURE 4. Chromatography of racemic lysine, alanine, serine, and leucine (A), and proline, aminobutyric acid, and threonine (B) on a glass microbore column (1 × 100 mm) with a polystyrene-type sorbent containing groups of *N′*-benzyl-(R)-diaminopropane-1,2 (2.7 mmol/g; d_p 7.5 + 2.5 μm). Eluent, 0.25 *M* sodium acetate + 1.5 m*M* copper acetate, pH 5.2, 1 mℓ/hr. Ultraviolet detection at 260 nm. (From Davankov, V. A. and Kurganov, A. A., *Chromatographia,* 13, 339, 1980. With permission.)

Several authors were concerned with the relatively low efficiency of columns packed with polystyrene-type chiral resins. This was, however, by no means a general drawback of the sorbents, but rather a matter of equipment and technique. In the first stages of developing LEC of optical isomers, it was considered to be more important to examine the possible application area of this new technique and to understand mechanisms of the chiral recognition of mobile ligands than to spend efforts on raising the plate number of the columns. The possibility of solving this problem also by means of the macronet isoporous-type matrices was shown by the example of the anion exchanger[71] containing 2.7 mmol/g of (R)-N^1-benzyl-1,2-propanediamine fixed to a polystyrene matrix of the degree of cross-linking of 5.5% and capable of taking up 1.80 g water per gram of resin. Glass microbore columns packed with resin particles of d_p = 7.5 ± 2.5 μm demonstrated good efficiency,[93] as illustrated by Figure 4.

Through a systematic study of copper-diamine-1,2 complexes, it has been found that at least two *N*-substituents in a *bis*-chelate are needed in order to achieve a rigid conformation of the chelate rings,[112-115] which obviously is an important prerequisite for enantioselectivity phenomena in these systems.[116] Actually, the chiral ligand of the above anion exchanger meets this requirement, since it has the structure of N^1,N^1-dibenzyl-propanediamine due to the additional bond to the polystyrene matrix:

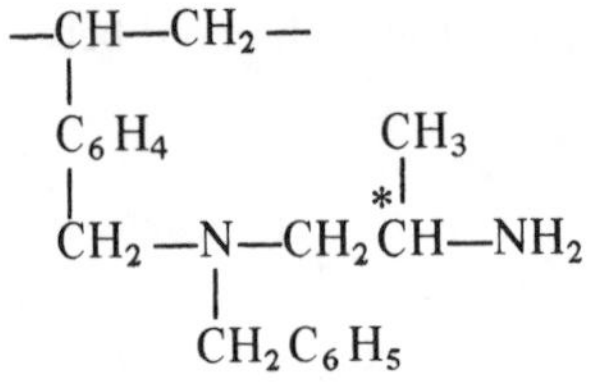

```
—CH—CH2—
 |
 C6H4          CH3
 |             *|
 CH2—N—CH2CH—NH2
     |
     CH2C6H5
```

As can be seen from Figure 4 and Table 5, not only is the ligand exchange equilibrium rapidly established with this type of resin, but its resolving power is also fairly good. Even alanine, an amino acid with the smallest molecular asymmetry, is easily resolved into enantiomers ($\alpha = 1.43$). Still greater selectivity factors are observed for other amino acids (Table 5). Alone, the basic molecules (lysine and ornithine) are excluded from the anion-exchanger phase without being resolved into enantiomers.

The main objective of Lefebvre et al.[74,75] by introducing a hydrophilic polyacrylamide-type matrix was to enhance the water uptake and efficiency of chiral ligand exchangers. By means of formaldehyde, the authors grafted L-proline onto cross-linked polyacrylamide beads of 10 to 20 μm in diameter to obtain a chiral resin of the following structure:

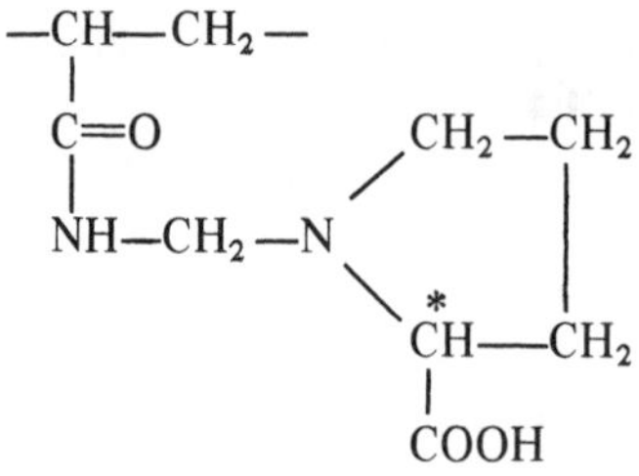

```
—CH—CH2—
 |
 C=O          CH2—CH2
 |           /      |
 NH—CH2—N           |
           \ *      |
            CH——CH2
            |
            COOH
```

Several amino acids could be resolved on the copper(II) form of the resin. Mixtures of DL-Val and DL-Phe,[74] as well as DL-Phe and DL-Trp[75] were separated into four peaks on columns of 30 and 5 cm in length, respectively. About a dozen other complexing metal ions have been tested, with the result that copper appears to be one of the more favored ones. However, the column efficiency with respect to complexed mobile ligands (valine) was about ten times lower than that for the dimethyl ether of ethylene glycol, which is not coordinated to copper. This points to a relatively low rate of ligand exchange in ternary sorption complexes. The column efficiency could be significantly enhanced by raising the column temperature to 50 to 60°C.[74] It should be noted here that with the macronet isoporous polystyrene-type sorbents, the temperature influence was observed to be much weaker.[95,117] It is remarkable that reversed elution order of bifunctional amino-acid enantiomers (D before L) has been found on the polyacrylamide-type resin as compared to the L-proline-containing polystyrene-type sorbent. Recently,[118] by means of low-molecular-weight models, it has been shown that L-isomers of valine and phenylalanine are indeed the preferred species in mixed-ligand complexes with *N*-acetylaminomethyl-L-proline, the latter simulating the repeating units of the polyacrylamide-type chiral resin.

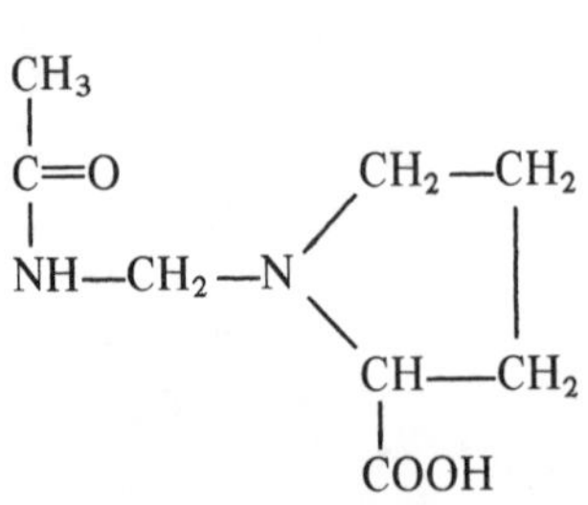

```
 CH3
 |
 C=O          CH2—CH2
 |           /      |
 NH—CH2—N           |
           \        |
            CH——CH2
            |
            COOH
```

Moreover, an acceptable agreement was observed between the ratio of stability constants $\beta_{CuLL'}$ and $\beta_{CuLD'}$, of two diastereomeric model complexes and the resolution selectivity values, α, on chromatography of racemic valine and phenylalanine: $\Delta \log \beta \approx \log \alpha$. However, no enantioselectivity was found in ternary model complexes with proline, though the latter was easily resolved by chromatography ($\alpha = k'_D/k'_L = 1.53$). Thus, low-molecular-weight models do not always adequately reflect the stereochemical situation in the resin phase. (We will discuss this statement at a later point.) Naturally, a comparison with a soluble linear polyacrylamide modified with L-proline in the same manner as was the cross-linked resin is more reliable.[119] The ability to distinguish enantiomers of valine and proline by forming mixed copper complexes with the soluble polymer ($\beta_D/\beta_L = 0.67$ and 1.25, respectively) was found to correspond well with the chromatographic resolution of these enantiomers using the cross-linked resin ($\alpha = 0.57$ and 1.25, respectively). However, polymeric models give but minimal advantages for stereochemical studies over insoluble resins.

Besides proline, which has become the most favored resolving ligand in various chiral ligand-exchanging systems, a series of other L-amino acids has been grafted onto polyacrylamide gels.[75] The highest resolving power was displayed by sorbents incorporating cyclic matrix-fixed ligands, proline, hydroxyproline, azetidine carboxylic acid, and pipecolic acid. None of them, however, show any enantioselectivity effects toward leucine.

Negative results were obtained in chromatographic tests with racemic proline, valine, and phenylanine on polyacrylamide gels modified by L-phenylalanine and L-phenylglycine.[75] Contrary to this, the Russian group[97] was really satisfied with their sorbent containing 1.4 mmol of L-phenylanine residues per gram polyacrylamide beads (Biogel® P-4, Serva®, d_p 64 μm). Enantioselectivity values of at least 1.25 to 1.30 were obtained in resolutions of all the 20 amino acids investigated, including the basic (Lyn, Orn), acidic (Asp, Glu) and the most difficult to resolve neutral amino acids (Ala, Met, Ser, Leu). Preparative-scale resolutions of the tritium-labeled DL-[^{3}H]-alanine[97] and DL-[^{3}H]-glutamic acid[120] on the copper(II) form of this resin were performed. L-enantiomers were always observed to eluete before the D-antipodes, opposite to the elution order on the L-proline-containing polyacrylamide.

Other chiral polymeric ligand exchangers prepared on the basis of hydrophilic TSK-Gel® 2000 PW gels[77] and polyhydroxyethylmethacrylate[76] were shown to possess the ability of resolving racemates of amino acids, but have been studied much less than the polystyrene- and polyacrylamide-type chiral resins.

Finally, a curious recent paper[121] should be mentioned claiming resolution of a series of racemic amino acids using LEC on the amino-copper form of a nonchiral polystyrene-based resin containing groups of iminodi (methane-phosphonic) acid:

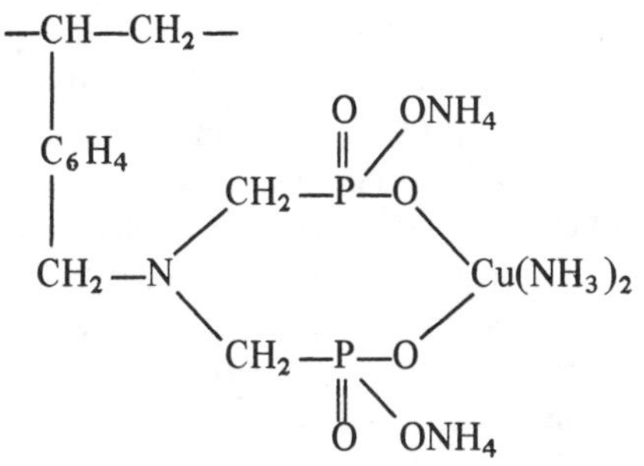

```
—CH—CH2—
 |
 |              O  ONH4
 |              ‖ /
C6H4            ‖/
 |        CH2—P—O
 |       /        \
CH2—N              Cu(NH3)2
       \          /
          CH2—P—O
              ‖\
              O  ONH4
```

Here, there is an obvious contradiction with the basic principle of stereochemistry stating that enantiomeric resolutions require involvement of chiral resolving agents.

B. Silica-Bonded Chiral Phases

Though chiral polymeric resins possess very high exchange capacity and excellent chemical stability, their pressure resistance is still inferior to that of porous silica. The latter parameter is very important for the high efficiency of column packings, since they must consist of small particles and consequently, are subjected to high pressures applied to the eluent (whose flow rate should be sufficiently high). Therefore, binding chiral ligands onto the surface of porous microparticulate silica was felt to be a timely and important task. In 1979, three different groups[122-124] reported the first results of synthesizing chiral silica bonded phases. Apparently, the chemical modification of a silica surface required a certain level of experience, so that these first results were not very impressive. Real success was achieved later.

The Russian group,[122] having advanced with the polystyrene-type resins, was at that time hardly able to achieve similar results with silica bonded analogues of the structure:

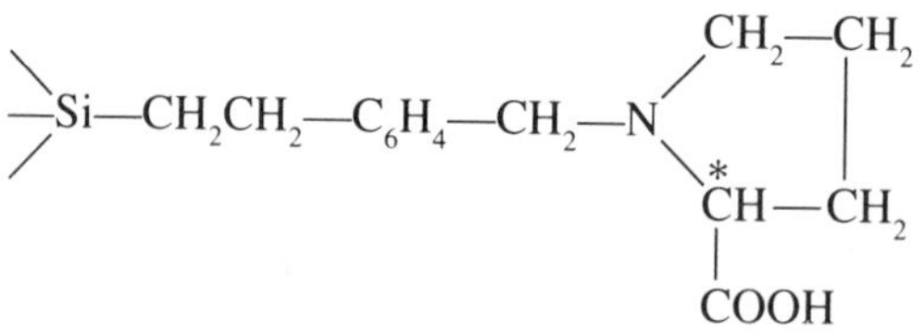

$$\geq Si-CH_2CH_2-C_6H_4-CH_2-N\langle\text{ring: }CH_2-CH_2-CH_2-\overset{*}{C}H(COOH)\rangle$$

Gübitz et al.[123] prepared a chiral bonded phase by treating the 10 μm LiChrosorb® Si 100 first with 3-glycidoxy-propyltrimethoxysilane in boiling benzene and then with sodium L-prolinate in dimethylformamide or methanol at room temperature to obtain fixed ligands of the following structure:

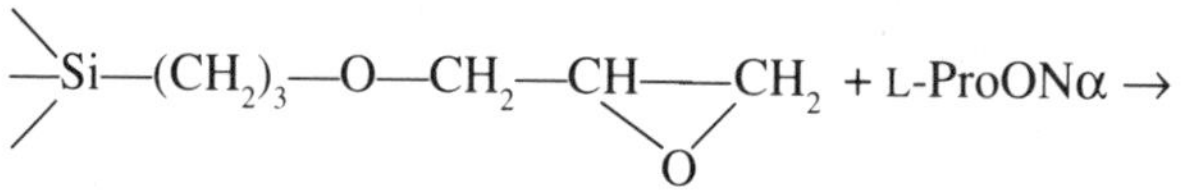

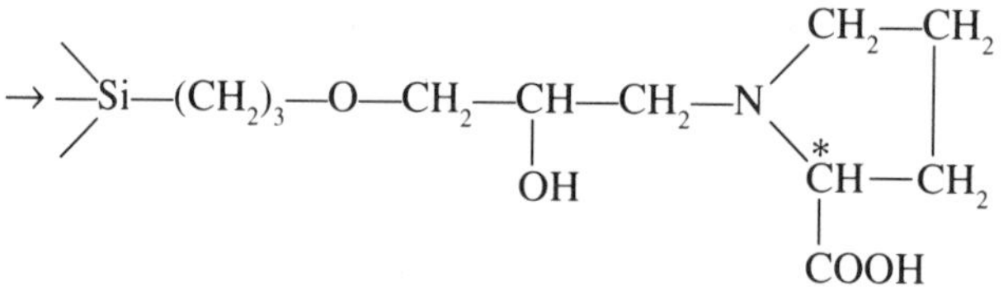

$$\geq Si-(CH_2)_3-O-CH_2-\underset{\backslash O/}{CH-CH_2} + \text{L-ProONa} \rightarrow$$

$$\rightarrow \geq Si-(CH_2)_3-O-CH_2-\underset{OH}{CH}-CH_2-N\langle\text{ring: }CH_2-CH_2-CH_2-\overset{*}{C}H(COOH)\rangle \quad 0.65\ \text{mmol/g or } 2\ \mu\text{mol/m}^2$$

For the copper(II) form of this sorbent, high selectivity with respect to enantiomers of tyrosine and tryptophan was observed, but a base line separation of two isomers still required 2 to 3 hr. Later,[125-127] chromatography conditions had been optimized so that the majority of amino acids (with the exception of alanine, leucine, glutamic acid, and cystine) could be resolved within a few minutes. Even mixtures of four to five racemic amino acids were resolved, almost to the base line (Figure 5), requiring 25 to 40 min.[126] The most important factors enhancing efficiency appeared to be the elevated temperature (60 to 80°C), the use of a neutral or even acidic phosphate buffer, and the relatively high ionic strength. Though the packing material showed a good stability in the tested pH range from 4 to 9, the lifetime of the columns was reduced by use at 80°C. On columns containing L-proline as a fixed ligand, the L-enantiomers of amino acids constantly appeared with higher k′ values, with the exception of proline. Up to 1.5 mg of amino acid racemate could be separated on an analytical column of 25 cm in length and 0.46 cm in inner diameter without significant change in k′ values and resolution.[125] The retention of the amino acids decreased with the rising ionic strength of the eluent, lowering the pH and addition of organic solvents to the mobile phase. Amino acids with aromatic α-substituents showed highest retention and enantioselectivity (α up to 3.5, Trp).

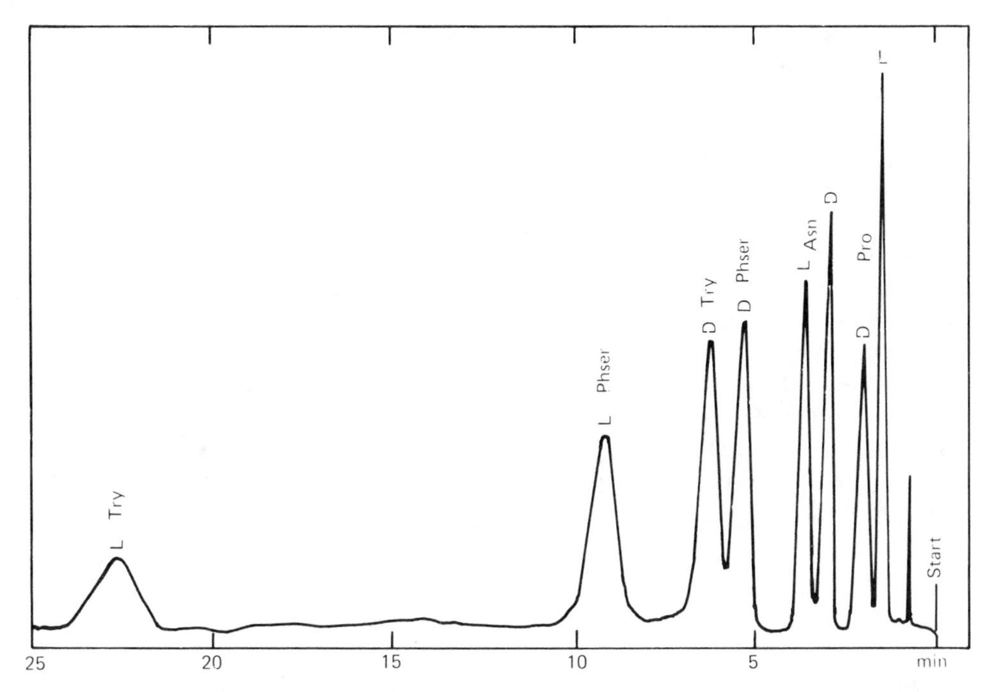

FIGURE 5. Separation of DL-proline, DL-asparagine, DL-phenylserine, and DL-tryptophan. Silica with L-proline as fixed ligand loaded with copper(II). Column, 4.6 × 250 mm; d_p 10 μm; eluent, 0.05 *M* KH_2PO_4, pH 4.6, 5 mℓ/min; temperature, 50°C. (From Gübitz, G., Jellenz, W., and Santi, W., *J. Chromatogr.*, 203, 377, 1981. With permission.)

Table 6
INFLUENCE OF THE SILICA FIXED LIGAND ON THE RESOLUTION SELECTIVITY OF AMINO ACIDS, $\alpha = k'_D/k'_L$;[126,127] MOBILE PHASE, 0.05 *M* KH_2PO_4, 0.1 m*M* COPPER(II), pH 4.5 TO 4.6, TEMPERATURE 25 TO 55°C

	Stationary ligand						
Racemic amino acid	**L-AzCA**	**L-Pro**	**L-Hyp**	**L-Pip**	**L-Val**	**L-His**	**L-Phe**
Valine	1.00	0.67	0.83	0.55	0.69	1.00	1.00
Proline	2.50	1.62	2.38	1.32	1.00	1.00	1.52
Threonine	0.85	0.62	0.69	0.61	0.80	1.00	1.14
Lysine	1.00	0.91	1.00	0.71	—	—	1.00
Aspartic acid	1.00	0.77	0.82	0.81	—	—	1.25
Histidine	—	0.56	0.43	0.64	0.50	1.00	—
Phenylalanine	0.60	0.34	0.43	0.44	0.59	1.25	1.00
Thyrosine	0.42	0.32	0.24	0.30	0.56	1.52	1.00
Tryptophan	0.40	0.29	0.23	0.39	0.54	1.32	1.00
Methionine	1.00	0.91	1.00	0.75	—	—	1.12

Copper(II), cobalt(II), nickel(II), and zinc(II) were tested as complexing ions. Acceptable results, however, were only obtained with copper.[126] Various amino acids, other than L-proline, were bonded as chiral fixed ligands to the 3-glycidoxypropyltrimethoxysilane-treated silica gel.[126-127] Cyclic amino acids as fixed ligands showed higher resolving ability than alicylic compounds, increasing in the series azetidine carboxylic acid < proline < hydroxyproline < pipecolic acid (Table 6).

More recently,[128] using a similar chiral bonded phase that contained L-hydroxyproline residues, Gübitz and co-workers successfully resolved a series of 2-hydroxy acid enantiomers (mandelic acid, its 4-hydroxy-, 3-hydroxy-, 3,4-dihydroxy-derivatives, 3-phenyllactic acid, atrolactic acid, and 2-hydroxycaproic acid). The eluent was pure water, $10^{-4}M$ in copper sulfate. The use of buffers in the mobile phase caused a decrease in retention time and resolution. The same was true when acetonitrile was added to the eluent. Though the efficiency of the system at room temperature was relatively low, high enantioselectivity values (α = 0.3 to 0.8) permitted a clear resolution in each enantiomeric pair and a simple detection of components at 223 nm. It is remarkable that on chiral bonded phases containing other types of resolving reagents (proline, valine, histidine, phenylalanine, azetidine carboxylic acid, pipecolic acid, propylenediamine, ephedrine, and tartaric acid), only slight or no separations of hydroxy acids were obtained. Substitution of copper(II) by other transition metal ions also resulted in the loss of resolution.

In the approach of Foucault et al.,[124] the carboxyl function of L-proline was used to form a link to the silica matrix. 3-Aminopropyltriethoxysilane was acylated by L-proline and then bonded to silica surface:

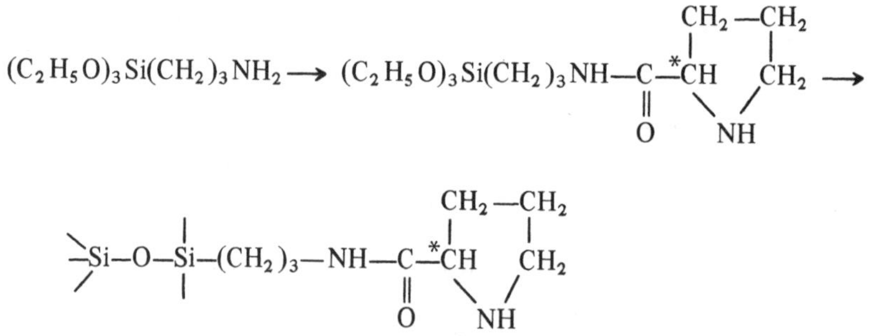

$$(C_2H_5O)_3Si(CH_2)_3NH_2 \longrightarrow (C_2H_5O)_3Si(CH_2)_3NH{-}\underset{\underset{O}{\|}}{C}{-}\overset{*}{C}H \; \begin{array}{c} CH_2{-}CH_2 \\ \text{(ring)} \\ NH \end{array} \; CH_2 \longrightarrow$$

$$\begin{array}{c}\diagdown\\ -Si\\ \diagup\end{array}{-}O{-}\overset{|}{\underset{|}{Si}}{-}(CH_2)_3{-}NH{-}\underset{\underset{O}{\|}}{C}{-}\overset{*}{C}H \; \begin{array}{c} CH_2{-}CH_2 \\ \text{(ring)} \\ NH \end{array} \; CH_2$$

Though involved in formation of amide bonds, L-proline moieties still possess the ability to coordinate copper(II) ions and bind herewith mobile amino acid ligands. The enantioselectivity of the sorbent proved to be sufficiently high, but because of the relatively low column efficiency in a water-acetonitrile (47:53)-0.17 *M* ammonia mobile phase, of the eight amino acids tested, only tryptophan, phenylalanine, and tyrosine were partially resolved with D-isomers emerging before the L-antipodes.

Lindner[129] examined this sorbent in more detail. He reacted 3-aminopropylsililated silica, LiChrosorb® NH_2, with hydroxysuccinimide ester of *N*-*t*-butyloxycarbonyl-L-proline and set free the amino group of the latter from the protecting *t*-BOC group by treatment with trifluoroacetic acid. Lindner, however, involved other sorbates into resolution; namely, 5-dimethylaminonaphthalene-1-sulfonyl (dansyl, Dns) amino acids:

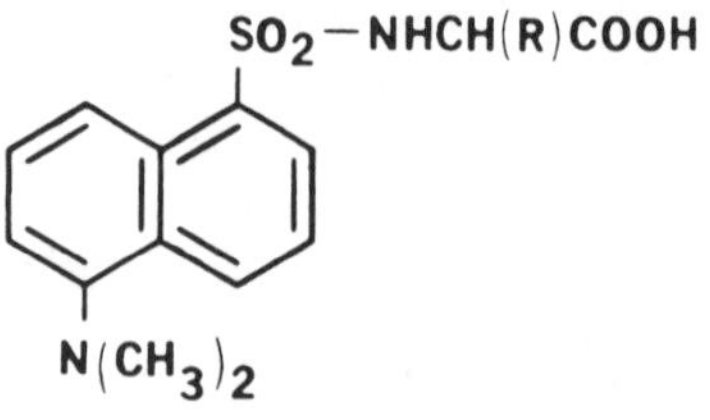

SO_2—NHCH(R)COOH

$N(CH_3)_2$

Using acetonitrile-0.1*M* ammonium acetate mixtures of pH 9.0, resolution of all dansyl-amino acids, except proline, was achieved with selectivity values ranging from 1.04 (Tyr) to 3.16 (Asp) and L-isomers eluting first in the case where the sorbent is being used in the copper(II) form. If cadmium(II) is the complexing metal ion, the elution order of enantiomers reverses; the overall resolutions are no longer as good, with α-values varying from 1.08 (alloisoleucine) to 1.61 (methionine). Nevertheless, in the latter case, 0.2% dansyl-D-methionine could be confidently detected in the L-product. On raising the temperature to 50°C, the column efficiency was observed to increase at the expense of selectivity. Substitution of acetonitrile in the eluent by methanol brings about a loss of selectivity.

Applying 1,1′-carbonyl-diimidazole as a condensing reagent was a more versatile and convenient method of binding *t*-BOC-protected chiral amino acids to the amino phase,[130] as compared with the use of hydroxysuccinimide esters of *N*-BOC amino acids. BOC-His(Tos) was coupled to the aminopropyl groups in this manner to give a chiral bonded phase of the following structure (after removing the protecting BOC group with trifluoroacetic acid and the tosyl group with hydroxybenzotriazole):

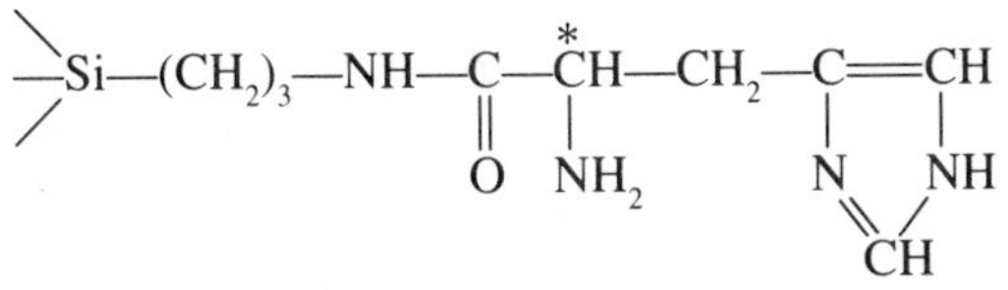

$$\geq Si{-}(CH_2)_3{-}NH{-}C({=}O){-}\overset{*}{C}H(NH_2){-}CH_2{-}C{=}CH{-}NH{-}CH{=}N \text{ (imidazole ring)}$$

In a 0.066 *M* phosphate buffer of pH 4.56, 10^{-4} *M* in copper sulfate, the above sorbent displayed enantioselectivity toward all tested amino acids with unprotected amino and carboxyl groups. The selectivity, $\alpha = k'_D/k'_L$, ranged from 1.03 (Asp) to 1.79 (Trp). Mandelic acid could also be resolved with α = 1.08. It is noteworthy that an acidic eluent was used in these experiments, which should improve the lifetime of the column packing, and a rapid increase in the retention of amino acids was observed with increasing pH.

The *t*-BOC-protected "intermediate" of the above proline-amide-bonded phase was also used as a self-consistent chiral sorbent, as well as its L-valine-containing analogue:[131,132]

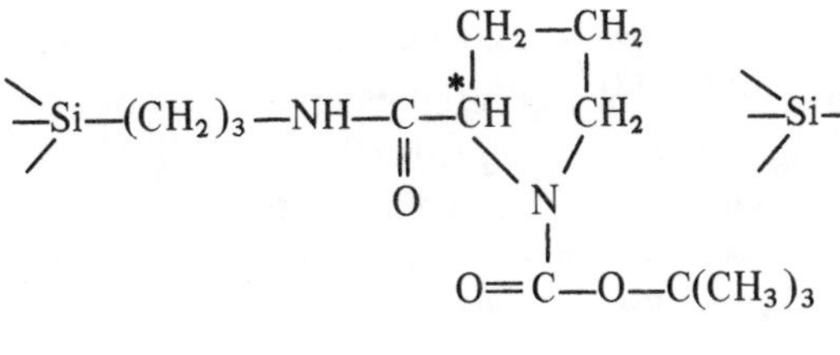

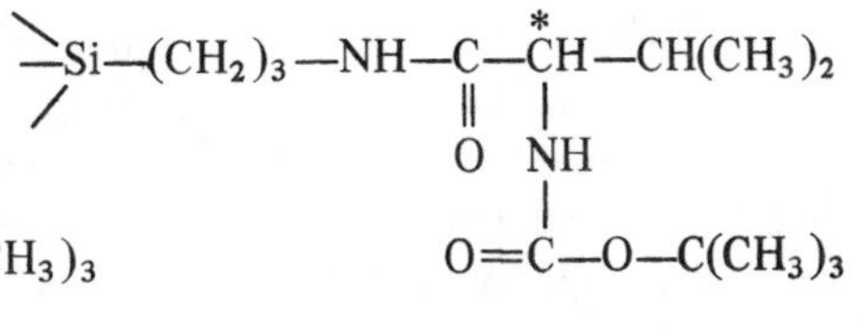

$$\geq Si{-}(CH_2)_3{-}NH{-}C({=}O){-}\overset{*}{C}H{-}CH_2{-}CH_2{-}CH_2{-}N(\text{ring}){-}C({=}O){-}O{-}C(CH_3)_3$$

$$\geq Si{-}(CH_2)_3{-}NH{-}C({=}O){-}\overset{*}{C}H(NH{-}C({=}O){-}O{-}C(CH_3)_3){-}CH(CH_3)_2$$

Both the carboxyl and amino functions of these ligands are converted into amide groups, and nevertheless, the sorbents act as ligand exchangers, bind copper(II) ions from a 30:70 acetonitrile-0.1 *M* ammonium acetate eluent of pH 6.9, and effectively resolve many racemic Dns-amino acids in this medium. Table 7 and Figure 6 give an idea of selectivities and efficiencies of columns packed with these sorbents.

A favorite way of obtaining enantioselective silica-bonded phases was considered to consist in covalently bonding optically active ligands via an akylene spacer to the surface of the support. Sugden et al.[133] reacted LiChrosorb® Si-60 (5 μm) with 3-chloropropyltrichlorosilane and then with L-proline in a boiling mixture of chloroform with methanol (85:15) in the presence of sodium iodide and diisopropylethylamine:

Table 7
RESOLUTION OF RACEMIC DANSYL-AMINO ACIDS ON CHIRAL-BONDED AMIDE PHASES CONTAINING RESIDUES OF BOC-L-PRO AND BOC-L-VAL; ELUENT, ACETONITRILE-0.1 *M* AMMONIUM ACETATE OF 30:70, pH 6.9, 10^{-5} *M* $CuSO_4$; TEMPERATURE, 55°C

	Si–C_3NH–Pro–BOC			Si–C_3NH–Val–BOC		
Dansyl-amino acid	k'_D	k'_L	α	k'_D	k'_L	α
Serine	1.49	1.10	1.35	4.97	3.03	1.64
Threonine	1.68	1.19	1.41	2.96	2.46	1.20
Leucine	3.29	2.71	1.21	4.60	4.23	1.08
Glutamic acid	1.40	1.07	1.31	3.58	1.98	1.80
Norcleucine	3.80	3.06	1.24	—	—	1.00
Methionine	3.00	2.26	1.33	5.93	5.01	1.18
Phenylalanine	4.56	3.71	1.23	6.78	6.35	1.07
Norvaline	2.71	2.18	1.24	—	—	1.00
Valine	2.26	1.95	1.16	—	—	1.00
Aminobutyric acid	2.12	1.71	1.24	—	—	1.00
Tryptophan	—	—	1.00	8.88	6.20	1.43

From Engelhardt, H. and Kromidas, S., *Naturwissenschaften*, 67, 353, 1980. With permission.

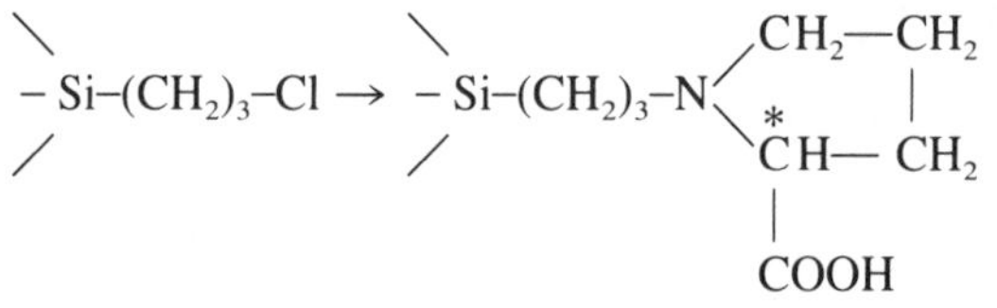

$$\geq Si-(CH_2)_3-Cl \rightarrow \geq Si-(CH_2)_3-N\langle\begin{matrix} CH_2-CH_2 \\ \overset{*}{C}H-CH_2 \\ | \\ COOH \end{matrix}$$

The packing material obtained was found to completely resolve racemates of proline and 3,3-dimethylproline and partially that of phenylalanine, using 1.0 m*M* copper acetate solution of pH 4.6 as eluent. Low efficiency and severe peak tailing were reported, so that histidine, alanine, and glutamic acid could not be resolved on a column 25 cm in length. For L-proline, the on-column limit of detection was found to be 400 ng at 240 nm where copper complexes exhibit an absorption maximum.

Chiral packings of the same chemical structure have been examined in detail by the groups of Unger and Davankov,[134] along with the other chiral bonded phases differing in the length and nature of the hydrocarbon spacer between the L-proline or L-hydroxyproline moieties and the surface of LiChrosorb® Si-100 (d_p = 10 μm):

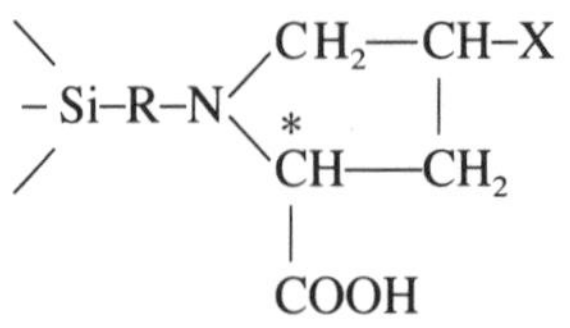

$$\geq Si-R-N\langle\begin{matrix} CH_2-CH-X \\ \overset{*}{C}H-CH_2 \\ | \\ COOH \end{matrix} \quad R = -CH_2-,\ -(CH_2)_3-,\ -(CH_2)_8-,\ -(CH_2)_2-C_6H_4-CH_2- \quad X = -H, -OH$$

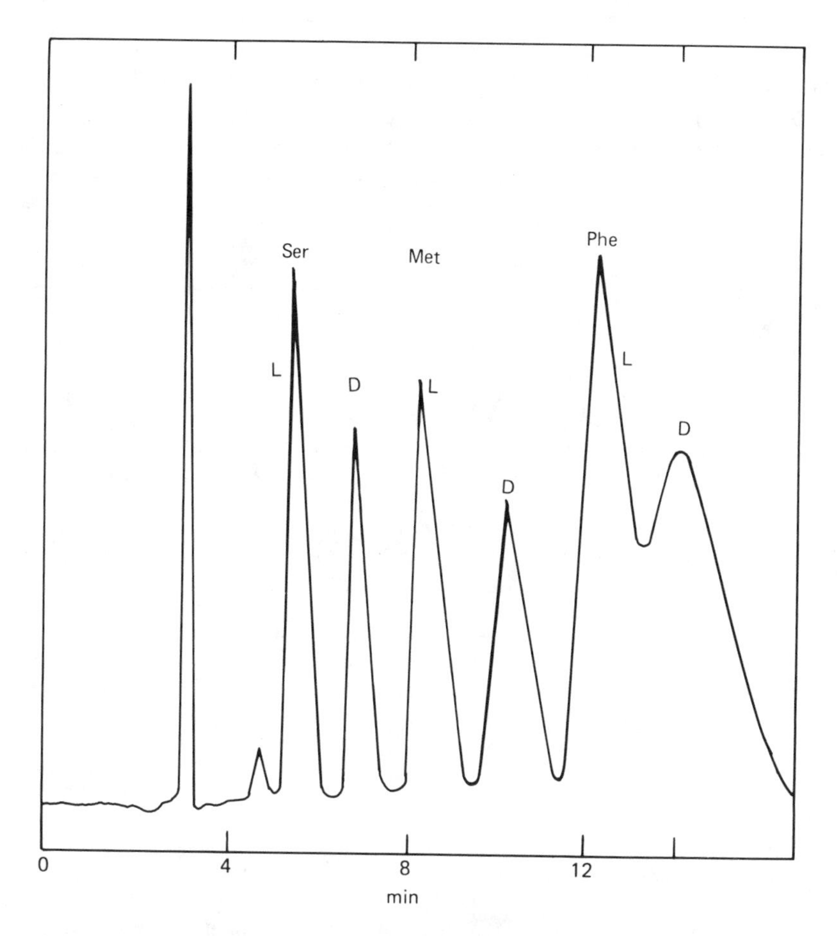

FIGURE 6. Resolution of dansyl derivatives of racemic serine, methionine and phenylalanine. *tert*-BOC-L-prolineamide bonded phase. Column, 4.1 × 300 mm; d_p 10 μm; eluent, 10^{-5} *M* $CuSO_4$, CH_3CN/O.1 *M* NH_4OAc 30/70, pH 6.9, 1 mℓ/min.; temperature, 55°C. (From Englehardt, H. and Kromidas, S., *Naturwissenschaften*, 67, 353, 1980. With permission.)

The starting sililating reagents were chloromethyl-trimethoxysilane, 3-chloropropyl-triethoxysilane, ω-bromooctyl-trichlorosilane, and 2-(*p*-chloromethylphenyl)-ethyldimethylchlorosilane, respectively. They are bonded to the silica surface in a trivial manner, and then the halogen atoms are substituted for L-proline or L-hydroxyproline methyl esters in the presence of sodium iodide. An alternative route is reacting chloroalkyltrialkoxysilane first with the amino acid methyl esters, and then bonding the chiral siloxane obtained to silica. Finally, combination of the above two reactions, the halogen to nitrogen exchange in the sililating agent, and bonding of the latter to silica, in one process, was also shown to give satisfactory results. Under optimized conditions, the concentration of surface functional groups of 3 μmol/m^2 can be easily achieved by any of the above synthetic routes. Removing of the protective ester groups proceeds automatically on treatment of the material with a cooper acetate solution. Herewith, the color of the sorbents gradually turns from green to blue, indicating formation of copper-amino acidato chelates.

The chiral-bonded phases offer good enantioselectivity for all α-amino acids. Table 8 shows examples of the separation of different enantiomeric pairs. Assessment of the variation of eluent composition and column temperature on the retention and chiral recognition of amino acid enantiomers[135-137] leads to a conclusion that a complex combination of hydrophobic, electrostatic, and complexation interactions with the support govern the behavior of

Table 8
RESOLUTION OF RACEMIC AMINO ACIDS ON CHIRAL SILICA-BONDED PHASES OF THE GENERAL FORMULA Si–R–L–Hyp, DIFFERING IN THE SPACER GROUP R; ELUENT, $10^{-2}M$ NH_4OAc 10^{-4} M $Cu(OAc)_2$, pH 5.0, TEMPERATURE 40°C

Amino acid	R = $-CH_2-$			R = $-(CH_2)_3-$			R = $-(CH_2)_8-$			R = $-(CH_2)_2C_6H_4CH_2-$		
	k'_L	k'_D	α	k'_L	k'_D	α	k'_L	k'_D	α	k'_L	k'_D	α
Asp	1.10	1.03	0.94	2.38	2.22	0.93	2.01	2.13	1.06	16.83	16.53	0.98
Glu	1.02	1.13	1.11	1.93	2.02	1.05	1.57	1.57	1.00	12.37	14.72	1.19
His	3.80	2.33	0.61	4.13	2.48	0.60	2.22	1.92	0.86	18.50	6.53	0.35
Ala	1.40	1.47	1.05	1.33	1.60	1.20	0.40	0.58	1.45	5.50	5.70	1.04
Asn	2.05	1.76	0.86	2.52	2.35	0.93	0.89	0.82	0.92	8.58	8.28	0.97
Ser	1.72	1.46	0.85	2.12	1.67	0.79	0.82	0.74	0.92	7.36	6.25	0.85
Pro	1.93	2.51	1.30	1.93	3.15	1.63	1.07	2.06	1.93	16.43	16.33	0.99
Thr	2.14	1.94	0.91	2.78	2.32	0.83	1.14	1.16	1.02	11.73	11.50	0.98
Val	2.10	2.12	1.01	3.00	2.52	0.84	2.81	2.71	0.96	14.38	10.80	0.75
Lys	2.77	2.61	0.94	1.67	1.83	1.10	0.07	0.06	0.86	2.05	2.23	1.09
Tyr	2.30	1.52	0.66	5.85	3.67	0.63	12.25	9.72	0.79	28.58	19.73	0.69
Met	2.87	2.82	0.98	4.12	3.90	0.95	4.65	4.56	0.98	19.18	16.85	0.88
Arg	3.40	3.44	1.01	2.58	2.78	1.08	0.31	0.39	1.26	3.80	3.50	0.92
Ile	2.45	2.52	1.02	4.25	3.82	0.90	5.86	5.94	1.01	21.95	14.43	0.41
Phe	3.35	2.87	0.86	8.32	5.78	0.69	17.60	15.45	0.88	33.18	29.10	0.88
Leu	2.83	3.02	1.07	3.50	3.45	0.99	4.80	5.14	1.07	20.46	14.43	0.71
Trp	4.69	2.69	0.57	16.12	9.37	0.58	—	—	—	100.70	28.87	0.29

From Roumeliotis, P., Unger, K. K., Kurganov, A. A., and Davankov, V. A., *Angew. Chem. Int. Ed. Engl.*, 21, 930, 1982. With permission.

the solutes. With the hydrophobicity of the sorbents rising in the above series with the growing length of the alkylene spacer (from methylene to propylene, octylene, and finally *p*-ethylenebenzyl), the retention of aromatic and heavy aliphatic amino acids gradually increases. Simultaneously, retention of these solutes becomes strongly dependent on the content of organic modifier in the eluent (methanol or acetonitrile additives diminish the retention of hydrophobic solutes). Contrary to this, retention of polar amino acids (such as aspartic acid, glutamic acid, lysine, histidine, and serine) remains largely unaffected by the addition of methanol to the eluent, and in the case of the least hydrophobic sorbent where L-hydroxyproline ligands are linked to silica via methylene bridges, organic modifiers of the eluent cause an increase in retention of polar amino acids. In these last systems, hydrophobic interactions play but a minimal role, and the retention is governed by polar and complexation interactions, both of which are favored by partially substituting water for organic solvents.

Formation of mixed-ligand sorption complexes with the silica-bonded chiral ligands is the most important mechanism of retention of the amino acid solutes. Copper amino acid complexes dissociate below pH 3.5. Accordingly, retention of all amino acids was found to gradually diminish by lowering the pH of the eluent from 7 to 4.

Purely electrostatic interactions with the negatively charged surface silanol groups are especially characteristic of sorbents with short spacers (C_1 and C_3) between the main sorption sites and the surface. Owing to this sort of interactions, positively charged species — arginine and lysine — appear to display the strongest retention of all amino acids. Contrary to this, they are scarcely retained on sorbents with octylene and ethylenebenzyl spacers which shield the surface silanol groups much more effectively than the methylene bridges.

Finally, addition of ammonium acetate to the eluent was observed to diminish the k′ values in the majority of systems. This can be understood in terms of ammonia molecules and acetate anions competing with the mobile amino acid ligands for the complexing sorption sites of the sorbents. Another effect of the salt additives is the diminishing of contributions from all electrostatic interactions.

We will discuss elsewhere the influence of the varying parameters of chromatography on the chiral recognition of enantiomers. In general, the enantioselectivity of silica bonded chiral phases concerned is not very high (see Table 8). Nevertheless, owing to the fairly good efficiency of columns, when operated at temperatures of 40°C and higher, mixtures of five to six racemic amino acids can be separated into the constituent enantiomers, as is illustrated by Figures 7 and 8.

The outlined development of silica-bonded chiral phases clearly reflects the main trend in modern liquid chromatography which consists of enhancing the efficiency of chromatographic systems. Applied to the LEC of enantiomers, this trend resulted in the real possibility of analyzing the enantiomeric composition of almost all chiral compounds capable of entering weak coordination interactions with transition metal ions. In our first publications[138] and the review on enantioselectivity effects in coordination compounds,[139] a suggestion was made that, contrary to the generally accepted ideas, enantioselectivity phenomena in labile chiral complexes are the rule rather than the exception. Therefore, with the efficiency of chromatographic columns increasing, practical use will be made of many new systems displaying enantioselectivity on a very low level. The higher the efficiency of the column packing, the wider its application range, since the efficiency largely compensates for the poor selectivity of the system.

We are now aware of the fact that the ligand exchange process in some labile complexes may be too slow[140] for the requirements of modern high-speed chromatography, but in the majority of cases, both selectivity and efficiency of columns operating according to the ligand exchange mechanisms can be influenced by varying the same parameters of the system that are under variation in other types of chromatography. Particularly important are the column temperature and the composition of the mobile phase (the value of pH, content and type of the organic modifier, content and type of inorganic or organic salts). Of course, of major importance is the performance of packing materials. Here again, the same optimization of the preparation of the chiral ligand exchangers that has been needed for all modern chromatographic sorbents is required.

In addition to well known mechanical, chemical, and structural parameters that should be met by a sorbent to produce highly efficient and selective columns, Karger and coworkers[141] point out the concentration of the chiral ligands in the chemically bonded phase as an important variable, strongly influencing the behavior of a silica-based ligand exchanger. The authors compare two types of chiral-bonded phases. The one was *N*-ω-(dimethylsiloxyl)-undecanoyl-L-valine bonded to 5 μm spherical silica. When loaded with copper(II), it provided high enantioselectivity in the separation of a number of racemic dansyl-amino acids,[141] as well as salicylaldehyde Schiff bases of primary amino alcohols including catecholamino alcohols.[142] In preparations of the second type of sorbents,[141] the monochlorosilane of the above undecanoyl-L-valine chiral ligand was cobonded, in varying concentration ratios, with another chlorosilane which can act as a dilutent with respect to the ligand exchange process. Thus, inert bonded groups were introduced such as butyl, decyl, and eicosyl, as well as 3-[2-(2-methoxyethoxy)ethoxy]-propyl:

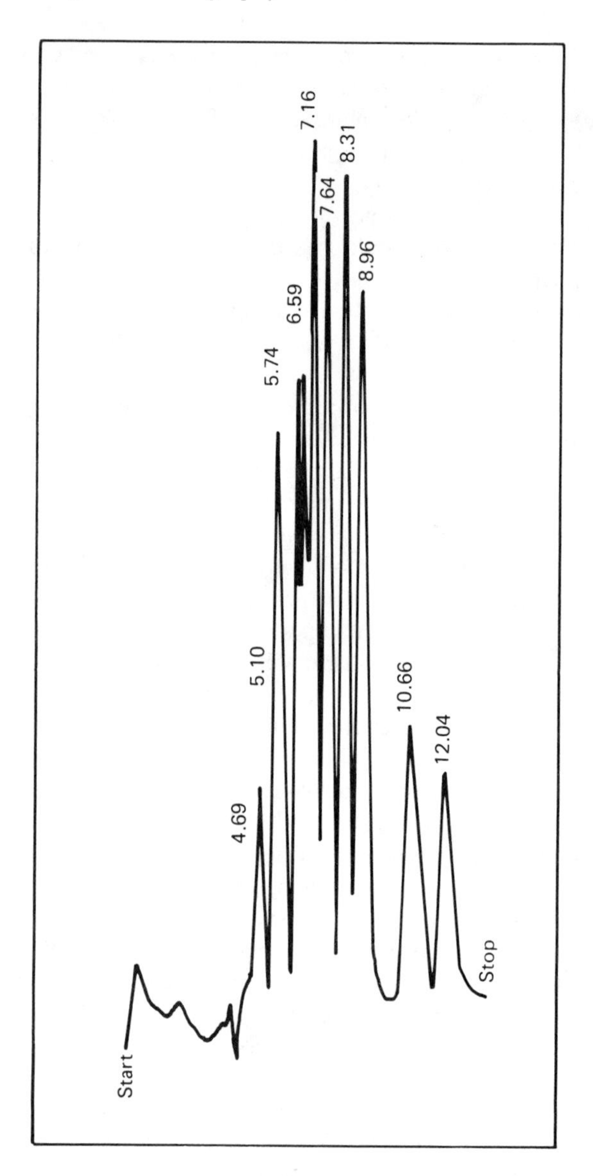

FIGURE 7. Separation of racemic amino acids on the *N*-propyl-L-hydroxyproline bonded phase; column, 4.0 × 250 mm; d_p 10 μm; eluent, 10^{-4} *M* $Cu(OAc)_2$, 0.01 *M* NH_4OAc, pH 4.3, 0.75 mℓ/min.; temperature, 50°C. Elution sequence: DL-Gln, L-Ala, D-Ala, D-Ser, L-Ser, D-Thr, L-Thr, D-Phe, L-Phe. (From Roumeliotis, P., Unger, K. K., Kurganov, A. A., and Davankov, V. A., J. Chromatogr., 255, 51, 1983. With permission.)

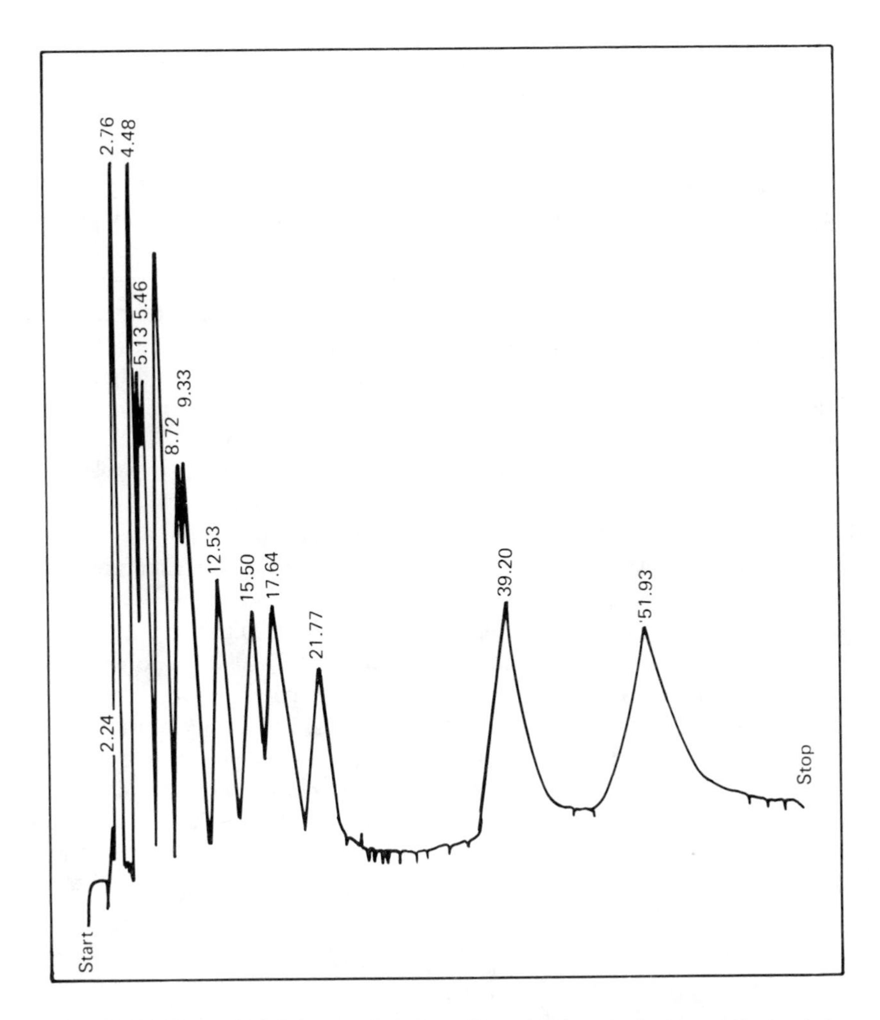

FIGURE 8. Separation of six racemic amino acids on the *N*-octyl-L-hydroxyproline bonded phase; column, 4.6 × 100 mm; d_p 10 μm; eluent, 10^{-4} *M* $Cu(OAc)_2$, 0.01 NH_4OAc, pH 4.8, methanol-water 30/70, 1.5 mℓ/min; temperature, 50°C. Elution sequence: DL-Arg, L-Pro, D-Ser, L-Ser, DPro, D-Gln, L-Gln, D-Tyr, L-Tyr, D-Phe, L-Phe, D-Trp, L-Trp. (From Roumeliotis, P., Kurganov, A. A., and Davankov, V. A., *J. Chromatogr.*, 266, 439, 1983. With permission.)

$$-Si(CH_3)_2-(CH_2)_{10}-C(=O)-NH-\overset{*}{C}H(COOH)-CH(CH_3)_2$$

$$-Si(CH_3)_2-R$$

$R = C_4H_9$

$C_{10}H_{21}$

$C_{20}H_{41}$

$(CH_2)_3O(CH_2CH_2O)_2CH_3$

These diluents, depending on their chemical structure and concentration, alter the environment surrounding the bonding ligand exchange site, thus altering the additional interactions of the mixed-ligand sorption complex with the interface layer. In several cases it was found that the retention and selectivity of D,L-Dns-amino acids was much greater on chiral phases diluted with decyl groups than on nondiluted phases, in spite of the fact that the amount of copper(II) loaded on the latter bonded phase was significantly greater.[141]

Of utmost importance is the chemical stability of silica-bonded chiral phases. Though the polystyrene-type ligand exchangers will probably remain superior in this respect, the lifetime of the majority of silica-based sorbents should be fairly long. Alone, chiral phases prepared from the 3-aminopropylsililated material could prove rather labile when operated at elevated temperatures and in the region above 5 to 6 pH. Due to the flexibility of the trimethylene spacer, the amino groups probably can approach the bonding sites of neighboring grafted groups acting as catalysts for their cleavage from the silica matrix.

A recent idea for enhancing the hydrolytic stability of silica-bonded phases is that of preparing a polymeric organic interface layer which would have several links to the matrix. A partial hydrolysis of these links would not necessarily lead to a loss of the bonded phase. The first realization of this idea was binding polystyrene chains containing residues of L-proline or L-hydroxyproline to the surface of microparticulate silica.[143]

According to one of the methods developed for chemical bonding short polystyrene chains of 100 to 250 monomer units to the surface of LiChrosorb® Si-100, a copolymer of styrene with small amounts of methylvinyldimethoxysilane was prepared. It was subjected to chloromethylation, then reacted with the silica silanol groups, and finally substituted with the chiral amino acid residues in the usual manner developed for the polystyrene-type sorbents. In the second method, chloromethylated polystyrene was reacted with a small amount of 3-aminopropyltriethoxysilane and then bonded to the silica surface. Thus, a strong differentiation in the length of the spacer between the polymer chain and the matrix surface was achieved:[143]

```
                                   CH2 - CH - X
                                  /         |
                         CH2 - N            |
                                  \ CH  -  CH2
                                     |
                                     COOH                 n = 90,   X = H
           |
 - CH2 CH - CH2 - CH - CH2 - (CH - CH2)n - CH -
                  |                        |              n = 14,   X = OH
                  Si - CH3
                 /   \
                O     O
             //////////

 - CH - CH2 - CH - CH2 - (CH - CH2)n - CH -
   |                               |
                                      CH2 - CH - X        n = 90,   X = OH
         CH2        CH2 - N          /          |
          |                \ 
         NH                 CH  -  CH2
          \                 |
          (CH2)3            COOH
          /
         Si -- OH
        /   \
       O     O
    //////////
```

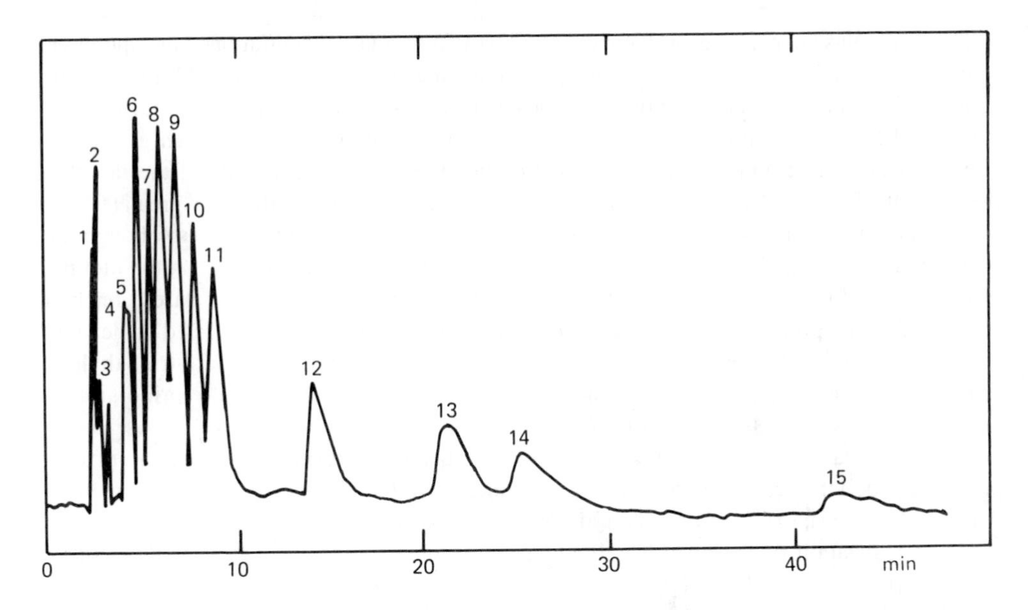

FIGURE 9. Resolution of racemic amino acids on L-hydroxyproline containing polystyrene bonded to silica gel. Column, 4.2 × 250 mm; d_p 10 μm; eluent, 10^{-4} *M* $Cu(OAc)_2$, 0.05 *M* $N(Bu)_4OAc$, pH 4.6, 1 mℓ/min.; temperature, 65°C. Elution sequence: 1 DL-Lys, 2 DL-Arg, 3 unknown, 4 D-Ala, 5 L-Ala, 6 D-Ser, 7 L-Ser, 8 D-His, 9 D-Thr, 10 L-Thr, 11 L-His, 12 D-Met, 13 L-Met, 14 D-Eth, 15 L-Eth. (From Kurganov, A. A., Tevlin, A. B., and Davankov, V. A., *J. Chromatogr.*, 261, 223, 1983. With permission.)

In addition to the chemical links, adsorption phenomena seem to provide a dense covering of the silica surface with the polystyrene chains. All attempts to block the remaining surface silanol groups by treating the sorbent with trimethylchlorosilane (either before or after the substitution of the polystyrene chloromethyl groups by the amino acid residues) resulted in an increase in the total carbon content of less than 1% (the carbon content of the sorbents amounted to 17 to 20%, corresponding to 1.4 to 1.6 mmol styrene units per 1 g of sorbent). The concentration of the polymer-fixed amino-acid ligands was approximately 0.6 m mol/g. As was expected, the polymeric-bonded phases showed good stability even at temperatures of 50 to 65°C.

Owing to the high surface density of styrene units that are partially unsubstituted, the above bonded phases display marked hydrophobic properties. Therefore, the retention of hydrophobic amino acids (Trp, Phe, Leu, Tyr) was found to be especially high, but falls drastically on adding 20 to 30% of an organic modifier to the aqueous eluent. It is remarkable that the typical basic amino acid lysine was observed to remain the least-retained solute in the 4.6- to 5.0-pH range investigated, where it certainly carries a positive charge and should actively interact with the surface silanol groups if these were not sterically blocked by the interface polystyrene layer.

Figure 9 shows that performance of the polymeric chiral-bonded phases does not yield to those of the monomerically bonded chiral phases and permits resolution of mixtures of several racemic amino acids. A quite unexpected result in the LEC with the above sorbents was that the chiral phase which was densely bonded to the matrix surface (short spacer groups and short distances between the two neighboring links to the surface) showed an inverted elution order of amino-acid enantiomers compared to those on polymeric phases bonded less densely to the silica surface. In the latter case, the elution order was D after L, which is also characteristic of polystyrene-type resins containing L-proline or L-hydroxyproline as fixed ligands. This phenomenon was explained (see below) by steric interference of the silica gel surface with the mixed-ligand sorption complexes formed in close proximity to the surface.

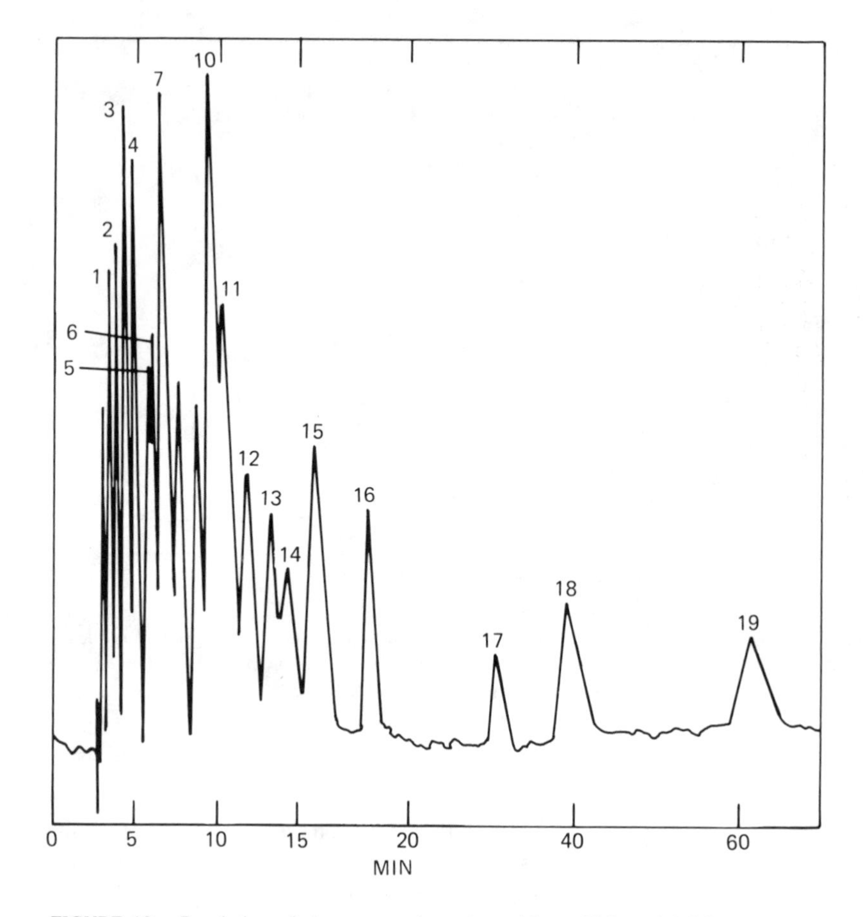

FIGURE 10. Resolution of eleven racemic amino acids on N'-benzyl-1,2(R)-propanediamine containing polystyrene bonded to silica gel. Column, 4.2 × 250 mm; d_p 10 μm; eluent, 10^{-4} *M* $Cu(OAc)_2$, 0.01 *M* NH_4OAc, water-CH_3CN 70/30, pH 4.0, 0.5 mℓ/min.; temperature, 75°C. Elution sequence: 1 D-Pro, 2 L-Pro, 3 D-Ala, 4 L-Ala, 5 D-Ser, 6 L-Ser, 7 D-Val, 8 D-Thr, 9 L-Thr, 10 D-Leu + D-Met + D-Ile, 11 L-Val, 12 D-Tyr, 13 L-Met, 14 L-Leu, 15 L-Tyr + L-Ile, 16 D-Phe, 17 L-Phe, 18 D-Trp, 19 L-Trp.

In closing the chapter on polymeric chiral-bonded phases, we would like to present Figure 10 as an example of chromatography of a mixture of 11 racemic amino acids on the silica-bonded polystyrene containing chiral ligands of the type N^1,N^1-dibenzyl-1,2-propanediamine, which again showed high efficiency and a broad application range.

In addition to the above-described numerous amino acid-type and N^1,N^1-dibenzyl-1,2-propanediamine-type chiral-bonded phases, the L(+)tartaric-acid-modified silica gel was recently synthesized:

$$\geq Si{-}(CH_2)_3{-}NH{-}CO{-}\overset{*}{C}H(OH){-}\overset{*}{C}H(OH){-}COOH$$

The synthesis was carried out[144] by bonding *N*-[3-(trimethylsilylpropyl)]-L(+)diacetyltartaric acid amide silica gel followed by cleavage of the protecting acetyl groups. It seems to be important that the carboxyl group of the bonded tartaric acid phase is free for the complexation with the copper(II) ions. It is only in this form that the chiral ligand exchanger acquires the ability to separate enantiomers of catecholamines:

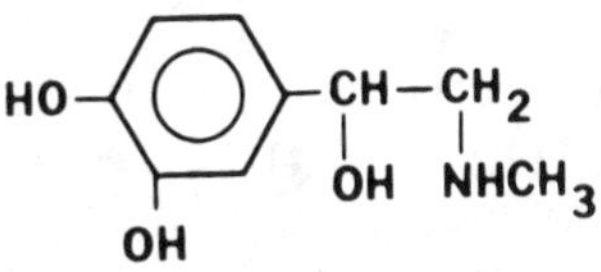

HO–C6H3(OH)–CH(OH)–CH2–NHCH3

adrenaline

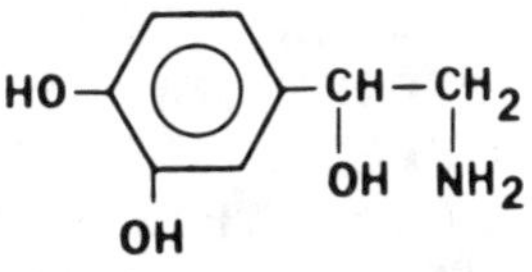

HO–C6H3(OH)–CH(OH)–CH2–NH2

noradrenaline

aromatic amino acids (phenylalanine, tryptophan, 3,4-dihydroxyphenylalanine, methyl-DOPA, 3-*O*-methyl-DOPA), and mandelic acid.[145] It is most remarkable that in an ammonium acetate buffer (pH 5.5 to 7.5) the D-enantiomers were observed to be more strongly retained than the corresponding L-enantiomers, whereas the opposite elution order was found in a phosphate buffer (pH 4.5 to 5.5). The last eluent provided higher column efficiency. Nickel(II) and cobalt(II) forms of the ligand exchanger gave no resolution of the enantiomers.

C. Coating of Silica With Chiral Ligands

One of the very early attempts to apply enantioselectivity effects in formation of labile mixed-ligand complexes to the chromatographic resolution of racemic compounds was made by Bernauer et al.[146-148]

These authors saturated ion-exchanging resins, Dowex® 1 × 2, with optically active anionic complexes and used the packings obtained in chromatography of racemates. In this manner, it was possible to detect a definite difference in the interaction of an anionic complex, iron^{III} (D-OPTA)$^-$, where D-OPTA is D-β-hydroxyethylpropylenediaminetriacetate, with enantiomeric forms of *N*-acetyl and *N*-benzoyl derivatives of alanine, phenylalanine, leucine, and methionine.[147] This difference resulted in a partial resolution of racemic solutes, with the highest optical purity of the eluate fractions amounting to 4.8 to 25.8%. The total sorption selectivity was α = 1.025 to 1.063. Since the measurements carried out with different solvents resulted in changing values of α, stereoselectivity was attributed to differences in solvation energies of the diastereomeric mixed-ligand complexes. Substantially higher selectivity effects were observed[148] with nickel, copper, and zinc complexes of D(−)-propylenediaminetetraacetate, M^{II}(D-PDTA)$^{2-}$, interacting with 1-phenylethylamine, with the values of α being 1.7, 1.6, and 1.4, respectively. In this instance, chromatography was easy to perform in pure water because phenylethylamine could not displace the doubly charged resolving complex from the anion exchanger. Even a relatively short column (37 cm) gives approximately 60% resolution of 1-phenylethylamine. Besides this practical result, the papers are interesting from two other aspects: (1) it has been demonstrated that an adsorptional modification of a nonchiral sorbent with a chiral resolving ligand or complex is a promising way for obtaining chiral packings, and (2) it has been shown for the first time that even unidentate ligands may be coordinated enantioselectively in chiral ternary structures:

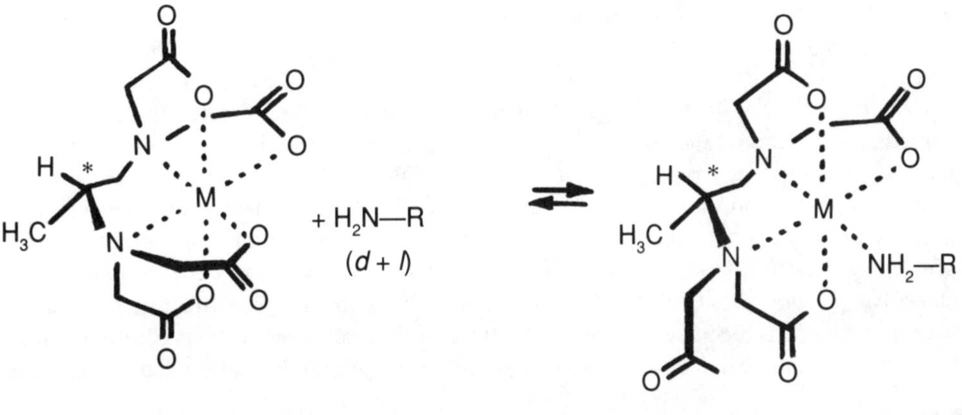

M(PDTA) complex + H_2N—R ($d + l$) ⇌ mixed complex with NH_2—R

Table 9
REPRESENTATIVE CAPACITY FACTORS k'_L AND k'_D AND ENANTIOSELECTIVITY α FOR AMINO ACIDS ON A C_{16}–L–HYP-COATED LICHROSORB® RP-18 COLUMN[a] AND A C_{10}–L–HIS-COATED ZORBAX® C_8 COLUMN[b149,150]

	C_{16}–L–Hyp			C_{10}–L–His		
Amino acid	k'_L	k'_D	α	k'_L	k'_D	α
Aspartic acid	0.08	0.10	1.17	0.86	0.63	0.73
Glutamic acid	0.13	0.17	1.33	2.39	2.39	1.00
Histidine	0.97	0.57	0.59	0.50	0.50	1.00
Alanine	0.58	0.91	1.56	2.48	3.55	1.35
Asparagine	0.66	0.66	1.00	0.63	0.63	1.00
Hydroxyproline	0.65	1.24	1.91	—	—	—
Glutamine	0.68	0.85	1.25	—	—	—
Serine	0.73	0.73	1.00	1.23	1.54	1.25
Proline	0.81	6.10	7.54	4.66	6.70	1.44
Citrulline	0.86	1.06	1.29	2.00	2.00	1.00
Threonine	0.83	0.85	1.02	1.03	1.33	1.29
Aminobutyric acid	1.00	2.49	2.49	7.20	11.60	1.61
Valine	1.63	7.42	4.55	19.33	35.33	1.83
Norvaline	2.33	7.94	3.41	—	—	—
Thyrosine	2.51	7.01	2.79	31.78	44.22	1.39
Methionine	2.54	5.06	1.99	34.00	43.00	1.26
Arginine	2.97	3.67	1.24	4.77	4.77	1.00
Leucine	5.01	17.82	3.56	—	—	—
Phenylalanine	10.33	38.30	3.71	—	—	—
Tryptophan	21.52	54.22	2.52	—	—	—

[a] Eluent, 10^{-4} *M* Cu(OAc) in methanol/water of 15:85, 25°C.
[b] Eluent, 10^{-4} *M* Cu(OAc) in water, 35°C.

With the same idea of transforming a commercially available nonchiral sorbent into a chiral ligand-exchanger, reversed-phase microparticulate silica, LiChrosorb® RP-18 has been dynamically coated with *N*-alkyl derivates of L-hydroxyproline.[149] The anchoring *N*-alkyl groups (where alkyl is n-C_7H_{15}–, n-$C_{10}H_{21}$–, and n-$C_{16}H_{33}$–) of L-hydroxyproline provide for a permanent adsorption of the resolving chiral agent on the hydrophobic interface layer of the reversed-phase packing material. These *N*-alkyl residues are supposed to be integrated in some way or another between the $C_{18}H_{37}$-chains modifying the surface of the support. With the precaution that the content of organic solvents in the eluent is kept below a certain level (to avoid eluting the chiral coating material), the columns were observed to remain stable and did not show any "bleeding" effects. As the hydrophilic amino acid part of *N*-alkyl-L-hydroxyproline is directed toward the polar mobile phase, it is accessible to formation of ternary complexes with copper(II) ions and mobile ligands. Enantioselectivity of the system was found to be very high, with α-values amounting up to 16.4 for DL-proline and 6.9 for DL-valine. With the exception of histidine, D-amino acids were stronger retained than L-enantiomers. Of the 26 different amino acids investigated (Table 9), only asparagine, serine, and ornithine, having polar side chains, could not be resolved on the 10-cm-long column used. As shown in Figure 11, a 10-cm-long column, coated with *N*-hexadecyl-L-hydroxyproline, is effective enough to perform base line separation of six racemic amino acids. The detection limit of a common 254-nm ultraviolet detector was determined to be about 10^{-10} mol, i.e., 10^{-8} to 10^{-9} g of amino acid having a capacity factor, k′, of about two. The data of Table 9 and Figure 11 imply that hydrophobic interactions contribute much to retention of mobile ligands in the above systems; tryptophan, phenylalanine, and norleucine

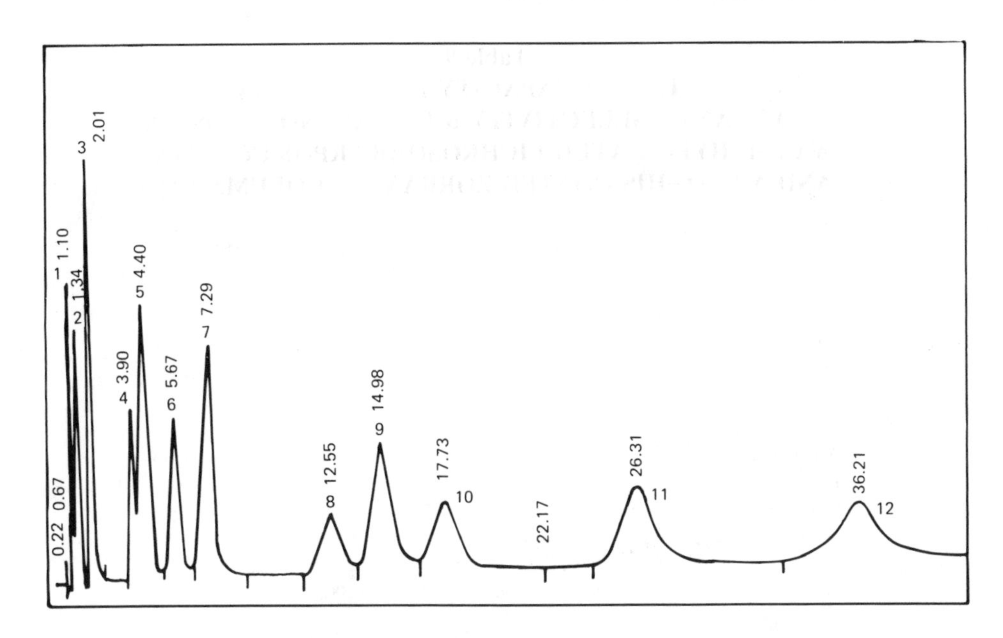

FIGURE 11. Resolution of six racemic amino acids on C_{16}-L-hydroxyproline coated LiChrosorb® RP-18 column. Column, 4.2 × 100 mm; d_p 5 μm; eluent, 10^{-4} *M* $Cu(AcO)_2$, methanol-water 15/85, pH 5.0, 2.0 mℓ/min; temperature, 20°C. Elution sequence: 1 L-Ala, 2 D-Ala, 3 L-Nva, 4 L-Leu, 5 L-Nle, 6 D-Nva, 7 L-Phe, 8 D-Leu, 9 L-Trp, 10 D-Nle, 11 D-Phe, 12 D-Trp. (From Davankov, V. A., Bochkov, A. S., Kurganov, A. A., Roumeliotis, P., and Unger, K. K., *Chromatographia*, 13, 677, 1980. With permission.)

appear to be the most retained species. Simultaneously, the α-values for these and other amino acids capable of entering distinct hydrophobic interactions with the reversed-phase material were observed to be extremely high, thus permitting preparative-scale resolutions.

Speaking about the analytical application of chiral coatings on a reversed-phase material, it is probably more convenient to use *N*-alkyl-L-histidine as resolving agent.[150] With *N*-decyl-L-histidine, enantioselectivity of amino acid resolutions does not exceed the value of α = 2, which favors a quantitative comparison of two peak areas and enhances the precision of determining the enantiomeric composition of amino acids. Of 23 racemates examined, 17 could be resolved on a 10-cm-long column in aqueous eluent containing nothing more than a trace of copper acetate (10^{-4} *M*). With respect to chromatographic parameters influencing retention and discrimination of amino acid enantiomers, *N*-decyl-L-histidine coatings were found to behave similarly to *N*-alkyl-L-hydroxyproline. Increasing the content of the organic modifier in the eluent and lowering the pH of the latter brings about a decrease in the k′ and α values, whereas the elevated temperature and copper content of the eluent decreases retention of amino acids, but not the resolution selectivity. The latter observation deserves a short comment because it is valid for all ligand-exchanging systems where retention of mobile ligands requires formation of ternary complexes. In accordance with the equilibrium [Cu(AA)(*N*-alkyl-L-His)] + $Cu(AcO)_2 \rightleftarrows [Cu(AA)]^+AcO^-$ + $[Cu(N\text{-alkyl-L-His})]^+AcO^-$, the rising concentration of $Cu(AcO)_2$ in the eluent favors the formation of the mixed-ligand (amino acidato) (acetato) copper complexes and facilitates the elution of amino acid solutes, AA.

The very simple technique of modifying column packing via adsorption of chiral ligands can be easily extended to various new systems. Even L-phenylalanine was found[151] to adsorb from aqueous solutions on a reversed-phase silica gel in quantities sufficient to transform the packing into a chiral ligand exchanger. Nice enantiomeric resolutions of a mixture of *p*-hydroxymandelic acid, *m*-hydroxymandelic acid, and mandelic acid were achieved on thus-treated μBondapak® C-18 10-μm column (540 × 4.6 mm) in an aqueous eluent which contained 5.10^{-4} *M* $CuSO_4$. Left-rotating enantiomers were observed to form stronger species in each diastereomeric pair of sorption complexes. The fact that unsubstituted mandelic acid isomers display longer retention times than hydroxy-substituted analogues implies that the phenylalanine-modified packing still preserves its overall hydrophobic nature. Similarly, adsorption of palladium(II)-*S*-ethyl-L-cysteine complex on the same reversed-phase material enhances retention of L-methionine and diminishes that of D-methionine on elution with a methanol-water mixture (40:60) adjusted to pH 3 with orthophosphoric acid.[152] Note that no metal ions were added to the eluent in this experiment.

An elegant preparative-scale resolution of enantiomers of 3-methylene-7-benzylidenebicyclo[3.3.1]nonane of the following formula

has been achieved on silica-gel impregnated with 7.5% silver(I) *d*-camphor-10-sulphonate.[153] In an apolar hexane-methanol (96:4) eluent, π-complexes were formed on the sorbent surface between silver ions and diene molecules.

Chiral polymeric coating could obviously have an advantage over low-molecular-weight coating materials in that the total adsorption energy of polymeric chains on the silica surface is rather high, thus precluding a danger of column "bleeding". Boue et al.[119] made use of this advantage of chiral chelating polymers. They reacted a linear soluble polyacrylamide ($\overline{M}_n$ = 23,000) with formaldehyde and L-proline in an aqueous alkaline solution to graft 0. to 4.2 mmol chiral ligands per gram of the polymer obtained (i.e., the extent of substitution varied between p = 0 and p = 0.90):

$$\begin{array}{llll}
-CH_2-\underset{|}{C}H- \; + \; CH_2O \; + & \quad CH_2-CH_2 & \quad -\underset{|}{C}H-CH_2 - & \underset{|}{C}H-CH_2- \\
\quad\; C{=}O & HN \qquad\quad | & \longrightarrow \quad C{=}O & C{=}O \\
\quad\;\; | & \quad\;\; \overset{*}{C}H-CH_2 & \qquad | & \;\; | \\
\quad\; NH_2 & \qquad | & \quad\; NH & NH \qquad CH_2-CH_2 \\
 & \quad\; COOH & \qquad | & \;\; | \\
 & & \quad CH_2OH & CH_2-N \qquad\quad | \\
 & & & \qquad\quad \overset{*}{C}H-CH_2 \\
 & & & \qquad\qquad | \\
 & & & \qquad\quad COOH
\end{array}$$

Spherosil® XOA 600 (5 to 7 μm) readily adsorbs this material from an aqueous solution in quantities up to 6 to 7 wt %. Herewith, however, the column efficiency was observed to fall from 7000 to 8000 plates claimed for a 10-cm column to 1200 plates with naphthalene eluted in *n*-hexane. For the elution of amino acids in water from the copper(II) form of the packing, the efficiency dropped to about 100 plates for 10 cm so that only single racemates could be resolved to the base line, in spite of the high enantioselectivity of the packing (α up to 2.45 for tryptophan eluted with 0.1 *M* KNO_3 solutions, 2.10^{-5} *M* in copper). It is interesting that the retention of amino acids was observed to change proportionally to the expression $p(1 - p)^2$, which represents the probability of finding isolated chiral grafts on the polyacrylamide chain. These isolated chiral ligands are thought to be responsible for the retention of the mobile ligands via formation of ternary sorption complexes, whereas the "vicinal" grafts probably form stationary complexes of the composition of 2:1 with copper(II) ions which could be too stable to participate in the ligand exchange. This assumption is verified fairly well by the observation that the k′ values reach a maximum at the extent of polymer substitution of $p = 0.33$ (about 2 mmol L-Pro per gram of polymer).

Another interesting observation with the above polymeric coating materials is that the retention of mobile ligands definitely drops with the amount of injected sample increasing, whereas the ratio of capacity factors for the two enantiomers, $\alpha = k'_D/k'_L$, is practically constant when the column is not overloaded.[119] A similar behavior was also reported[76] for chiral resins prepared by grafting L-proline to poly(2,3-epoxypropyl methacrylate). For other chiral ligand-exchanging packings, including those prepared by coating an inert matrix with chiral amino acid derivatives, both the k′ and α values were observed to remain fairly independent of the sample size.

Closing the consideration of chiral complexing stationary phases in the LEC of enantiomers, we would like to emphasize that polymeric resins, in particular, those based on macronet isoporous polystyrene, will predominate in preparative-scale resolutions, due to the excellent chemical stability and high exchange capacity of these resins. The favored application area of chemically and adsorptionally modified silica packings is the enantiomeric analysis of complex mixtures where the column efficiency is of utmost importance.

After the analytical potentials of silica packings coated with chiral ligand have been generally recognized, it appeared logical to extend this technique to thin layer chromatography (TLC). Indeed, this has recently been done with reversed-phase TLC plates in two different ways.

In experiments of the German group,[154] RP 18-TLC plates were immersed for 1 min in a 0.25% copper acetate solution (methanol-water of 1:9), dried in the air, then immersed into a 0.8% methanolic solution of (2*S*, 4*R*, 2′*RS*)-4-hydroxy-1-(2′-hydroxydodecyl)-proline and allowed to dry again. After this the plate was ready for chromatography of amino acid enantiomers, as the sorbent surface now incorporated copper complexes of L-hydroxyproline bearing hydrophobic 2-hydroxy-dodecyl chain at the nitrogen atom. Several amino acids could be resolved into enantiomers (Ile, Phe, Tyr, Trp, Pro, Gln, and 3-thiazolidine-4-carboxylic acid) on development with appropriate methanol-water-acentonitrile mixtures and visualization of spots with a 0.1% ninhydrin reagent. As could be expected from regularities observed in column experiments with similar chiral resolving systems,[149] D-isomers of amino acids (with the exception of tyrosine) were found to show smaller R_f values than L-antipodes.

Chiral plates prepared in the above described manner are now commercially available as Chiral-Plate® for TLC, Machery Nagel, Dueren. The application scope of the technique has been extended[155] to nonnatural amino acids, *N*-methylated amino acids, *N*-formyl amino acids, and other derivatives of amino acids. Even dipeptides and a lacton derivative were resolved into enantiomers (Table 10).

Weinstein[156] immersed RP TLC plates into a 4×10^{-3} *M* solution of copper-*bis*-dimethyl-L-alaninate in acetonitrile containing 2.5% water. After drying, the plates easily resolved dansyl derivatives of all common amino acids using 0.3 *M* sodium acetate buffer (pH 7) mixed with 25 to 40% acetonitrile. Resolutions were registered under UV irradiation.

To achieve resolution of mixtures of several dansyl-amino acids, a two-dimensional RP-TLC technique was applied.[157] In the first dimension, the Dns-amino acids were separated in a nonchiral mode using a gradient elution with aqueous sodium acetate buffers and increasing concentration of acetonitrile. For the second dimension, the plates were sprayed with a copper-*bis*-dipropyl-L-alaninate solution (4 m*M*) in water and acetonitrile (5:95) and developed with the same mixture (plus 0.3 *M* sodium acetate) using a temperature gradient. Thus, it has been possible to resolve five racemic solutes, Dns-derivatives of aspartic acid, serine, methionine, alanine, and phenylalanine, into individual enantiomers.

The resolution of enantiomers in both systems was usually so good that the respective

Table 10
ENANTIOMERIC RESOLUTION OF RACEMATES BY TLC (DEVELOPMENT DISTANCE 13 cm, SATURATED CHAMBER, CHIRAL-PLATE®)

Racemate	R_f value (configuration)		Eluent[a]
Valine	0.54(D)	0.62(L)	A
Methionine	0.54(D)	0.59(L)	A
allo-Isoleucine	0.51(D)	0.61(L)	A
Norleucine	0.53(D)	0.62(L)	A
2-Aminobutyric acid	0.48	0.52	A
O-Benzylserine	0.54(D)	0.65(L)	A
3-Chloralanine	0.57	0.64	A
S-(2-Chlorobenzyl)-cysteine	0.45	0.58	A
S-(3-Thiabutyl)-cysteine	0.53	0.64	A
S-(2-Thiapropyl)-cysteine	0.53	0.64	A
cis-4-Hydroxyproline	0.41(L)	0.59(D)	A
Phenylglycine	0.57	0.67	A
3-Cyclopentylalanine	0.46	0.56	A
Homophenylalanine	0.49(D)	0.58(L)	A
4-Methoxyphenylalanine	0.52	0.64	A
4-Aminophenylalanine	0.33	0.47	A
4-Bromophenylalanine	0.44	0.58	A
4-Chlorophenylalanine	0.46	0.59	A
2-Fluorophenylalanine	0.55	0.61	A
4-Jodophenylalanine	0.45(D)	0.61(L)	A
4-Nitrophenylalanine	0.52	0.61	A
O-Benzyltyrosine	0.48(D)	0.64(L)	A
3-Fluorotyrosine	0.64	0.71	A
4-Methyltryptophan	0.50	0.58	A
5-Methyltryptophan	0.52	0.63	A
6-Methyltryptophan	0.52	0.64	A
7-Methyltryptophan	0.51	0.64	A
5-Bromotryptophan	0.46	0.58	A
5-Methoxytryptophan	0.55	0.66	A
2-(1-Methylcyclopropyl)-glycin	0.49	0.57	A
N-Methylphenylalanine	0.50(D)	0.61(L)	A
N-Formyl-*tert*.-leucine	0.48(+)	0.61(−)	A
3-Amino-3,5,5-trimethyl-butyrolactone · HCl	0.50	0.59	A
N-Glycylphenylalanine	0.51(L)	0.57(D)	B

[a] A: methanol/water/acetonitrile = 50:50:200(vvv); B: methanol/water/acetonitrile = 50:50:30(vvv).

From Günther, K., Schickedanz, M., and Martens, J., *Naturwissenschaften*, 72, 149, 1985. With permission.

antipodes could be determined at trace levels of about 1%.[157] It should be noted, however, that the hydrophobicity of the *N,N*-dimethyl-L-alanine as well as *N,N*-di-*n*-propyl-L-alanine is not sufficient to make the chiral resolving ligand stick onto RP silica. The chiral copper complex certainly moved with the eluent front along the TLC plate. Therefore, the system is closely related to another modification of LEC technique which was called "chiral eluent method" and should be dealt with in the next section.

III. CHIRAL COMPLEXES IN THE MOBILE PHASE

The first chiral chelating packing materials became commercially available in 1983 with the production of silica gels modified with L-proline or L-valine by Serva®, West Germany. Before this, LEC of enantiomers was the privilege of a limited number of research groups having a certain level of experience in macromolecular chemistry or silica gel modification, since experience is needed to produce chiral sorbents giving acceptable performance. To overcome this handicap, the addition of chiral metal complexes to the eluent was sugested in 1979 as a novel approach to LEC of enantiomers.[158,159] This method of "chiral eluent" relates to an earlier suggestion by Cram and co-workers[160] to resolve racemic amino esters on Celite® or alumina by adding chiral crown ether to the eluent. Thus, the method permits one to resolve racemic compounds using conventional achiral column packings. The resolving chiral reagent, an optically active transition metal complex, is continuously introduced into the chromatographic column by means of the mobile phase. Again, resolution is based on enantioselectivity of formation of ternary metal complexes that involve one of the solute enantiomers to be resolved and the resolving chiral ligand. However, ternary diastereomeric complexes can be located now both on the sorbent surface and/or in the mobile phase.

One can easily notice that a chromatographic system operating in accordance with the chiral eluent method consumes the chiral resolving reagent in quantities by far exceeding the amount of enantiomers to be separated. Moreover, excess of chiral resolving agent accompanies the separated D- and L-isomers as they come out of the column.

Therefore, unless small amounts of very expensive, optically active compounds have to be obtained, a preparative application of the method is out of the question.

On the contrary, analytical applications of the method seem very promising and versatile since highly efficient achiral analytical columns can be used in combination with various available optically active chelating compounds. Moreover, the ease of substituting the resolving chiral agent for its enantiomeric form, and then for a racemic form, gives the unique possibility of a safe identification of chiral components in complex mixtures with achiral compounds.[159,161,162] The above operations will alter positions of peaks belonging to chiral components, whereas the retention of achiral mobile ligands is independent of configuration of the chelate additive. However, some possible detection problems caused by the presence of a chiral background in the effluent should be taken into account. The enantiomers resolved in the column enter the detector cell in the form of ternary complexes with the chiral resolving ligand. These complexes are diastereomeric and differ therefore in the molecular extinction as well as other properties. Separate calibration for each of the enantiomers is required.

Tables 11 and 12 give an overview of chiral metal chelate additives and other main components of the chromatographic systems described in the literature and successfully tested in the resolution of enantiomeric amino acids and derivatives therefrom, respectively. Both tables relate to reversed-phase systems, which are especially popular, owing to high efficiency and versatility of these systems; here, aqueous or aqueous-organic eluents are used (which are very suitable for underivatized amino acids, as well as the majority of their derivatives) in combination with highly hydrophobic sorbents, in particular, reversed-phase macroporous silica. Its surface has been made hydrophobic through chemically bonding alkyl chains. In accordance with the common experience with reversed-phase system, it is not advisable to use packings with short-bonded alkyl chains in the chiral eluent technique.[162] Octyl and octadecyl radicals provide better covering of the silica surface and give equally good results.[162,182] In this case, retention of solutes dissolved in the polar mobile phase is governed by hydrophobic interactions with the hydrocarbonaceous interphase layer.

Concerning the retention mechanisms of amino acid enantiomers under resolution according to the chiral eluent technique, three different extreme situations should be regarded as important:[93] (1) the sorbent retains the resolving chiral agent and its complexes, (2) the sorbent retains the enantiomeric ligands to be separated, and (3) the sorbent retains ternary mixed-ligand complexes. Though in practice a combination of these mechanisms can occur, this is not necessarily the case, as the electrostatic charges and other properties of the above species can be substantially different. We have to bear in mind these differences in order to understand the influence of varying chromatography parameters on the retention and resolution of amino acids.

Table 11
RESOLUTION OF UNDERIVATIZED AMINO ACIDS USING CHIRAL ELUENT TECHNIQUE IN REVERSED-PHASE SYSTEMS

Chiral resolving ligand	Metal ion	Racemates resolved, maximum selectivity	Elution order of enantiomers	Experimental conditions
L-proline[161]	Cu^{2+}	19 amino acids; α = 6.5 (Val)	D, L, except His, Asp, Thr	4—8 m*M* Cu(L-Pro)$_2$, pH ≤5 for Glu, pH 7—8 of Asp, Thr
L-phenylalanine[162]	Cu^{2+}	Trp, α-methyl-DOPA, 5-hydroxytryptophan	D, L	4 m*M* Cu(L-Phe)$_2$, 20% MeOH, 20—25°C
L-phenylalanine[163]	Cu^{2+}	α-methyl-DOPA	D, L	4 m*M* Cu(L-Phe)$_2$, MeOH/H_2O of 1/5
L-phenylalanine[164]	Cu^{2+}	21 amino acids, α = 5.2 (Pro)	D, L, except Thr, Ser, His, Asn	1 m*M* Cu(L-Phe)$_2$, 5 m*M* NH_4OAc, pH 4.5
L-phenylalanine[190]	Cu^{2+}	*p*-ethylphenylalanine	D, L	4 m*M* Cu(L-Phe)$_2$, 45% MeOH
N,N-dipropyl-L-alanine[165-167]	Cu^{2+}	19 amino acids; α = 2.7 (Ileu)	D, L	4 m*M* Cu(Pr_2-L-Ala)$_2$, 0—14% MeCN
N,N-dialkyl-L-alanine and L-valine, Alk = Me, Et, Pr, But, Pent[167]	Cu^{2+}	6 amino acids	D, L, except Val (for Et_2Ala and Et_2Val) and His (for Me_2Ala and Me_2Val)	4 m*M* Cu(Alk_2-L-Ala)$_2$ or Cu(Alk_2-L-Val)$_2$
N,N-dimethyl-L-valine[168]	Cu^{2+}	8 α-methylamino acids		4 m*M* Cu(Me_2-L-Val)$_2$, pH 5.5—7.0
N,N-dimethyl-L-leucine[20,202]	Cu^{2+}	Ala, Val, Ser, Thr, Asp, Asn, Glu, Lys, His, Arg	D, L	4 m*M* Cu(Me_2-L-Leu)$_2$, H_2O, pH 5.3
N,N-dimethyl-D-phenylglycine[20]	Cu^{2+}	Ala, Val, Ser, Lys, Arg	L, D	4 m*M* Cu(Me_2-D-Phgly)$_2$, H_2O pH 5.1—5.5
N,N-dimethyl-L-methionine-sulfoxide[20]	Cu^{2+}	Ala, Val, Glu, Tyr, Abu, His	D, L, except His	4 m*M* Cu(Me_2-L-MetSO)$_2$, H_2O, pH 5.4
L-pipecolic acid[20]	Cu^{2+}	Ala, Val, *t*-Leu, Tyr, Abu	D, L	4 m*M* Cu(L-Pip)$_2$, H_2O, pH 5.0
N-methyl-L-proline[20]	Cu^{2+}	Ala, Val, *t*-Leu, Tyr, Abu	D, L	4 m*M* Cu(Me-L-Pro)$_2$, H_2O, pH 5.0—5.7
N-benzyl-L-proline[202]	Cu^{2+}	Val, Pro, Hyp, aHyp; α = 3.7 (Pro)	L, D, except aHyp	1 m*M* Cu(Bzl-L-Pro)$_2$, H_2O
N-methyl-L-valine, L-prolyl-glycine, L-prolyl-L-valine, L-prolyl-L-thyrosine[161]	Cu^{2+}, Zn^{2+}, Co^{2+}, and Hg^{2+}	Amino acids		
L-aspartyl-L-phenylalanine methyl ester (Aspartame)[169,184,191]	Cu^{2+}	Tyr, Phe, Trp, DOPA 7 aliphatic amino acids; α = 2 (Ileu)	L, D	1 m*M* Cu (Aspartame)$^+$, 0—10% MeCN, 32°C
L-aspartyl-L-phenylalanine methyl ester (Aspartame)[169]	Zn^{2+}	Tyr, Phe, Trp, α = 1.9 (Phe)	L, D	0.5 m*M* Zn (Aspartame)$^+$
L-aspartyl-cyclohexylamide[170-173]	Cu^{2+}	11 amino acids; α = 3.9 (Pro)	L, D, except His	0.1—1.0 m*M* Cu(L-AspNH-cHex)$_2$, 5% MeCN for Trp
L-aspartyl-cyclohexylamide[171]	Ni^{2+}	Pro, Val, Tyr; α = 3.9 (Pro)	L, D	0.3 m*M* Ni(L-AspNH-cHex)$_2$

Table 11 (continued)
RESOLUTION OF UNDERIVATIZED AMINO ACIDS USING CHIRAL ELUENT TECHNIQUE IN REVERSED-PHASE SYSTEMS

Chiral resolving ligand	Metal ion	Racemates resolved, maximum selectivity	Elution order of enantiomers	Experimental conditions
L-aspartyl-ethylamide, -butylamide, -hexylamide, -octylamide[171]	Cu^{2+}	Pro, Tyr, Val, Leu, Met, DOPA	L, D	0.3 m*M* Cu(L-AspNH-Alk)$_2$, phosphate buffer, pH 5.0
			L, D	
N-tosyl-L-phenylalanine[174-176]	Cu^{2+}	18 amino acids, except Gln; α = 3.9 (Pro)	L, D for neutral amino acids; D, L for Ser, Asp, Asn, Glu, His, Lys, Arg	0.5 m*M* Cu(Tos-L-Phe)$_2$, 0—15% MeCN, pH6, 30°C
N-tosyl-D-phenylglycine[176,177]	Cu^{2+}	19 amino acids; α = 2.6 (PhGly)	D, L for neutral amino acids; L, D for Ser, Asp, Asn, Glu, His, Lys, Arg, Gln	0.5 m*M* Cu(Tos-D-PhGly)$_2$ 0—15% MeCN, pH 6
N-methyl-L-phenylalanine[164]	Cu^{2+}	19 amino acids; α = 2.9 (Pro)	D, L except Thr, Ser, His	1 m*M* Cu(Me-L-Phe)$_2$; pH 4.5
N,*N*-dimethyl-L-phenylalanine[164]	Cu^{2+}	23 amino acids; α = 4.8 (*tert*-Leu)	D, L except His	1 m*M* Cu(Me$_2$-L-Phe)$_2$, pH 4.5
N,*N*,*N*′,*N*′-tetramethyl-(R)-propanediamine[178]	Cu^{2+}	PhSer, Phe, Tyr, Trp, DOPA; α = 1.14 (Trp)	L, D	0.5 m*M* Cu(Me$_4$pn)$_2^{2+}$, 15% MeCN

Let us consider first the influence of the pH of the eluent. This parameter governs the dissociation of the amino acid species, and consequently, their ability to coordinate metal ions. Chelation of metal ions is favored at higher pH values, and since the chiral recognition in the LEC requires the formation of ternary mixed-ligand complexes, the enantiomeric resolution, $\alpha = k'_D/k'_L$, always improves on raising the pH of the chiral eluent.[158,171,174,179,187,188] However, reversed-phase silica packings are unstable in alkaline solutions and it is advisable to lower the pH of the eluent as far as possible, while still remaining in the pH range that provides a sufficient extent of ternary complex formation. The optimal pH range may lay between 4.5 and 6.0 if underivatized amino acids are resolved using strongly chelating chiral additives to the eluent, such as L-proline, L-phenylalanine, or L-aspartyl-alkylamide. In the latter case, amino acid resolutions were observed to occur even at pH values as low as 3.5.[173] If dansyl-amino acids are subjected to chromatography, the optimal pH value may lie above seven because the sulfonamide fragment $-SO_2NH-$ requires deprotonation in order to coordinate a transition metal ion. When using zinc(II) and nickel(II) as complex-forming metal ions in combination with 2-(L)-isopropyl-4-octyl-diethylene-triamine[158,179] and L-proline-octylamide,[185] respectively, producing chiral chelate additives of the following structures:

Table 12
RESOLUTION OF DANSYL DERIVATIVES OF AMINO ACIDS (di-Dns DERIVATIVES OF HIS, ORN, LYS, AND TYR) AND DIPEPTIDES USING THE CHIRAL ELUENT TECHNIQUE IN REVERSED-PHASE SYSTEMS

Chiral resolving ligand	Metal ion	Racemates resolved, maximum selectivity	Elution order of enantiomers	Experimental conditions
2-(L)-alkyl-4-octyl-diethylenetriamine	Zn^{2+} or Cd^{2+}	25 amino acids but Pro; α = 2.5 (Ser)	L, D, except Asp for R = iPro; Ser and Thr for R = Et; Leu for R = iBu	0.65 m*M* Zn-chelate, 0.17 *M* NH_4OAc, pH 9, 35% MeCN, 30°C
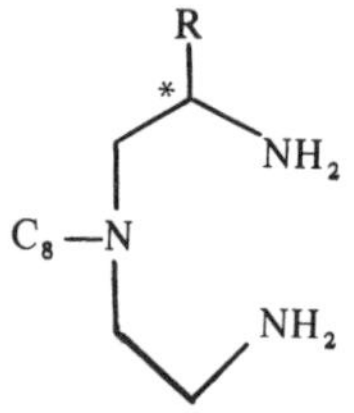 R = Et, iPro, iBu[158,179]				
2-(L)-isopropyl-4-octyl-diethylenetriamine[179]	Hg^{2+}	Ser, Thr, Ala, Ileu, allo-Ileu	D, L	0.8 m*M* Hg-chelate, 0.19*M* NH_4OAc, pH 9.0—9.2, 40% MeCN, 50°C
	Ni^{2+}	Thr, Cys	D, L	0.8 m*M* Ni-chelate, 0.19 *M* NH_4OAc, pH 9, 35% MeCN, 30°C
	Ni^{2+} or Zn^{2+}	10 glycil-containing dipeptides	L, D for Zn D, L for Ni	0.8 m*M* Me-chelate, 0.17 *M* NH_4OAc, pH 9, 35% MeCN, 30°C
L-proline[180-183]	Cu^{2+}	13 neutral amino acids; α = 1.9 (Trp)	L, D except Ser	2.5 m*M* Cu(L-Pro)$_2$, 5 m*M* NH_4OAc, pH 7, 15—20% MeCN
L-prolyl-octylamide[185]	Ni^{2+}	25 amino acids; α = 3.5 (Ser)	D, L except Asp, $CySO_3H$	4 m*M* Ni(L-ProNH-Oct)$_2$, 87.5 m*M* NH_4OAc, pH 9.2, 60% MeOH
L-arginine[181]	Cu^{2+}	10 aliphatic and aromatic amino acids; α = 1.26 (Trp, Nleu)	D, L	2.5 m*M* Cu(L-Arg)$_2$, 25 m*M* NH_4OAc, pH 7.5, 20% MeCN
L-histidine[188]	Cu^{2+}	12 amino acids; α = 2.4 (Nleu)	L, D, except Trp, Ser, Thr, Phe	2.5 m*M* Cu(L-His)$_2$, 25 m*M* NH_4OAc, pH 7.0, 10—17% MeCN
L-histidinemethyl ester[183,184,188,189]	Cu^{2+}	16 amino acids; α = 1.5 (Nleu) except Pro, Hyp, Ala, Thr, Cys	L, D, except Asp, Phe, Trp, thyroxine	2.5 m*M* Cu(L-His-OMe)$_2$, 25 m*M* NH_4OAc, pH 7.0, 20% MeCN
N,N-dipropyl-L-alanine[186]	Cu^{2+}	Common amino acids except Tyr	D, L	1 m*M* Cu(Me$_2$-L-Ala)$_2$, 0.3 *M* NaOAc, pH 7.0, 23.5—40% MeCN gradient
	Cu^{2+}	9 amino acids; α = 2.5 (Asp)	D, L	2 m*M* Cu-chelate, 0.3 *M* NaOAc, pH 7.5, 27% MeCN

Table 12 (continued)

RESOLUTION OF DANSYL DERIVATIVES OF AMINO ACIDS (di-Dns DERIVATIVES OF HIS, ORN, LYS, AND TYR) AND DIPEPTIDES USING THE CHIRAL ELUENT TECHNIQUE IN REVERSED-PHASE SYSTEMS

Chiral resolving ligand	Metal ion	Racemates resolved, maximum selectivity	Elution order of enantiomers	Experimental conditions
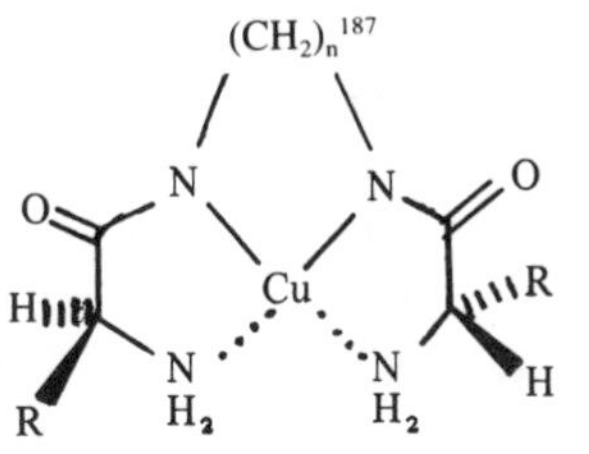 $(CH_2)_n$ [187] — N, N, O, O, Cu, H, R, R, H, N H_2, N H_2				
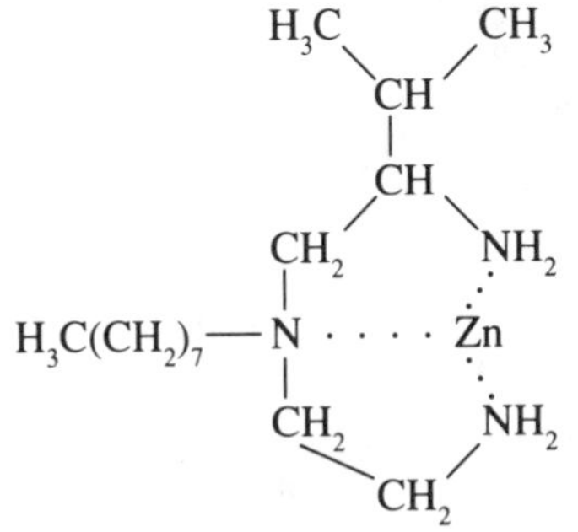 H_3C, CH_3, CH, CH, CH_2, NH_2, $H_3C(CH_2)_7$—N · · · · Zn, CH_2, NH_2, CH_2				
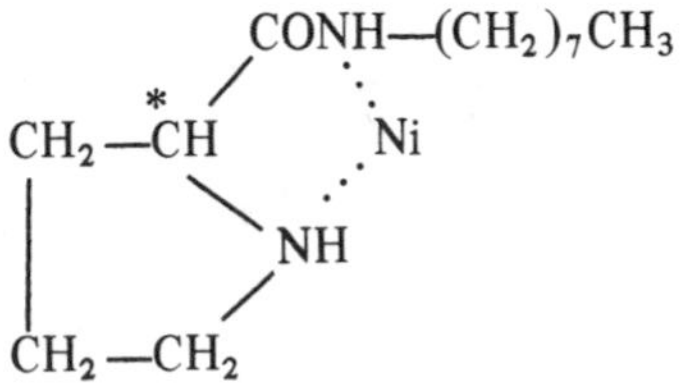 CONH—$(CH_2)_7CH_3$, CH_2—$\overset{*}{C}H$, Ni, NH, CH_2—CH_2				

Tapuhi and co-workers[185] were bound to raise the pH value of the eluent from 9.0 to 9.2, since the coordination ability of zinc and nickel with dansyl-amino acids and the above chiral additives is less than that of copper(II) ions. (A short precolumn was needed to enhance the lifetime of the reversed-phase material in the main column.)

Whereas with the chemically bonded chiral phases, the increasing extent of ternary complex formation on raising the pH of the mobile phase automatically enhanced the retention of the enantiomeric solutes, raising the pH value of a chiral eluent can sometimes result in lowering the solute retention. Obviously, such a situation indicates that the ternary mixed-ligand complexes are less hydrophobic than the enantiomeric solutes alone. This situation was observed on chromatography of hydrophobic dansyl-amino acids in eluents containing hydrophilic L-prolinato- or L-histidinato-copper.[181,188] Here, retention of Dns-amino acids was found to fall on raising the pH of the eluent from 5 to 7. With hydrophobic chiral additives like L-phenylalanine, *N*-methyl-L-phenylalanine, *N,N*-dimethyl-L-phenylalanine,[164] L-aspartyl-cyclohexylamide,[171] *N*-tosyl-L-phenylalanine,[174] or chiral alkyl-diethylenetriamine,[158,179] retention of enantiomeric amino acids and dansyl-amino acids increases in accordance with the extent of the formation of ternary complexes.

Similar to the influence of pH on the resolution selectivity, the impact of the chiral additive concentration is unambiguous. As the amount of the resolving complex increases, the probability of formation of ternary complexes approaches unity, and the α-values gradually reach a certain maximal level.[170,174,181,183] Contrary to this, the retention of the enantiomers may vary in both directions, depending on the relative hydrophobicity of all species standing at equilibrium.

In its turn, the retention of enantiomers under resolution in a reversed-phase system unambiguously depends on the amount of organic modifiers in the eluent, which usually are acetonitrile, methanol, or tetrahydrofuran. Via adsorption on the hydrocarbonaceous interface layer, these solvents make the stationary phase less hydrophobic, thus speeding up the elution of all types of solutes, including free enantiomers and their ternary complexes.[164,169,174,188] Herewith, the resolution selectivity remains unaffected or decreases insignificantly. Therefore, the addition of organic solvents to the eluent is a very useful instrument for adjusting the capacity factors, k′, of enantiomers of interest to a desired level. With hydrophilic chiral additives, like copper-*bis*-L-prolinate and hydrophilic amino acid solutes, pure water or water-rich eluents are the appropriate media. With hydrophobic chiral chelate additives like L-proline-octylamide or alkyl-octyl-diethylenetriamine and hydrophobic dansyl-amino acid solutes, the amount of acetonitrile or methanol required for the analysis time to be sufficiently short may be as high as 35 to 60%.[179,185]

In the case of hydrophobic chelate additives, relatively small solutes, and moderate concentrations of organic modifiers in the eluent, saturation of the column packing with the chiral resolving ligand occurs so that chiral eluent is no longer needed, and the column preserves its chiral resolving ability for long periods of time with aqueous or aqueous-organic eluents containing nothing more than small amounts of the appropriate transition metal salt. Such possibilities were shown for L-aspartyl-octyl and dodecylamide,[173] L-proline-octylamide,[185] *N*-tosyl-L-phenylalanine, and *N*-tosyl-D-phenylglycine.[176] Herewith, a close relation of the chiral eluent method becomes evident to the previously discussed applications of the permanently coated silica sorbents (with chiral ligands) or other types of chiral ligand exchangers.

Due to the use of reversed-phase packing materials having a well optimized porous structure, resolutions of amino acids, dansyl-amino acids, and dansyl-dipeptides, according to the method of chiral eluents, are distinguished by a relatively high efficiency. This is evident from Figures 12 to 14 showing some examples of enantiomeric resolutions of the above types of racemic solutes. The column efficiency is, however, no better than that of good chemically bonded chiral ligand-exchanging phases. Anyway, it is still impossible to resolve on a short analytical column about 50 individual enantiomers, which would be required for a complete enantiomeric analysis of a mixture of all common protein amino acids.

To solve the above problem, several approaches have been tried. Weinstein et al.[165,166] suggested the initial amino acid mixture to be separated first into three fractions using a conventional cation-exchanging resin and volatile aqueous pyridine buffers. After evaporating the eluent, the amino acids within the groups are separated into the individual enantiomers on a reversed-phase column with *N,N*-dipropyl-L-alanine and copper acetate in the aqueous mobile phase. Subnanomole sensitivity is achieved by postcolumn amino acid derivatization with *o*-phthalaldehyde and subsequent fluorometric detection. Proline, having a secondary amino group unable to form Schiff-bases with *o*-phthalaldehyde, is reacted with ninhydrine prior to the photometric detection at 440 nm.

Nimura and co-workers[177] prefer copper complex of *N*-(*p*-toluenesulfonyl)-D-phenylglycine as a chiral additive in the mobile phase for the resolution of underivatized amino acids on a reversed-phase packing. They suggested an elegant two-column system (Figure 15) that permits simultaneous resolution of common protein amino acids by a column-switching technique and gradient elution with acetonitrile. Two eluents were needed. Eluent A consisted of 75 mg of Na_2CO_3, 125 mg of $CuSO_4 \cdot 5H_2O$, 10 mℓ of a 100 m*M* solution of Tos-D-phenylglycine, and water to make the volume up to 1 ℓ. The second eluent, B, was prepared by adding three volumes of acetonitrile to seven volumes of the first chiral mobile phase. Amino acids injected into the chromatograph were first resolved with eluent A on the two columns, C-1 and C-2, connected in succession. After the faster moving, relatively polar amino acids (aspartic acid, glutamic acid, serine, threonine, and alanine) had passed through C-1 into the second column, valve V-1 was switched to the position in the second step which results in final isocratic resolution of the above solutes. Finally, in the third step,

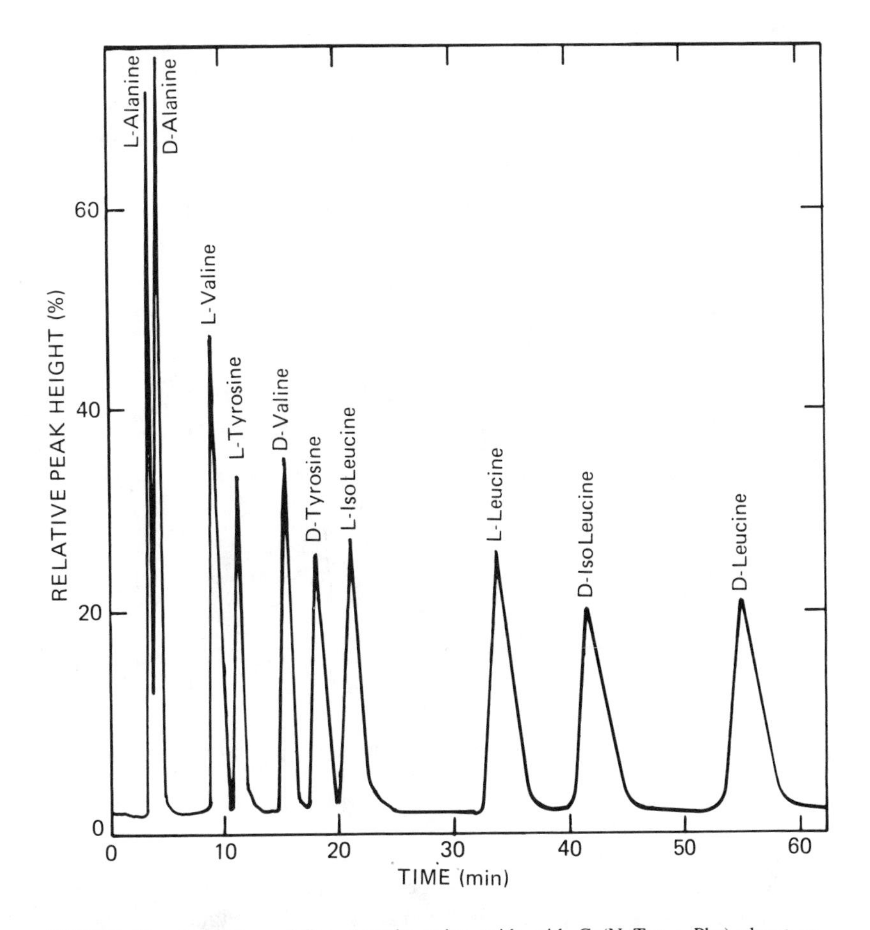

FIGURE 12. Separation of five racemic amino acids with Cu(N–Tos–L–Phe) eluent on Develosil C8, d_p 5 μm. Column, 4.0 × 100 mm; eluent, 0.5 m*M* $Cu(TosPhe)_2$ in water acetonitrile 90/10, pH 6.0, 1.0 mℓ/min; temperature, 30°C. (From Nimura, N., Suzuki, T., Kashahara, Y., and Kinoshita, T., *Anal. Chem.*, 53, 1380, 1981. With permission.)

strongly retained, less polar amino acids were resolved in the shorter column C-1, using a linear gradient to mobile phase B. The resulting chromatogram, as monitored fluorometrically after reaction with *o*-phthalaldehyde, is presented in Figure 16.

Karger and co-workers[185] suggested that mixtures of dansyl-amino acids should be separated into individual racemic components using conventional reversed-phase conditions and a linear methanol gradient, and that some of the eluted peaks should be subjected to a subsequent enantiomeric resolution on a second reversed-phase column percolated with a L-prolyl-*n*-dodecylamide-nickel complex solution. In this combined system, quantitative amino acid analysis is thus performed in the first part, and the enantiomeric ratio of some amino acids of interest is determined by directing them into the second module of the system. As a more dilute solute sample is injected into the second column, detection is performed at two different sensitivity levels, namely, using ultraviolet and fluorescence detections after the first and second columns, respectively. The detection limit was estimated to lie in the picomole-subpicomole region, with the maximal ratio of two enantiomers of 5000:1 still permitting enantiomeric analysis.[179]

Weinstein and Weiner[186] succeeded in developing a gradient program for resolution of almost all dansyl derivatives of amino acids on a single reversed-phase column 24 cm in length. The eluent contained 1 m*M* copper-bis-*N,N*-dipropyl-L-alanine, 0.3 *M* sodium acetate, and an increasing proportion — from 23.5 to 40% — of acetonitrile and was adjusted

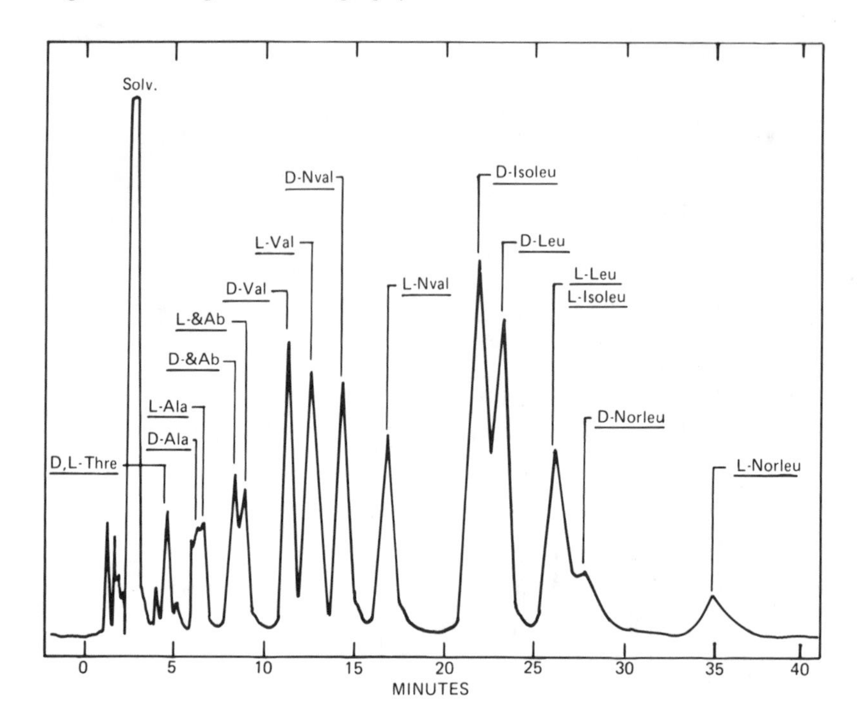

FIGURE 13. Separation of racemic dansyl-amino acids with Cu(L-Arg) on Nucleosil 5 μm C. Column, 2.6 × 250 mm. Eluent, 2.5 m*M* $Cu(Arg)_2$, 0.025 *M* NH_4OAc, water acetonitrile 80/20, pH 7.5, 2.0 mℓ/min (From Lam, S., Chow, F., and Karmen, A., *J. Chromatogr.*, 199, 295, 1980. With permission.)

to pH 7.0 with glacial acetic acid. As can be seen from Figure 17, the elution sequence of Dns-amino acids differs from that of unsubstituted amino acids in Figure 16. Dansyl-tyrosine was observed to rapidly oxidize in the presence of copper.

Though the overwhelming majority of enantiomeric resolutions in accordance with the chiral eluent technique has been made with reversed-phase packings, there is no reason why normal-phase systems (Table 13) should not operate with comparable selectivity and efficiency. Indeed, Oelrich et al.[162] successfully resolved thyroxine and triidothyronine, amino acids of the following structures,

on silica gel LiChrosorb® Si 60 in the presence of 0.2 m*M* *bis*-(L-prolinato) copper(II) in a nonpolar eluent consisting of hexane/*n*-propanol/water of 60/37.5/2.5. L-enantiomers were observed to be retained longer, just as in a reversed-phase system containing the same chiral chelate additive.

The latter fact seems to corroborate the frequently discussed suggestion that the enantiomeric resolution using the chiral eluent technique results from the enantioselectivity of formation of ternary mixed-ligand complexes. In each pair of diastereomeric ternary complexes formed from two enantiomeric solutes, one complex could possess higher stability than the other. The stabler diastereomer was thought to correspond to the longer-retained enantiomer of the solute.[170,171,172,180,182,183,188,191] In this case, reversed- and normal-phase systems should produce identical elution orders of the solute enantiomers.

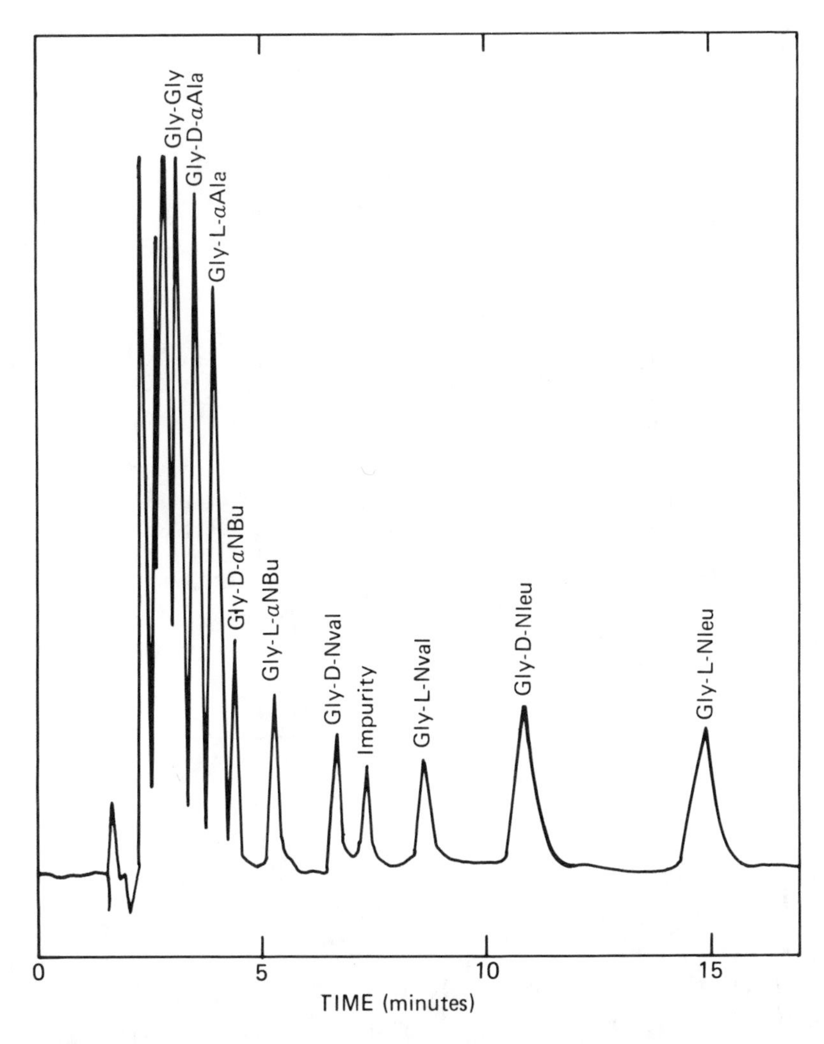

FIGURE 14. Separation of dansyl-glycyl-DL-amino acid dipeptides with Ni(2-iso-propyl-4-octyl-diethylenetriamine) on Hypersil 5 μmC_8. Column, 4.6 × 150 mm; eluent, 0.8 m*M* C_3-C_8-dien-Ni, 0.19 *M* NH_4OAc, water acetonitrile 65/35, pH 9.0, 1.0 mℓ/min.; temperature, 30°C. (From Lindner, W., LePage, J. N., Davies, G., Seitz, D. E., and Karger, B. L., *J. Chromatogr.*, 185, 323, 1979. With permission.)

However, Kurganov and Davankov[178] have demonstrated an interesting example of a total inversion of the elution order of amino-acid enantiomers on passing from a reversed- to normal-phase system. In both cases identical chiral chelate additives were used — copper(II) complex with *N,N,N',N'*-tetramethyl-(R)-propanediamine-1,2. This example unambiguously implies that the enantioselectivity observed in a chromatographic system that operates in a chiral eluent has little to do with the enantioselectivity of ternary complexes in solutions. Instead, enantioselectivity of ternary sorption complexes is important. This means that the mode of interaction of the diastereomeric ternary complexes with the sorbent surface plays the decisive role. These interactions obviously differ in the normal-phase system from those in the reversed phase, which results in an inverted sign of the overall enantioselectivity effect. A total of 14 amino acids could be resolved on a packed microbore column of 10 cm in length using reversed-phase conditions and copper-tetramethylpropanediamine complex as the chiral additive. Mixtures of four to five racemates can be quickly resolved; e.g., a mixture of Asp, Trp, Abu, Pro, and Thr. Unmodified silica, contrary to the reversed-

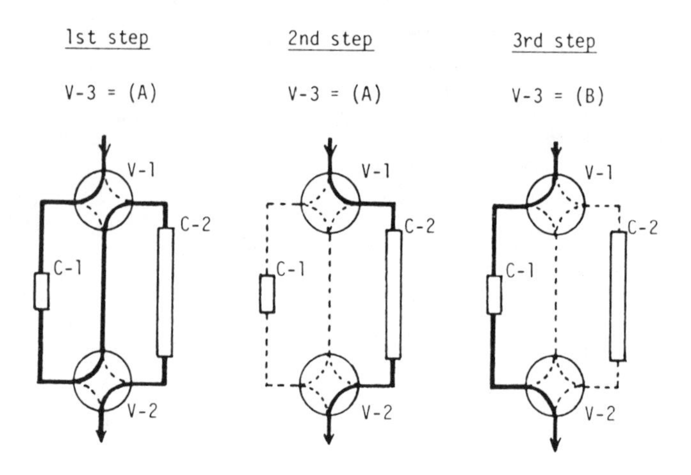

FIGURE 15. Mode of column switching for the simultaneous analysis of free D, L-amino acids with the Tos-D-PhG/Cu(II) eluent system. C-1 column, ERC-ODS, 50 × 6 mm. C-2 column, ERC-ODS, 200 × 6 mm. The solid line shows the flow of the chiral mobile phase. (From Nimura, N., Toyama, A., and Kinoshita, T., *J. Chromatogr.*, 316, 547, 1984. With permission.)

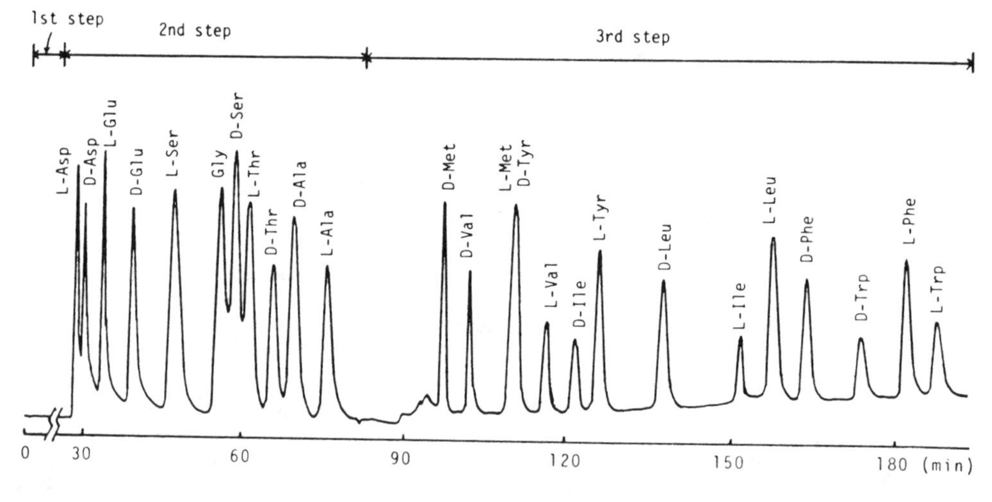

FIGURE 16. Chromatogram of the simultaneous resolution of free D,L-amino acids with the Tos-D-PhG/Cu(II) eluent system and column-switching technique shown in Figure 15. Injection, 100 to 250 pmol of each amino acid. Fluorescence detection. (From Nimura, N., Toyama, A., and Kinoshita, T., *J. Chromatogr.*, 316, 547, 1984. With permission.)

phase sorbents, retains polar threonine more strongly than hydrophobic tryptophan. Aspartic acid emerges first owing to electrostatic repulsions from the partially dissociated surface silanol groups.

There is probably a single chiral eluent system described thus far where no intensive interactions between the ternary mixed-ligand complexes and the sorbent occur. This is a combination of swollen sulfonated polystyrene-type cation-exchange resin and an aqueous 4 m*M* Cu(L-Pro)$_2$ solution as the chiral eluent. Hare and Gil-Av[159] observed resolution of 19 racemic amino acids in this system with the D-enantiomers being the stronger-retained species. The resolution enantioselectivity did not exceed the value of α = 1.28 (DL-Tyr), which is much less than the enantioselectivity of other above-described chiral eluent systems,

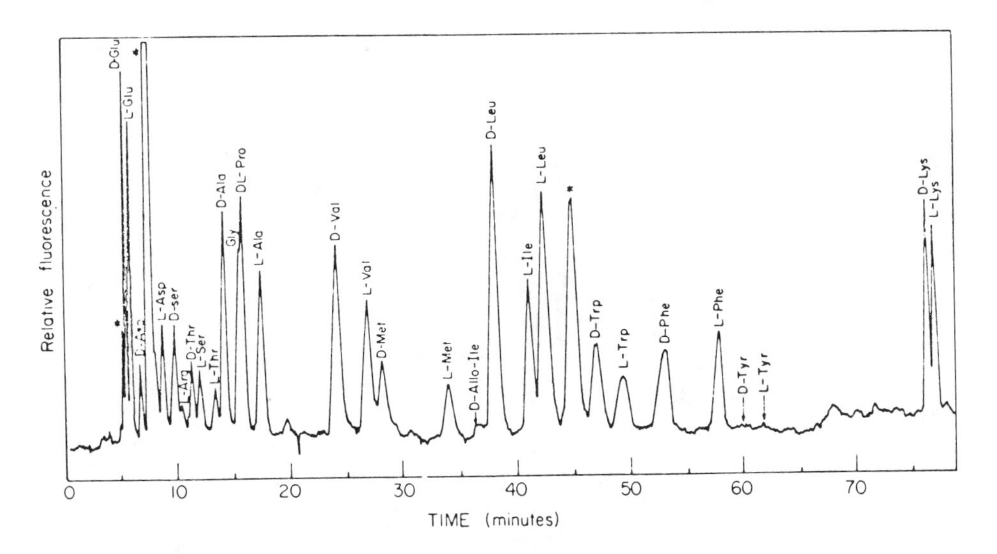

FIGURE 17. Chromatogram showing the separation of almost all the amino acids present in a protein hydrolysate analyzed as their fluorescent Dns derivatives. Dns derivatization side products are indicated by asterisks. The large peak appearing after 7 min is Dns-OH and the peak at 45 min is Dns-amide. Arrows denote the elution positions of compounds not present in the mixture (Dns-D-*allo*-isoleucine), or decomposed during analysis (Dns-D- and Dns-L-tyrosine). Elution with the *N,N*-dipropyl-L-alanine/Cu(II) system and acetonitrile gradient (from 23.5 to 40%). Column, 24 × 0.46 cm. Nucleosil C_{18}, dp 5 μm. Flow rate, 0.8 mℓ/min. (From Weinstein, S. and Weiner, S., *J. Chromatogr.*, 303, 244, 1984. With permission.)

Table 13
RESOLUTION OF UNDERIVATIZED AMINO ACIDS USING CHIRAL ELUENT TECHNIQUE IN STRAIGHT-PHASE SYSTEMS

Chiral resolving ligand	Metal ion	Racemates revolved, maximum selectivity	Elution order of enantiomers	Experimental conditions
L-proline[159]	Cu^{2+}	19 amino acids, except Asp, Glu, Ala, Gln; α = 1.28 (Tyr)	L, D	Sulfonated polystyrene cation exchanger, 4—8 m*M* Cu(L-Pro)$_2$, 0.05—1.0 *M* NaOAc, pH 5.5, 75°C
L-proline[162]	Cu^{2+}	Triiodothyronine, thyroxine; α = 1.6	D, L	LiChrosorb® Si 60, 0.2 m*M* Cu(L-Pro)$_2$, 45°C, hexane/*n*-propanol/water of 60:37.5:2.5
N,N,N′,N′-tetramethyl-(R)-propanediamine[178]	Cu^{2+}	14 amino acids; α = 1.4 (Asp)	D, L, except Asp, Glu	LiChrosorb® Si 100, 0.5 m*M* Cu(Me$_4$pn)$_2^{2+}$, 5 m*M* Me$_4$pn, 90% MeCN

including those containing L-prolinato-copper as the chiral additive. This fact and the opposite sign of enantioselectivity as compared with the silica-based packings, both reversed and normal phase, points to some important peculiarities of the system. This could be the absence of immediate interactions between the negatively charged resin phase and neutral ternary mixed-ligand complexes. In this case, the enantioselectivity of complex formation in the liquid phase could be the governing factor, whereas the retention of amino acids could

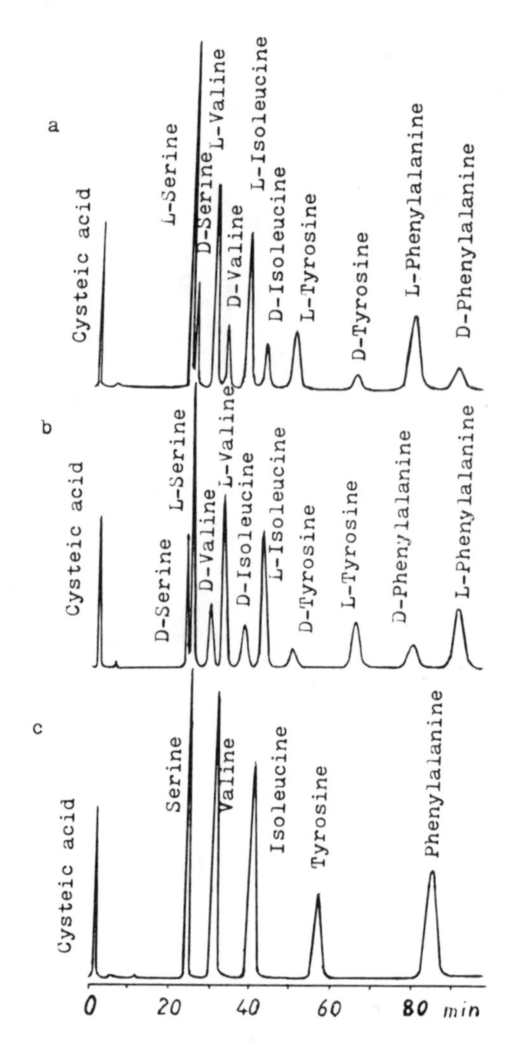

FIGURE 18. Effect of the chirality of the eluant on the separation of D- and L-amino acid enantiomers by ligand-exchange chromatography. Identical portions of five pairs of amino acid enantiomers, each consisting of 0.375 nmole of L form and 0.125 nmole D form, were injected in each run. Sodium acetate buffer (0.05 *N*, pH 5.5) containing 4×10^{-3} *M* $CuSO_4$ and 8×10^{-3} *M* proline was used as eluant. The chirality of the proline ligand was L, D, and DL. The column was 12 × 0.2 cm packed with sulfonated polystyrene-type DC 4a resin. The eluant flow rate was 10 mℓ/hr, the reagent (*o*-phthalaldehyde) flow rate was 10 mℓ/hr, and the column temperature was 75°C. (a) L-Proline effected the separation of all five pairs of enantiomers with the L-enantiomers eluting before the corresponding D-enantiomers; (b) D-proline reversed the order of elution; (c) with racemic proline, no resolution occurred. The amino acids eluted halfway between the corresponding enantiomeric peaks in (a) and (b). (From Hare, P. E. and Gil-Av, E., *Science*, 204, 1227, 1979. With permission.)

proceed in the form of protonated species or positively charged mono complexes $Cu(AA)^+$. The difference in the retention of two enantiomers should correspond then to the difference in their involvement into ternary complex formation. One of the very first examples of employing chiral eluents in LEC, the paper[159] by Hare and Gil-Av, still remains an extremely impressive demonstration of the great potential of this new technique (see Figure 18).

Meanwhile, examples of practical applications of enantiomeric analysis using the chiral eluent mode of LEC have been published. Thus, in the presence of copper-L-aspartame, varying proportions and concentrations of D- and L-isomers of pipecolic acid were detected in the urines of patients with hyperpipecolatemia.[184] Amino acid profiling of cerebrospinal fluid was carried out with copper(II) L-histidine methyl ester system. The high sensitivity of fluorescence detection of Dns-derivatives permitted the profile of amino-acid composition to be safely registered starting with 50 μℓ of sample of cerebrospinal fluid which only contains about 10 μmol/ℓ per amino acid compared to 100 μmol/ℓ in serum. Though D-isomers could not be detected for certain, several samples had easily discernible, different compositions, particularly from patients with bacterial meningitis or with metastatic cancer of the breast.[183,184]

It is only natural that the method of chiral eluent is generally applicable to various other derivatives of amino acids, as well as members of other classes of organic compounds. Thus, using copper complexes of *N,N*-dimethyl-L-valine, a series of α-methyl α-amino acids was shown to resolve into enantiomers.[168] In eluents containing copper complexes of L-proline, L-valine, or aspartame, with selectivities of 1.03, 1.11, and 1.28, respectively, enantiomers have been separated of a potentially sweet, sucrose-like homoserinedehydro-chalcone conjugate of the formula:[192]

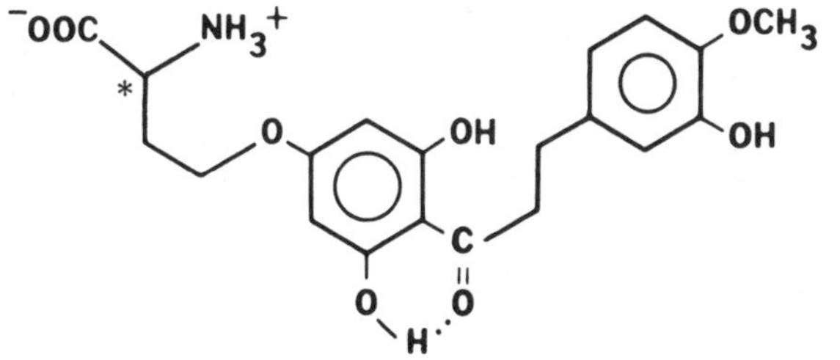

9-(3,4-Dihydroxybutyl)guanine:

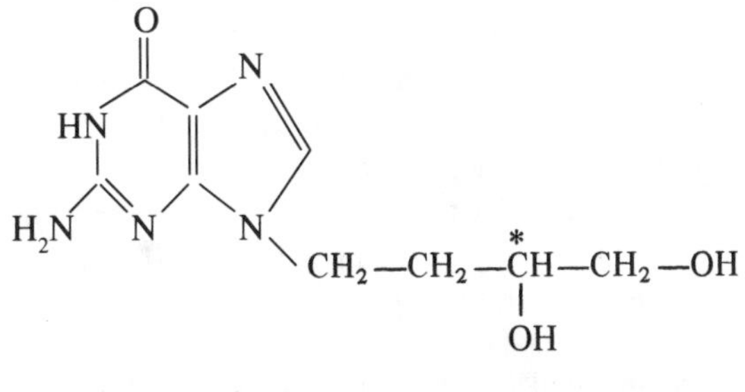

which two enantiomers exhibit antiherpes activity of different magnitudes, were resolved using aqueous 3 m*M* solution of copper-*bis*-L-phenylalaninate.[193] The detection limit of the *S*-form in the later-eluting R-form was approximately 0.3%.

Mandelic acid was observed to resolve into enantiomers in eluents containing copper complexes of *N,N,N',N'*-tetramethyl-(R)-propanediamine-1,2[178] or L-phenylalanine,[162,164] again on reversed-phase packings. The last system,[151,164] as well as copper complexes with *N*-methyl- , *N,N*-dimethyl- , and *N*-ethyl-*L*-valine,[164] are also applicable to enantiomers of *p*-hydroxy-[151] and *m*-hydroxymandelic acid,[151,164] 4-hydroxy-3-methoxymandelic acid, tropic acid, and atrolactic acid.[164] A more detailed study into chromatography of α-hydroxy acids[194] in reversed-phase systems with chiral additives of copper(II) complexes with di-*n*-propyl-L-alanine, or still better, dimethyl-L-valine showed the approach to be generally successful. Of 2-hydroxy substituted aliphatic acids, derivatives of propionic (lactic acid), butyric, 2-methylbutyric, valeric, isovaleric, isocaproic, and octanoic acids were resolved. On the contrary, species containing additional β-hydroxy or β-carboxylic groups (glyceric, tartaric, malic acids) elute in the form of a single peak. Of mandelic acid analogues, the following aromatic hydroxy acids are resolved: 2-methylmandelic, 4-hydroxy-3-methoxymandelic, 3-

hydroxy-4-methoxymandelic, atrolactic, 3-phenyllactic. Selectivity values can amount to three or more (mandelic, hydroxyisovaleric acids). Similar to amino acids, L-isomers of hydroxy acids were observed to be retained more strongly in the column.[164,194] This implies the same, *trans*(carboxy)-structure of heteroligand copper [(amino acid)(hydroxy acid)] complexes.

IV. MECHANISM OF CHIRAL RECOGNITION OF ENANTIOMERS

Resolution of two enantiomers in LEC is always due to the enantioselectivity in formation of their ternary mixed-ligand complexes with the complexing metal ion and the chiral resolving ligand, the latter being bonded to the insoluble sorbent matrix or added to the mobile phase. With L-hydroxyproline taken as a typical resolving ligand, the two possible mixed-ligand copper complexes (involving L- or D-enantiomers of an α-amino acid under resolution) are diastereomeric, as shown in the following scheme:

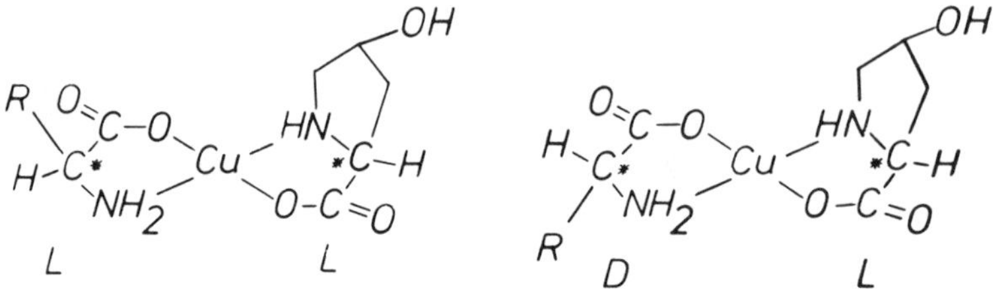

These two labile diastereomeric complexes can possess different stabilities if the resolving agent recognizes the configuration of the amino acid ligand. According to the three-point interaction model, at least three contacts should exist between the two ligands within the mixed-ligand structure in order to bring about the desired chiral recognition.

Two contact points are evident from the above scheme. These are the copper-mediated interactions between the electron-donating nitrogen and oxygen atoms of the constituent amino acid ligands. The third contact point must be that between the substituents at the asymmetric α-carbon atoms. In order to consider the required conditions and nature of this third interaction, the structure of copper complexes with α-amino acids and *N*-substituted amino acids should be examined in more detail.

N-substituted amino acids and particular *N*-benzyl amino acids can be regarded as appropriate low-molecular-weight models for fixed ligands of chiral polystyrene-type resins described above. A searching examination of copper complexes of different bifunctional amino acids containing *N*-benzyl and *N*-methyl substituents has shown[195,196] that the complexes can form four different structures standing at equilibrium in solution:

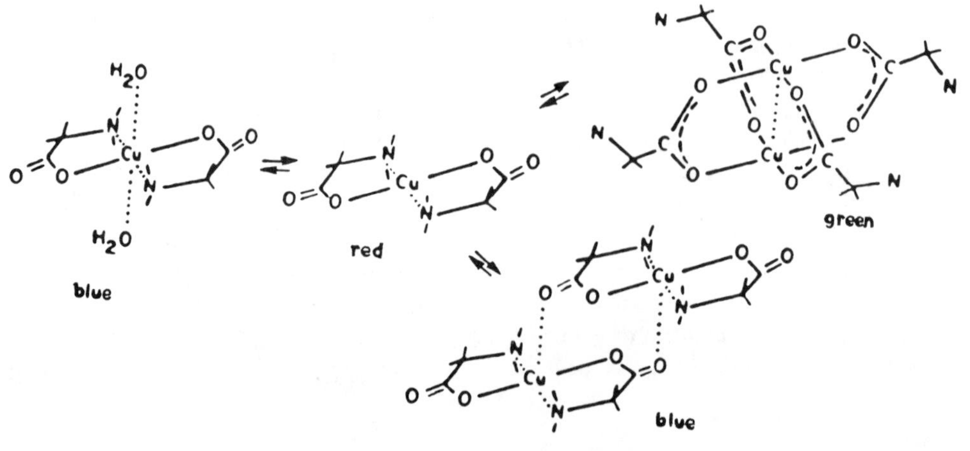

The position of this equilibrium depends on the number and size of substituents at the *N*- and α-C-atoms, the nature of the solvent, and the temperature and concentration of the solution. For steric reasons, formation of the last two structures in the sorbent phase is rather unlikely. Because of the low concentration of components in the mobile phase of a chromatographic system, these structures cannot predominate in solution. The first two structures alone should be considered in LEC.

One is the well-known, blue-colored, distorted octahedral structure with two solvent molecules in the extended axial positions. The carboxylic and amino groups of the two ligands are situated in the main coordination square in the *trans* position to each other. Though *cis-bis*(amino acidato) copper often is suggested[126,127] to explain some enantioselectivity phenomena and probably exists to a certain degree in the polymeric resin, there is no evidence for the *cis* structure formation. In fact, two negatively charged oxygen atoms should tend to occupy *trans* positions which are more removed from each other as compared with *cis* positions. Similarly, nitrogen atoms, each of them carrying two substituents (hydrogen atoms or alkyl groups), should prefer to be situated *trans* in order to minimize the nonbonded interactions.

The second important structure of *bis*(amino acidato) copper differs in that the axial positions of the coordination sphere remain vacant. They are shielded from solvent molecules by alkyl substituents at nitrogen and α-carbon atoms of the ligands. These four-coordinate, square planar structures display an unusual red color. They are also formed in solvents unable to coordinate in axial positions; for instance, when proline-containing, polystyrene-type resins are used in their copper(II) form in nonpolar media.

In both the above six- and four-coordinate copper complexes, the side groups attached to the asymmetric carbons atoms of the amino acid ligands appeared to be situated too far from each other to produce the required third interaction point between the ligands. Therefore, it has been generally accepted that no enantioselectivity can arise in copper complexes with bidentate amino acid ligands.[197] Nevertheless, by fixation of a bidentate chiral amino acid, L-proline (or L-hydroxyproline) onto a cross-linked polystyrene matrix, a ligand exchanger was obtained which proved to display enantioselectivity toward racemates of most of the amino acids.[138] An explanation of this phenomenon was found in the suggestion that certain elements of the nonchiral polystyrene matrix, namely, *N*-benzyl groups of the resin-fixed L-proline moieties, effectively enhance interactions with the α-R group of the mobile amino acid ligand in the ternary copper complex. And indeed, aromatic rings of *N*-benzyl-L-proline were found to approach the axial positions of copper(II) ions. However, this favorable conformation is accessible for both the ligands only in the meso complex (*N*-benzyl-L-prolinato)(*N*-benzyl-D-prolinato)copper, [CuDL], where as in the equally paired complex [CuLL], one ligand is forced into a distorted conformation.[102] Formation of diastereomeric *bis*-complexes in this system can be represented by the following scheme:

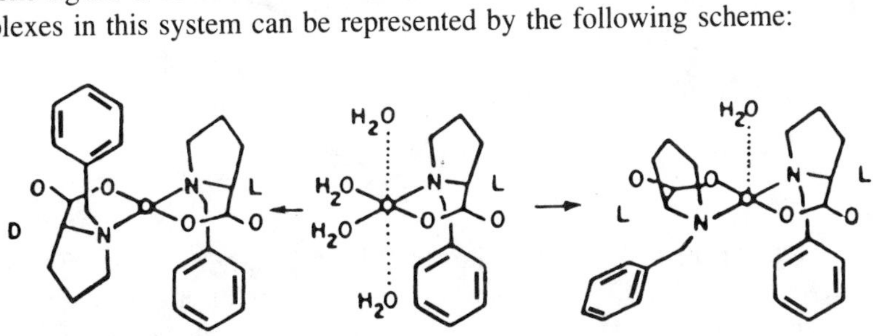

A detailed study of coordination equilibria[98-103] showed the meso structure of *bis*(*N*-benzylprolinato)copper [CuDL] to be considerably more stable than the equally paired complex [CuLL]. In the latter structure, however, one of the axial positions can be occupied by a solvent molecule so that the [CuLL] complex exists in solution to a certain degree in the five-coordination blue form.[103,198] In conformity with this, the enthalpy gain during the formation of the equally paired complex turns out to be higher than in the case of meso complex formation. At the same time, the entropy changes favor the formation of a more symmetrical, red-colored meso complex. Due to the fact that the entropy contribution throughout the studied temperature range predominates over the enthalpy contribution, the meso complex proves to be by 4.8 ± 0.8 kJ/mol more stable than the equally paired complex, and this superiority somewhat increases with the rise in temperature.[103]

It can be assumed that the resolving chiral agent in the L-proline-containing polystyrene-type resin acquires in a sorption complex the favored conformation of *N*-benzyl-L-proline. In this case, the hydrophobic side group R of a sorbed D-amino acid appears in the vicinity of the hydrophobic fragments of the resin matrix. Contrary to this, the hydrophobic R-groups of L-amino acids would interfere with water molecules coordinated in the upper axial positions[89] as shown below:

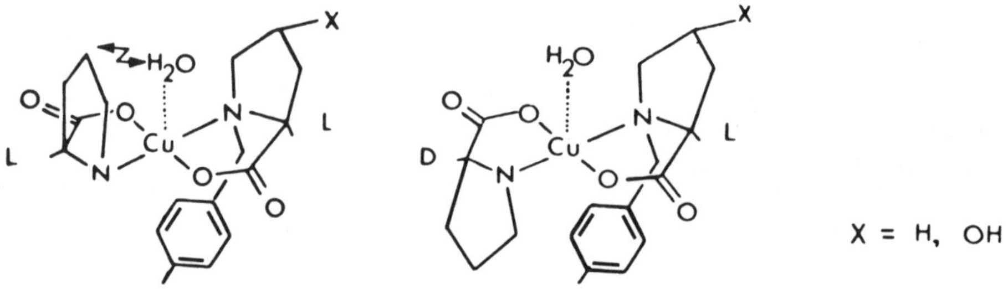

X = H, OH

In this particular system, the polystyrene matrix of the resin and the coordinated water molecules apparently provide the third interaction sites (enhancing retention of D-amino acids and diminishing retention of L-enantiomers) which are required for the chiral recognition of the DL-enantiomeric pair of an amino acid.

Contrary to the bidentate mobile ligands, typically tridentate amino acids like histidine, allo-hydroxyproline, aspartic acid, glutamic acid, and ornithine capable of displacing the water molecule from the axial position, are sorbed preferentially, providing that they belong to the L-configurational series. As shown below, L-enantiomers of trifunctional amino acids are the stronger retained species:

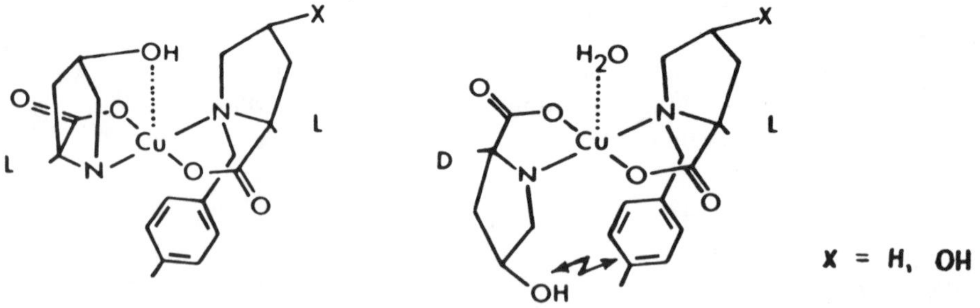

X = H, OH

Polystyrene-type resins containing cyclic chiral ligands, L-proline, L-hydroxyproline, or L-azetidine carboxylic acid, display a common pattern of the discriminating mechanism of amino acid enantiomers; elution order is L ahead of D for bifunctional ligands, and D ahead of L for trifunctional ones.

With the L-allo-hydroxyproline-incorporated resin, both the bifunctional and trifunctional amino acids act as bidentate mobile ligands so that L-enantiomers are always retained less strongly:

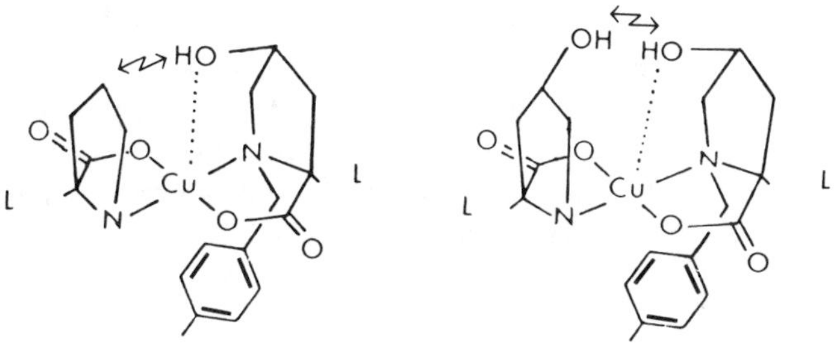

Having been first introduced as effective resolving agents for α-amino acid enantiomers, copper complexes of L-proline and L-hydroxyproline have been frequently used both as active sorption sites of chiral resins and as chiral additives to the mobile phase. Surprisingly, the above agents have often been observed to change their preferential affinity for D- or L-enantiomers of amino acids, depending on the chromatographic mode and the experimental conditions. Thus, in contrast to the polystyrene-type resins, chiral sorbents prepared by grafting L-proline (or L-hydroxyproline) onto the matrix of the poly(2,3-epoxypropyl methacrylate)[76] or polyacrylamide-type[75] display a higher affinity toward the L-enantiomers. Here, the donating groups –OH or =C=O, respectively, descending from the initial sorbent matrix, most probably occupy the lower axial positions of the metal coordination sphere.[199] They are unable to enter hydrophobic interactions with the second ligand that could stabilize the ternary structure formed. On the contrary, they would effectively hinder the rotation of the R-group (around the Cα-Cβ axis) of the sorbed D-amino acids, thus diminishing the retention of the latter as compared with L-enantiomers:[76,118,119,199]

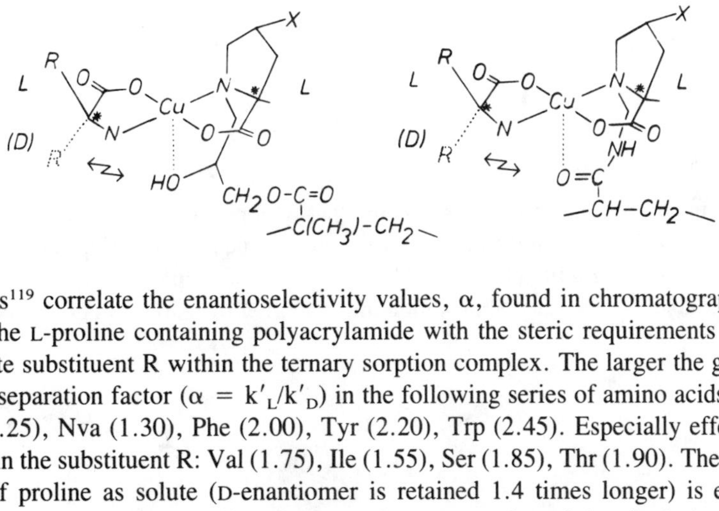

The authors[119] correlate the enantioselectivity values, α, found in chromatographic experiments on the L-proline containing polyacrylamide with the steric requirements for rotation of the solute substituent R within the ternary sorption complex. The larger the group R, the higher the separation factor ($\alpha = k'_L/k'_D$) in the following series of amino acids: Ala (circa 1), Abu (1.25), Nva (1.30), Phe (2.00), Tyr (2.20), Trp (2.45). Especially effective is the branching in the substituent R: Val (1.75), Ile (1.55), Ser (1.85), Thr (1.90). The exceptional behavior of proline as solute (D-enantiomer is retained 1.4 times longer) is explained in terms of the inability of its cyclic substituent R to rotate around the single bond to the α-carbon atom.

The above examples imply that the nature of additional interactions between sorbed enantiomers and some fragments of the nonchiral matrix of the sorbent, taking place within the structure of the ternary sorption complex, often determines the sign of the enantioselectivity of the whole system. Moreover, these additional interactions seem to be a necessity for the very existence of enantioselectivity in many systems comprising complexes of bidentate ligands. They complete up to three the number of interaction sites between the two chiral moieties of a ternary complex.

The decisive role of additional interactions between a mixed-ligand sorption complex and a nonchiral sorbent for the chiral recognition of the solute enantiomers is especially evident in experiments under reversed-phase conditions according to the chiral eluent method. Here, hydrophobic interactions occur between the sorbent surface and hydrocarbon groups of the diastereomeric ternary complexes involving the chiral component of the eluent and the enantiomers under resolution. Due to these hydrophobic interactions, the mixed-ligand complexes of D-solutes are additionally stabilized* and retained stronger when *N*-alkyl-L-hydroxyproline,[149] *N*-benzyl-L-proline,[202] or *N*-decyl-L-histidine[150] is the resolving agent in the eluent:

It is easy to see from the following scheme that the opposite elution order of amino acid enantiomers, L after D, should be expected in chiral eluents containing L-proline,[161] *N,N*-dimethyl-L-leucine,[202] L-phenylalanine,[162,164] or *N,N*-di-(*n*-propyl)-L-alanine[165] as resolving agents (see Table 14). Here, complexation and retention of L-solutes is enhanced:

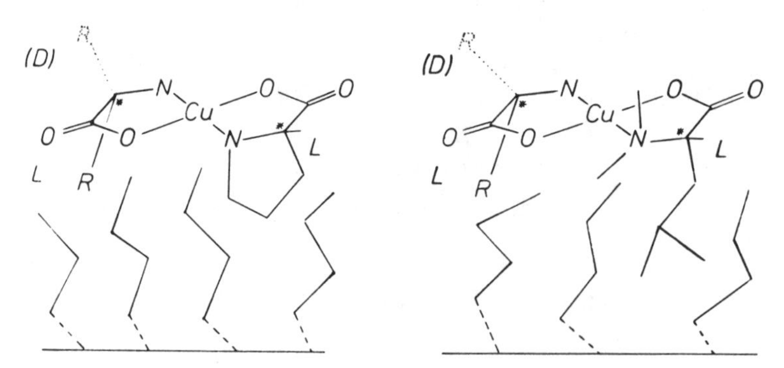

* Suggestion of an additional stabilization through adsorption on a hydrophobic surface of ternary complexes Cu(D-amino acid)(*N*-alkyl-L-hydroxyproline) can be corroborated by the recent finding[200,201] that these diastereomers are indeed the more stable species in the bulk solution (in butanol, pentanol, or octanol), but the enantioselectivity in solution is much less than that manifested in the reversed-phase chromatographic system.

Table 14
RESOLUTION SELECTIVITY, $\alpha = k'_D/k'_L$, OF TYPICAL RACEMIC AMINO ACIDS ON REVERSED-PHASE SILICA PACKINGS DYNAMICALLY MODIFIED WITH COPPER COMPLEXES OF CHIRAL LIGANDS

Chiral modifier	Ala	Abu	Val	Leu	Tyr	Phe	Trp	Pro
N-C_{16}-L-hydroxyproline[149]	1.6	—	2.5	3.6	2.8	3.7	2.5	7.5
N-Bzl-L-proline[202]	2.0	1.9	2.0	1.8	2.0	—	—	2.6
N-C_{10}-L-histidine[150]	1.4	1.6	1.8	1.5	1.4	1.7	1.3	1.4
L-proline[161]	0.3	0.2	0.2	0.4	0.4	0.5	—	—
N,N-Me_2-L-leucine[202]	0.8	0.5	0.3	0.3	0.3	0.3	—	0.2
N,N-(n-Pro)$_2$-L-alanine[165]	0.7	0.3	0.4	0.4	0.8	0.7	0.9	0.4
L-Aspartylhexylamide[171]	1.0	—	1.7	1.8	1.4	—	—	4.4

Thus, in the *bis*(L-prolinato)copper-containing chiral eluent, the L-isomer of valine is retained 6.5 times longer on a reversed phase silica than is D-valine.[161] This value of enantioselectivity would correspond to the predominance of the Cu(L-Pro)(L-Val) sorption complex over the Cu(L-Pro)(D-Val) sorption complex by more than 4 kJ/mol or, expressed in stability constants of the two complexes, by $\log \beta_{DL} - \log \beta_{LL} = 0.81$. It is most remarkable that the two diastereometric complexes in solution are known to be equally stable, at least within the following limits: $\Delta\log\beta = 0.11 \pm 0.29$[203] or $\Delta\log\beta = 0.083 \pm 0.008$.[204] This example is very important for emphasizing the fact that the adsorption at the hydrocarbonaceous interface layer of a reversed-phase sorbent can entirely change the sign of the enantioselectivity effect in ternary complexes.

Along the same lines of a simultaneous interaction of alkyl groups R′ and R of a D-amino acid ligand and L-aspartyl-alkylamide, respectively, with the hydrophobic sorbent surface, the stronger retention of D-amino acid enantiomers should be explained in chiral eluent systems based on copper-L-aspartyl amides:[170-173]

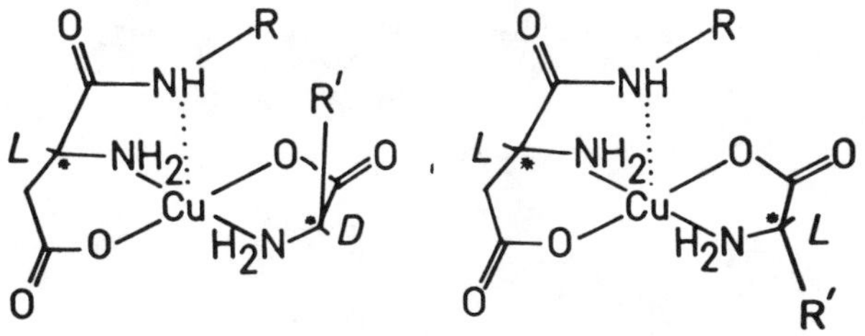

R = Et, But, Hex, Oct, cHex

It is only natural that the elution order of optical isomers of amino acids containing polar functions in their side substituents R differs from that of isomers of simple aliphatic and aromatic amino acids. In all the above discussed systems with hydrophobic reversed-phase sorbents, the polar substituents would tend to be oriented towards the polar mobile phase, not the hydrophobic stationary phase. This is why histidine, serine, aspartic acid, and threonine often show an opposite elution sequence as compared with other solutes.

Similar to hydrophobic interactions with a reversed-phase silica, the polar interaction, as well as hydrogen bond formation, can be responsible for the resolution of racemic thyroxine and triiodothyronine on unmodified silica in the presence of L-prolinato-copper in an unpolar eluent.[162] In this case, the coordination plane of the mixed-ligand complex should approach the polar surface of the silica whereas the hydrophobic parts of both L-proline and the above amino acids should be directed toward the eluent. As follows from the scheme below, coordination and stronger retention of L-amino acids would be the favored situation, which is consistent with the experiment.

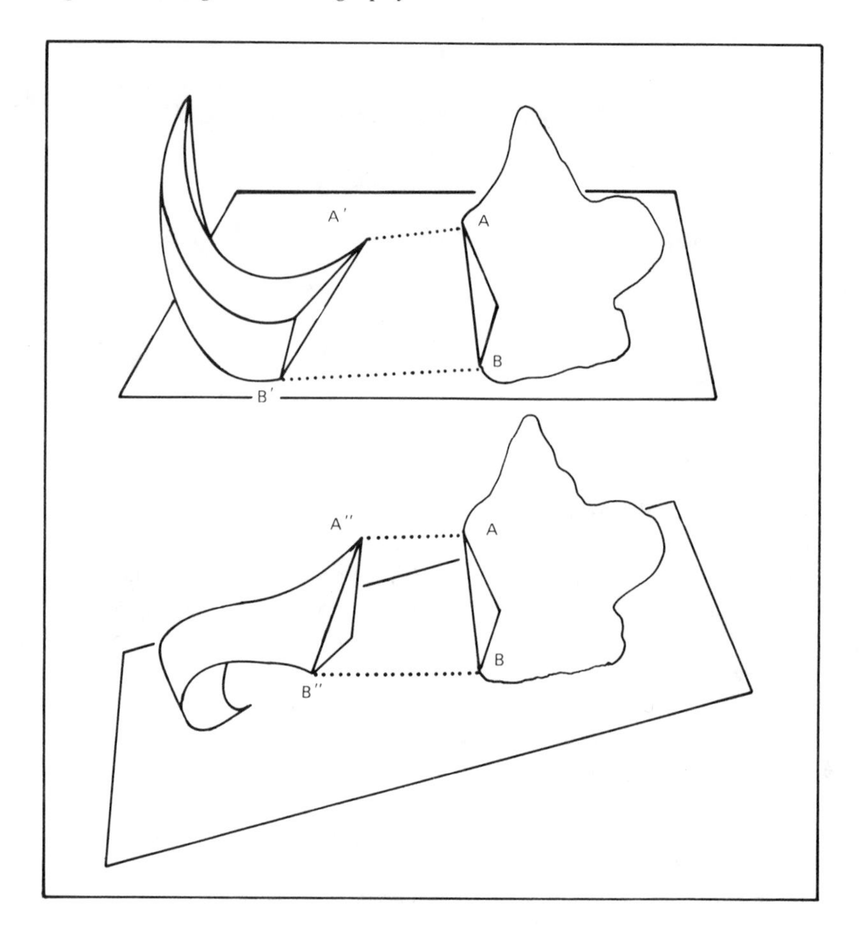

FIGURE 19. A two-point interaction between two chiral structures is sufficient for their mutual chiral recognition, provided each of these structures simultaneously contacts a third, nonchiral species, in particular, a solid surface. (From Davankov, V. A. and Kurganov, A. A., *Chromatographia,* 17, 686, 1983. With permission.)

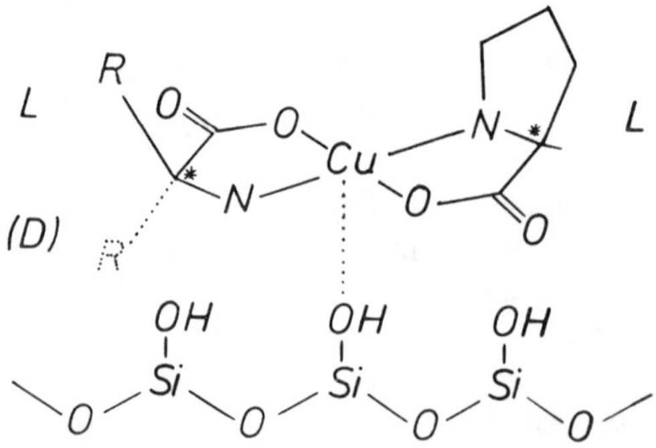

The above considerations of the essential role of certain fragments of a nonchiral matrix or of the surface of a sorbent in the mutual recognition of chiral configurations of two bidentate ligands in a ternary complex have led Davankov and Kurganov[202] to an important general statement that a two-point interaction between two chiral structures is sufficient for their mutual chiral recognition, provided that each of these structures is simultaneously in contact with a third nonchiral species, in particular, a solid surface. This idea is illustrated in Figure 19, which shows how a nonchiral surface contributes to chiral recognition of two enantiomers by a chiral resolving agent (compare with Figure 1).

This viewpoint logically explains the fact that enantioselectivity phenomena manifest themselves much more strongly in heterogeneous chromatographic systems operating in accordance with ligand exchange than in homogeneous solutions of simple bidentate ligand complexes. There is, however, one chromatographic system described where diastereomeric ternary complexes do not immediately interact with the sorbent. This happens on ion-exchange chromatography of racemic amino acids on sulfonated polystyrene cation exchangers in the presence of L-prolinato-copper in the eluent.[159] Here, all complexed and free amino acid species are partitioned between the resin gel and the mobile phase according to their charges. The enantiomeric resolution most probably arises from the true enantioselectivity of complexation in solution, which determines the involvement of the enantiomers in the formation of ternary complexes with L-proline and, consequently, the distribution ratio. It is important to emphasize that the selectivity factors in this system were observed to be very low. (The mutual chiral recognition of two bidentate ligands in a dissolved complex proceeds most probably via their interaction with solvent molecules coordinated in axial positions of the copper ion. Participation of solvent molecules on the processes of chiral recognition can be corroborated by the fact of strong dependence of the sign and magnitude of enantioselectivity[205] on the solvent nature.)

Unfortunately, it is not always possible to exactly define the nature of interactions between the solute molecules, chiral resolving agents of the system, and its nonchiral components. Therefore, it is hardly possible at present to forecast the influence of changing experimental conditions on the total enantioselectivity of the system. Nevertheless, starting from the idea that hydrophobic interactions enhance retention of D-enantiomers of aliphatic and aromatic amino acids on the L-hydroxyproline-containing polystyrene-type resin, authors[206] succeeded in improving resolution of alanine and aminobutyric acid by making the matrix more hydrophobic. This was done by alkylation of the polystyrene phenyl rings with *n*-butyl, *tert*-butyl or cyclohexyl chloride, or by acylation with cyclohexylpropionyl chloride. (However, resolution of other solutes, especially hydrophilic ones, was largely spoiled.) On the other hand, the inversion of the enantioselectivity sign of the system appeared quite unexpected after the L-hydroxyproline-modified polystyrene chains were densely grafted onto the surface of a porous silica.[143] A similar inversion is also characteristic of chiral phases prepared by bonding *N*-(*p*-ethylenebenzyl)-L-hydroxyproline to the silica surface.[134,137] The anomalous behavior of the latter two systems is probably due to the steric hindrance in the sorption of D-isomers arising from the rigid silica surface which appears immediately after the benzyl groups of the L-proline moieties.

Similarly, steric factors and polar interactions would probably hinder complexation of D-amino acids to L-hydroxyproline ligands bonded on the silica surface via *N*-methylene bridges (n = 1) as shown below:

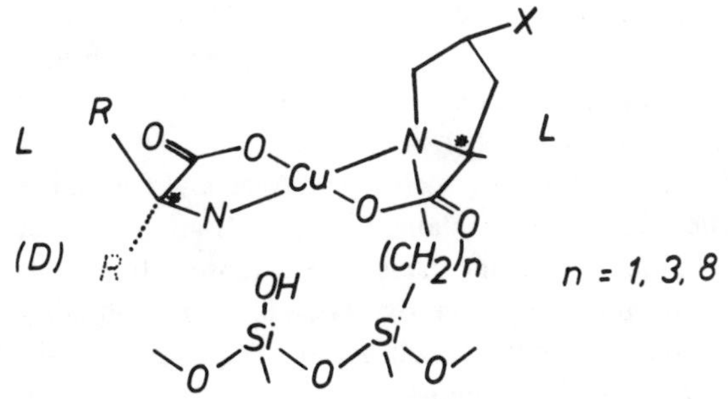

n = 1, 3, 8

In the case of sorbents with longer and more flexible trimethylene and octamethylene (n = 3, 8) spacer groups, steric factors would gradually diminish and hydrophobic interactions would emerge between the spacer groups and side substituents R of the mobile ligand. Under conditions favoring hydrophobic interactions, such sorbents begin to display a higher affinity to the D-isomers of hydrophobic amino acids.[134-136]

In this connection it is interesting to note that Feibush and co-workers[141] did not find any significant role of the micro environment of the undecanoyl-L-valine moieties bonded to the silica surface with respect to chiral recognition of dansyl-amino acid enantiomers in ternary copper(II) complexes. In their experiments, the above chiral resolving ligands were "diluted" with C_4, C_{10}, and C_{20} alkyl chains or propyloxy-2-(2′-methoxyethoxy)ethane groups. However, enantioselectivity of the system depended on the concentration of the chiral fixed ligands. At present, it is difficult to discuss the nature of enantioselectivity effects in the mixed-ligand amino acid/dansyl-amino acid complexes as practically nothing is known about their structure. Nevertheless, as will be shown in Section V of this chapter, the microenvironment of sorption sites in the series of chiral phases considered still exerts a definite influence on the resolution enantioselectivity.

Besides the structure and properties of the nonchiral sorbent matrix, components of the mobile phase sometimes influence the enantioselectivity of the complex formation. Thus, pyridine, when used as the displacing ligand in the eluent, was observed to diminish enantioselectivity of resolution of DL-proline on the L-proline-containing polystyrene resin.[117] Similarly, high concentrations of ammonia negatively influence the resolution selectivity of racemic histidine, proline, tryptophan,[86] and leucine[95] on the resin containing L-hydroxyproline. This indicates that the coordination of amine molecules in the axial positions of diastereomeric sorption complexes reduces the difference in their stabilities.

The above outlined ideas on the active participation of nonchiral matrix fragments or sorbent surface, as well as solvent molecules, on the process of chiral recognition of solute enantiomers within the diastereomeric mixed-ligand sorption complexes imply that it is extremely difficult to design low-molecular-weight models for these sorption complexes which would adequately simulate all the important interactions between their components. More or less suitable models have been suggested for some simple gel-type resins prepared from chloromethylated or chlorosulphonated polystyrene and from polyacrylamide, but no homogeneous system could ever represent the inversion of the elution order of amino acid enantiomers in a *N,N,N′,N′*-tetramethyl-(R)-propanediamine-copper-containing chiral eluent on replacing a reversed-phase sorbent for an unmodified silica.[178]

To support this pessimistic statement, we would like to cite another impressive example showing diversity and complexity of factors governing enantioselectivity of a chromatographic system.

Chiral resolution is generally supposed to be reciprocal in that if a chiral resolving agent, (+)A, can distinguish between (+)B and (−)B, then reagent (+)B should be able to distinguish (+)A from (−)A. Moreover, on an achiral reversed-phase sorbent, resolution selectivity of the diastereomeric pair of mixed-ligand complexes (+)A-(+)B and (+)A-(−)B in the (+)A-containing eluent is expected to coincide with the resolution selectivity of complexes (+)A-(+)B and (−)A-(+)B in the (+)B-containing eluent. To experimentally examine the above theoretical consideration, racemic proline, hydroxyproline, allo-hydroxyproline, and a series of their *N*-methyl and *N*-benzyl derivatives have been resolved in aqueous acetonitrile eluents that contained copper(II) complexes of a succession of optically active isomers of the same amino acids.[207] Enantioselectivity values for corresponding diastereomeric pairs were found to differ significantly from each other. Even cases of total reversal of elution sequences could be observed. Thus, if L-allo-hydroxyproline-copper complex was taken as chiral additives to the eluent, the elution order of *N*-benzyl-allo-hydroxyproline isomers was L after D. On the contrary, when *N*-benzyl-L-allo-hydroxyproline was to serve as chiral resolving agent in the system, the elution order of allo-hydroxyproline enantiomers was D after L. To put it the other way around, (+)A-(+)B complex was found to be the more strongly retained diastereomer in the first system, whereas in the second it was retained more weakly than the (+)A-(−)B diastereomer. Since concentration of chiral additives in the mobile phase was kept rather low, 10^{-3} *M*, no change in relative stabilities of the two diastereomeric ternary complexes in the mobile phase could have happened. However, it is logical to expect a significant change in the composition and following sorption properties of the sorbent interface layer, since after equilibration with two chiral mobile phases considered, it should contain much higher quantities of Bzl-L-aHyp than L-aHyp. Obviously, this difference in the stationary phase composition turns out to be sufficient for the inversion of the enantioselectivity sign of the chromatographic system. Likewise, a small change in the chiral additive concentration or content of organic modifier in the mobile phase may result in changing the enantioselectivity of the column. No low-molecular model complexes can follow and simulate all the peculiarities of a multicomponent heterogenic system.

V. THERMODYNAMICS OF FORMATION OF DIASTEREOMERIC SORPTION COMPLEXES

Chromatographic systems comprising chiral chelating sorbents and chiral eluents should be considered separately. In the first case, enantiomeric solutes partition between two phases and diastereomeric ternary complexes form in the sorbent phase alone. The second case is more complicated in that ternary complexes exist in both the mobile and stationary phases, strongly differing from each other in their structure, stability, and enantioselectivity.

The most simple situation is a silica-type rigid sorbent with a very low concentration of surface-bonded chiral stationary ligands (S). In this case, each sorbed metal ion (M) can coordinate just one stationary ligand to form a 1:1 complex (S–M). This situation is realized in ''diluted'' chiral bonded phases[141] where 0.13-μmol/m² *N*-undecanoyl-L-valine ligands are bonded among 3.0-μmol/m² decyl groups. Retention of a mobile ligand (A) can involve its distribution between the mobile and stationary phases:

$$A_m \overset{K}{\rightleftarrows} A_s \tag{1}$$

as well as formation of ternary sorption complexes:

$$A_m + (S{-}M)_s \overset{K_{S-M-A}}{\rightleftarrows} (S{-}M{-}A)_s \tag{2}$$

where the subscripts m and s refer to the mobile and stationary phases, respectively, and charges of all components are omitted. The capacity factor k′ of the solute A is given through Equation 3:

$$k' = \phi(K + K_{S-M-A}[S{-}M]_s) = k'_o + \phi K_{S-M-A}[S{-}M]_s \tag{3}$$

where ϕ is the phase ratio.

Enantioselectivity value ($\overline{\alpha}$) of the system evaluted as

$$\overline{\alpha} = \frac{k'_D}{k'_L} = \frac{k'_o + \phi K^{D}_{S-M-A}[S{-}M]_s}{k'_o + \phi K^{L}_{S-M-A}[S{-}M]_s} \tag{4}$$

varies with varying chromatography conditions in accordance with changing contributions from the nonselective partition (Equation 1) and enantioselective complexation (Equation 2). The corrected value (α) of the complexation enantioselectivity, however, should characterize the system more correctly:

$$\alpha = \frac{k'_D - k'_o}{k'_L - k'_o} = \frac{K^D_{S-M-A}}{K^L_{S-M-A}} \tag{5}$$

as it represents the ratio of stability constants of the two diastereomeric sorption complexes, $S\text{–}M\text{–}A^D$ and $S\text{–}M\text{–}A^L$, and is related to the difference in the free energy of formation of this diastereomeric pair through Equation 6:

$$\delta\Delta G = -RT\ln \alpha \tag{6}$$

(In order to obtain this important parameter, one has to know k'_o, i.e., the solute retention in the metal-free system.)

In case the surface concentration of bonded chiral ligands is high, for instance 2.4 μmol/m² of *N*-undecanoyl-L-valine groups,[141] stationary ligands metal complexes S–M–S of the composition 2:1 are predominantly formed in the stationary phase:

$$S + M_s + S \overset{k_{S-M-S}}{\rightleftarrows} (S\text{–}M\text{–}S)_s \tag{7}$$

Stability constant of these stationary complexes is

$$K_{S-M-S} = \frac{[S\text{–}M\text{–}S]_s}{[S]^2 \cdot [M]_s} = K_{S-M} \frac{[S\text{–}M\text{–}S]_s}{[S] \cdot [S\text{–}M]_s} \tag{8}$$

The concentration of the stationary mono complex, $(S\text{–}M)_s$ (which alone is in position to bind mobile ligands A) is then:

$$[S\text{–}M]_s = \frac{K_{S-M}}{K_{S-M-S}} \cdot \frac{[S\text{–}M\text{–}S]_s}{[S]} \tag{9}$$

Substitution of Equation 9 into Equation 3 yields for the retention of the mobile ligand:

$$k' = k'_o + \phi K_{S-M-A} \cdot \frac{K_{S-M}}{K_{S-M-S}} \cdot \frac{[S\text{–}M\text{–}S]_s}{[S]} \tag{10}$$

The experimental comparison of the above-mentioned diluted and concentrated *N*-undecanoyl-L-valine bonded phases[141] unexpectedly showed a much weaker retention of solutes (dansyl-amino acids) on the concentrated phase. Apparently, the equilibrium concentration of the active sorption sites (i.e., stationary monocomplexes S–M), is higher in the case of diluted phase where these complexes cannot be transformed into less active *bis*-structures S–M–S.

The same authors[141] also report lower resolution selectivity for the concentrated phase with respect to racemic dansyl-amino acids. This tendency remained even after the data presented in the paper were corrected for the nonselective contribution to the net retention of enantiomers arising from the usual phase partition in accordance with Equations 1, 3, and 5. Since the ratio of stability constants of the diastereomeric sorption complexes determines the value of enantioselectivity of both concentrated and diluted phases (Equation 5), the only explanation of the inconsistency of the α-values is the marked difference in the microenvironment of sorption complexes in the above diluted and concentrated phases. The latter is rich in strongly polar donating groups, whereas the former is definitely hydrophobic, which obviously results in some structural changes of the sorption complexes.

Boue et al.[119] also noticed that, curiously, the capacity factor k′ of a given amino acid solute does not continuously increase with the number of sorption sites in the column, i.e., with the exchange capacity of the sorbent. The latter was prepared by grafting L-proline onto an acrylamide polymer by means of formaldehyde. A maximum of retention was observed for the degree of substitution of the polymer by proline ligands of $p = 0.33$ (exchange capacity of the resin about 2 meq/g). The authors assume that two "vicinal" proline grafts form with a copper ion a 2:1 complex, S–M–S, which is too stable to participate in the ligand exchange and that only isolated grafts forming 1:1 complexes, S–M, are responsible for the retention and separation of amino acids. If p is the probability of finding a graft on a given polymer repeat unit, then (1-p) is the probability of finding an ungrafted unit. Consequently, the probability of finding a grafted unit with two free vicinal positions is $p(1\text{-}p)^2$. The retention of amino acid enantiomers was found to vary with the degree of polymer substitution by the L-proline residues parallel to the above extreme function. It should be noted, however, that in a soft gel-type exchanger, stable *bis*-complexes S–M–S can form not only between two vicinal grafts, but also between grafts situated on different polymeric chains. Nevertheless, the maximum affinity of the mobile ligands toward the chiral resin can be expected to manifest itself at moderate, nonmaximum concentrations of the resin-fixed ligands.

Another important parameter influencing the retention of mobile ligands is the concentration of complexing metal ions in the stationary phase. French authors[75,119] observed the capacity factors for the L-proline containing polyacrylamide resin to grow rapidly with the extent, r, of saturation of the resin by copper(II) ions, almost proportional to the function $r/(1 - r)$, where $r = 1$ at a complete formation of 2:1 stationary complexes, S–M–S, in the resin phase. Davankov et al.[88] have found the solute retention k′ to be an exponential function of r as shown in Figure 20. The resin examined was a polystyrene gel containing 3.44 mmol/g residues of L-hydroxyproline.

Thermodynamics of ligand exchange was studied in more detail using a polystyrene-type chiral resin containing residues of L-proline.[208,209] Here, by examining the distribution of complexing metal ions (copper) and mobile ligands (L-proline) between the swollen resin phase and solution, an attempt was made to estimate stability constants of both the stationary *bis*-complex S–M–S and mixed-ligand sorption complex S–M–A under various experimental conditions. The equilibria considered with were

$$S + M_s + S \overset{K_{S-M-S}}{\rightleftarrows} S\text{–}M\text{–}S \qquad K_{S-M-S} = \frac{[S\text{–}M\text{–}S]_s}{[M]_s \cdot [S]_s^2} \tag{11}$$

$$S + M_s + A_s \overset{K_{S-M-A}}{\rightleftarrows} S\text{–}M\text{–}A \qquad K_{S-M-A} = \frac{[S\text{–}M\text{–}A]_s}{[M]_s \cdot [S]_s\,[A]_s} \tag{12}$$

$$S\text{–}M\text{–}S + A_s \overset{K_e}{\rightleftarrows} S\text{–}M\text{–}A + S \qquad K_e = \frac{K_{S-M-A}}{K_{S-M-S}} \tag{13}$$

It is important to remember that in order to obtain thermodynamic constants of the above equilibria, it is necessary to use activities (or, at least, concentrations) of all components within the resin phase. For chromatography purposes, however, it is sufficient to know the phase distribution of copper(II) and L-proline under varying experimental conditions, and as a first approach, we can assume the concentrations of noncomplexed copper and proline moieties in the resin phase to be equal to those in the mobile phase. The results thus obtained are presented in Figure 21. They lead to these five important conclusions.

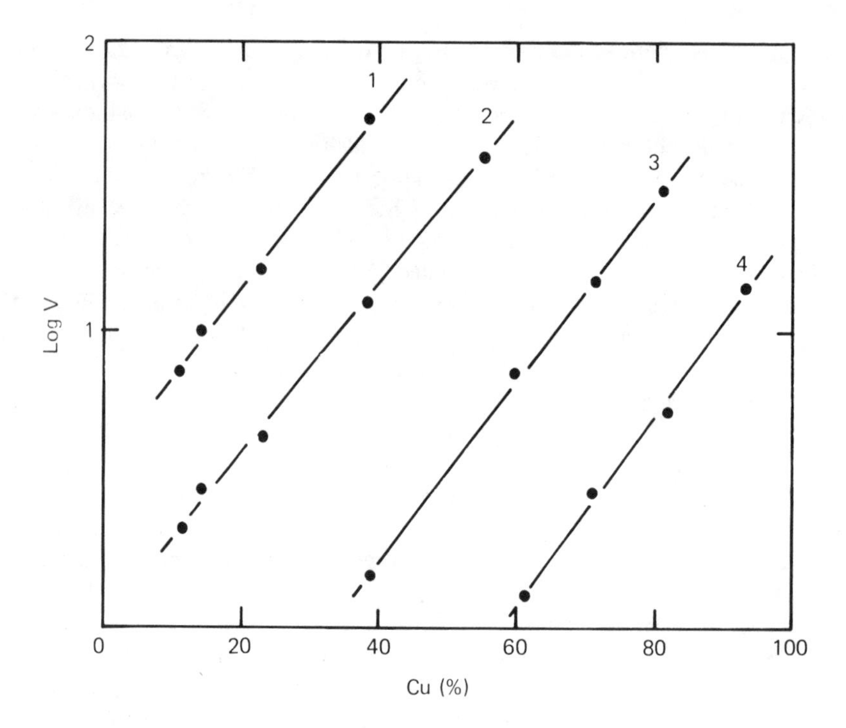

FIGURE 20. Influence of the degree of saturation of the L-hydroxyproline-containing resin (3.44 mmol/g; degree of cross-linking, 11%) with copper(II) ions on the mobile ligands retention volumes (measured in multiples of the column void volume); eluent, 0.1 *M* NH_4OH. 1 D-Trp, 2 L-Trp, 3 D-Pro, 4 L-Pro. (From Davankov, V. A., Zolotarev, Yu. A., and Tevlin, A. V., *Bioorg. Khim.*, 4, 1164, 1978. With permission.)

1. Stability of stationary *bis*-complexes S–M–S rapidly falls with the rising degree of saturation of the chiral resin with copper ions. This is a fundamental peculiarity of chelating resins forming stationary complexes in the composition 2:1 with metal ions, as was first pointed out by Hering.[210] Indeed, the first metal ions entering the resin phase can coordinate the fixed ligand pairs favorably located in space, thus forming stable *bis*- chelates S–M–S. Each succeeding portion of the metal ions can find only less favorably distributed fixed ligand pairs. To be involved in complexation, these ligands must move into positions corresponding to the geometry of the *bis*-complex, with the deformation energy of the relevant polymer chains being consumed at the expense of stability of the *bis*-complex to be formed. It is clear that stability constants of *bis*-complexes S–M–S in the sorbent phase will always depend on a series of additional factors, besides the extent of the complexation process. The most important factors should be (a) the concentration of the matrix-fixed chelating ligands, (b) the rigidity of the sorbent matrix, (c) the length and flexibility of the spacer group between the ligand and matrix, and (d) the geometry of the *bis*-chelate. The experimentally determined stability constants of stationary *bis*-complexes in the sorbent phase can be either above or below the stability constant of a soluble low-molecular-weight model complex, depending on the structure of the sorbent and experimental conditions.
2. The affinity of the proline-containing resin to the copper(II) ions is especially high in alkaline media (where the stationary amino acid-type ligands assume the anionic form needed for complexation) and in the absence of electrolytes like KCl. This results from the Donnan distribution favoring the concentration of the positively charged copper ions in the negatively charged resin phase.
3. The decrease in the stability constants of the ternary sorption complexes S–M–A with

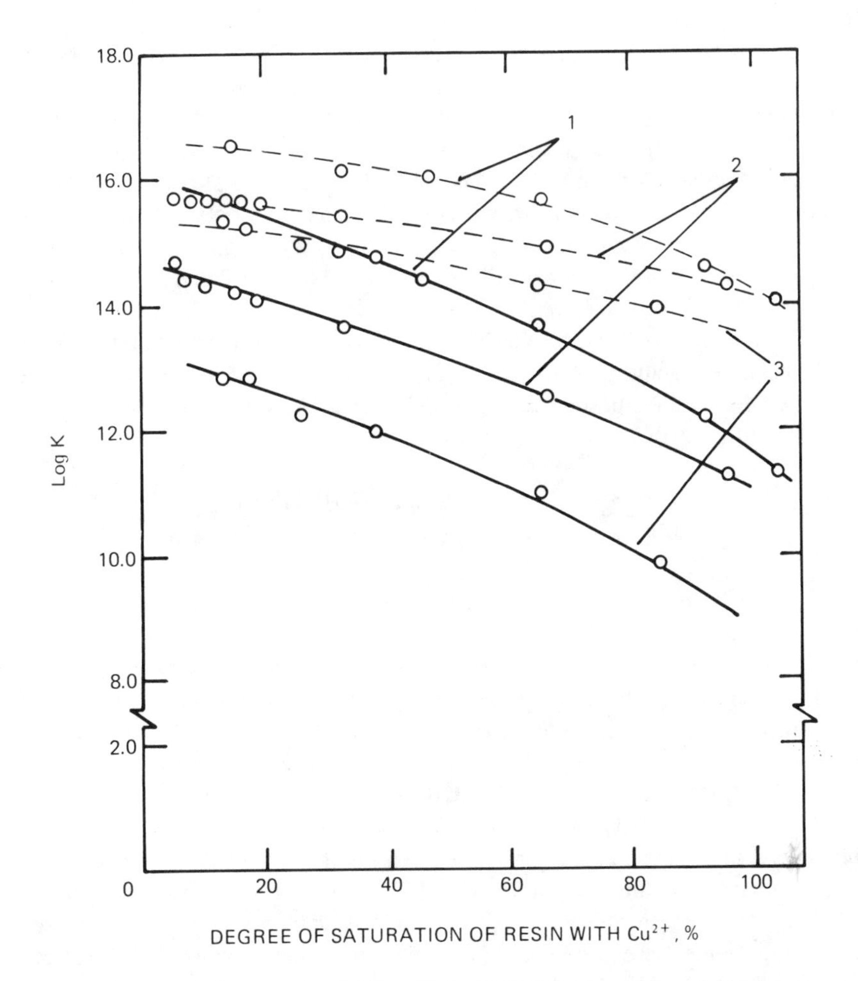

FIGURE 21. Apparent stability constants of stationary *bis*-complexes S–M–S (———) and mixed ligand sorption complexes S–M–A (- - -) over a wide range of saturation with copper(II) ions of the L-proline-containing resin (2.71 mmol/g; degree of cross-linking, 10%). 1 pH > 11.0; 2 pH > 11.0, 1 *M* KCl; 3 pH > 8.0, 1 *M* KCl. (From Zolotarev, Yu. A., Kurganov, A. A., Semechkin, A. V., and Davankov, V. A., *Talanta* 25, 499, 1978. With permission.)

rising copper content of the resin is definitely less dramatic than the decrease in stability of stationary complexes S–M–S. The mixed-ligand structure S–M–A includes only one matrix-fixed ligand. The second, or mobile, ligand, A, chelates the central metal ion without causing any strain in the polymeric matrix, provided that enough space is available for the bulky complex that is formed.

4. Formation of sorption complexes S–M–A is again favored in alkaline media and in the absence of KCl. These phenomena should merely reflect the rising concentration of the copper ion in the resin phase.
5. Stability constants of sorption complexes S–M–A in the system under examination throughout are higher than those of initial stationary complexes S–M–S. This implies a high affinity of the resin to the mobile L-proline ligands. Herewith, the difference between stabilities of the two *bis*-complexes steadily increases with the rising copper content in the resin. This should result in a rise of solute capacity factors, which is in full agreement with the experiment (Figure 20).

In order to evaluate the true, not the apparent, stability constants of stationary and sorption complexes, the following additional information is needed: (1) the dissociation constants of both stationary and mobile ligands, (2) the equilibrium water content in the resin phase, (3) the distribution coefficient of the double positively charged nonchelated copper ions between the mobile ligand solution and the negatively charged resin phase, and (4) the corresponding distribution coefficient for the mobile ligand in its anionic form. The dissociation constants of the carboxy and amino functions of the L-proline-containing polystyrene-type resin (macronet isoporous matrix, cross-linked to the extent of 11%; exchange capacity, 2.7 mmol/g; swelling ability, 180% w/w at pH 9.7) were found[208] to be 2.2 and 9.5, respectively, (n = 1.8), which corresponds fairly well with the dissociation constants[99] of *N*-benzyl-L-proline: 2.12 ± 0.05 and 9.77 ± 0.05. Similarly, when involving into calculation the above listed additional quantities, stability constants of stationary *bis*-complexes S–M–S and sorption complexes S–M–A with L-proline as the mobile ligand A were found[208,209] to correspond well with stabilities of the low-molecular-weight models:[99,104]

$$\log \beta = 12.4 \pm 0.1 \text{ for Cu(N-Bzl-L-Pro)}_2$$
$$\log \beta = 14.9 \pm 0.2 \text{ for Cu(N-Bzl-L-Pro)(L-Pro)}$$
$$\log \beta = 11.4 \text{ to } 12.5 \text{ for S-M-S}$$
$$\log \beta = 14.5 \text{ to } 15.1 \text{ for S-M-A}$$

(the last two values were estimated under various experimental conditions). In the case of sorption of DL-proline under static conditions, the value of enantioselectivity of 1.9 kJ/mol[3,211] was found to be consistent with the chromatographic result[58] 3.5 kJ/mol and the investigation results of the diastereomeric model complexes, Cu(*N*-Bzl-L-Pro)(L-Pro) and Cu(*N*-Bzl-L-Pro)(D-Pro), in solution[104] 3.8 ± 0.8 kJ/mol. However, one should not overvalue the correctness of comparisons of the polymeric and model systems, especially with respect to stability constants of the stationary *bis*-complexes S–M–S and of enantioselectivity values.

The last comments in this section are aimed at chromatographic systems operating with a chiral eluent. They are much more complicated in that more components and more equilibria should be dealt with. There is no general description of such systems thus far. Grushka et al.[173] made a great simplification by omitting complexation process between the resolving chiral additives (R) and metal ions and considering formation of sorption sites in the stationary phase as a simple phase distribution of chiral metal chelates R–M:

$$k_{R-M} = \frac{[R-M]_s}{[R-M]_m} \tag{14}$$

In this case, retention of the solute A can be described as

$$k' = k_o' + \phi \cdot K_{(R-M-A)_s}[R-M]_s = k_o' + \phi \cdot K_{(R-M-A)_s} k_{R-M}[R-M]_m \tag{15}$$

To a certain extent, Equation 15 explains the observation that retention of amino acid A increases on rising the concentration of copper-aspartylcyclohexylamide in the eluent in a reversed-phase system.[173] However, we have shown earlier that this happens only in the case where the hydrophobicity of the chiral additive R is higher than that of the solute A, which indeed, is valid for the above experiment. The authors[173] also estimated apparent values of $K_{(R-M-A)_s}$, as well as their temperature dependence. Some thermodynamic quantities obtained therefrom, however, should characterize the total chromatographic system, rather than the individual sorption complex $(R-M-A)_s$.

VI. LIGAND EXCHANGE IN THE OUTER COORDINATION SPHERE OF STABLE COMPLEXES

Spatial arrangement of ligands in the coordination sphere of transition metal ions has always been a subject of intensive investigation. Since important conclusions on the symmetry of a complex can be made from the presence or absence of optical isomerism, numerous attempts have been made to resolve, at least partially, stable complexes into enantiomers. Chromatography proved to be especially productive in this direction. It has long been noticed (see References 6 through 8) that the separation of optical isomers of kinetically inert octahedral complexes of cobalt(III), chromium(III), and rhodium(III) on naturally occurring asymmetric sorbents like cellulose, starch, and lactose (the latter in nonaqueous media) takes place more readily than does the resolution of conventional organic racemic compounds. The majority of these experiments have been done without paying much attention to the mechanism of chiral recognition. In the light of present knowledge, however, it is logical to ascribe such readiness of resolution of inert complexes to ligand exchange in their external coordination sphere.

The external coordination sphere (i.e., solvation shell) of inert complexes is known to possess a highly regular structure. Its ligands are bonded via hydrogen bonds or dipole interactions to the donating hetero atoms of the inner-sphere ligands fixed in the main coordination positions of the central metal atom. For instance, it has been found that methanol molecules bind to amino and carboxylate groups coordinated to copper(II) in *bis*(*N*-benzylprolinato) copper long before the axial positions of the complex are occupied by methanol.[100,101] Herewith, the binding energy of ligands in the external coordination sphere may amount to several kilojoules per mole (or more than 10 kJ/mol for polydentate ligands), which is convenient for a chromatographic process to be based upon. Ligand exchange in the external coordination sphere is a fast process, thus allowing a great number of kinetically inert complexes to be involved in LEC.

The resolution of racemic complexes on cellulose and other carbohydrate-type sorbents can be supposed to result from the circumstance that the asymmetrically distributed hydroxy groups of these sorbents fit into the external coordination sphere of one enantiomer of the complex more readily than into the external coordination sphere of its anitpode. Hereby, solvent molecules are replaced from the sphere by ligands belonging to the chiral sorbent, which causes retention of the complex in the chromatographic column.

As early as 1954, Krebs and co-workers[212,213] resolved into enantiomers a series of octahedral complexes (e.g., *tris*-glycinatocobalt[III]) using starch as the chiral sorbent. The terms "LEC" and "enantioselectivity" did not even exist at that time. It is remarkable that successful resolutions of racemic complexes according to the methods of "chiral eluent" and "chiral coating" have also been described long before these methods were recognized by chromatography specialists. Thus, Yoshikawa and Yamasaki,[214] in 1970, described the use of (+)-tartrate solutions in chromatography of racemic cobalt(III)-diamine and cobalt(III)-triamine complexes $[Co(en)_3]^{3+}$, $[Co(dien)_2]^{3+}$, and $[Co(tn)_3]^{3+}$ on cation-exchanging dextran gels or cellulose. In 1961, Piper[215] coated Al_2O_3 with (+)-tartaric acid to resolve *tris*-acetylacetonates of chromium(III) and cobalt(III). Moreover, an attempt is described to combine the action of both a chiral eluent and a chiral sorbent:[216] resolution of $[Co(en)_3]^{3+}$ on a (−)-tartrate-containing Sephadex® packing was observed to improve on using sodium (+)-tartrate solutions as the eluent instead of nonchiral salt solutions.

Table 15 represents a list of typical chiral complexes resolved on the naturally occurring and chemically modified carbohydrates. Low-molecular-weight sugars, being soluble in water, have to be used in nonpolar organic media. Starch and cellulose, when applied in aqueous or aqueous-organic eluents, resolve both neutral and electrically charged complexes. Carboxymethylated and phosphorylated cellulose are used for chromatography of cationic

Abbreviations to Tables 15, 16, and 17

acac	acetylacetonate
actp	3-(2-aminoethylthio)propionato, $H_2N(CH_2)_2S(CH_2)_2COO^-$
i-abu	α-amino isobutyrato
bipy	2,2′-bipyridine
bzac	benzoylacetonate
chxn	*trans*-1,2-cyclohexanediamine
cydta	*trans*-1,2-cyclohexanediaminetetraacetate
dien	diethylenetriamine
edda	ethylenediamine-*N,N′*-diacetate
edta	ethylenediaminetetraacetate
en	ethylenediamine
glygly	glycilglycinato
ida	iminodiacetate
mal	malonate
medien	*N*-methyldiethylenetriamine
ox	oxalate
phen	1,10-phenanthroline
pn	1,2-propanediamine
py	pyridine
sep	1,3,6,8,13,16,19-octaazabicyclo[6.6.6]eicosane
tdta	1,4-butanediaminetetraacetate
tmdata	trimethylenediaminetetraacetate
tn	1,3-propanediamine
trien	triethylenetetramine

Table 15
LIST OF COMPLEXES RESOLVED INTO ENANTIOMERS ON CHIRAL CARBOHYDRATES

Complex	Ref.
Starch	
$Co(en)_3^{3+}$	212,213,218
$Cr(en)_3^{3+}$	213,218
$Co(gly)_3$	212,213,219,220
$Co(ox)_3^{3-}$	213
$Cr(ox)_3^{3-}$	213,218
$Co(en)_2(gly)^{2+}$	213,218
$Co(en)_2(NO_2)_2^+$	213,218
$Co(en)_2(X)_2^+$ where $(X)_2 = (NO_3)_2$, (CO_3), (ox), $(NO_2)(Cl)$	218
$Co(glygly)_2^-$	220
$Co(i\text{-}abu)_3$	221
Co(edda)(gly)	222
$Co(S_2COCH_2CH_2SO_3)_3^{3-}$ (dithiocarbonate chelate, S,S-ring)	213,218
$Co(S_2C\text{–}N(CH_3)\text{–}C_6H_4OH)_3$ (dithiocarbamate chelate, S,S-ring)	213

Table 15 (continued)
LIST OF COMPLEXES RESOLVED INTO ENANTIOMERS ON CHIRAL CARBOHYDRATES

Starch

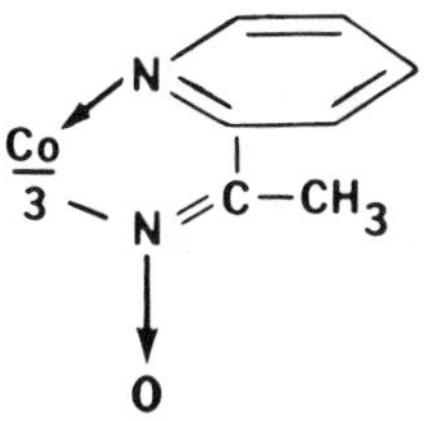	213,218
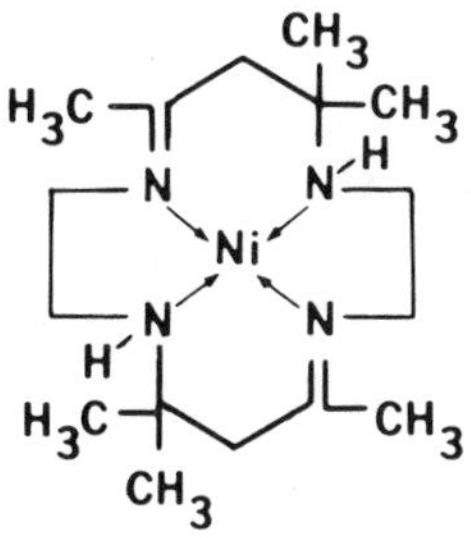	223
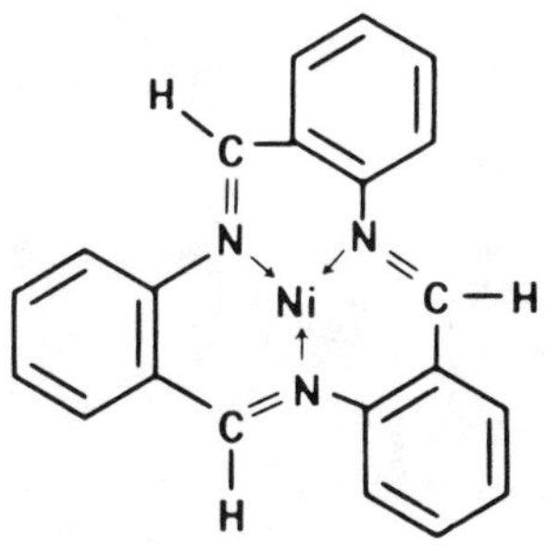	224

D(+)-Lactose

$Cr(acac)_3$	225—228
$Co(acac)_3$	225—230
$Ru(acac)_3$	227
$Rh(acac)_3$	226,227,229
$Y(acac)_3$ and $Gd(acac)_3$	231
Co(3-Br-acac)$_3$	229
$Co(bzac)_3$	227
$Cr(bzac)_3$	232
Co(H_3CC=CHCCHCH$_3$)$_3$ with O under the first C and =O under the second C, and CH$_3$ on the CH	233
Cu(RC=CHCR′)$_2$ and Ni(RC=CHCR′)$_2$ (O under the first C, NR″ under the second C) where R and R′ and CH_3 or C_6H_5, R″ is H, C_6H_5, or 2,6-$C_6H_3(CH_3)_2$	234

Table 15 (continued)
LIST OF COMPLEXES RESOLVED INTO ENANTIOMERS ON CHIRAL CARBOHYDRATES

D(+)-Lactose

Complex	Ref.
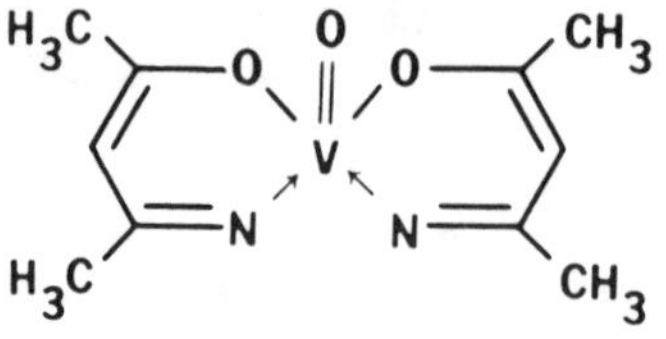	235
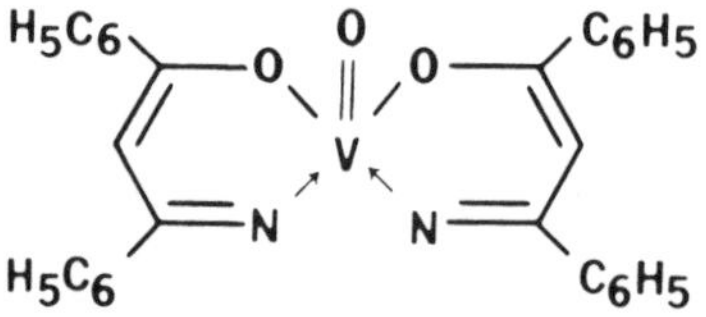	236

D(+)-Sorbitol

Complex	Ref.
Co(acac)$_3^{3-}$	237
Cr(acac)$_3^{3-}$	237
Co(bzac)$_3^{3-}$	237

D(+)-Manitol

Complex	Ref.
Co(acac)$_3^{3-}$	237
Cr(acac)$_3^{3-}$	237
Co(bzac)$_3^{3-}$	237

D(+)-Glucose

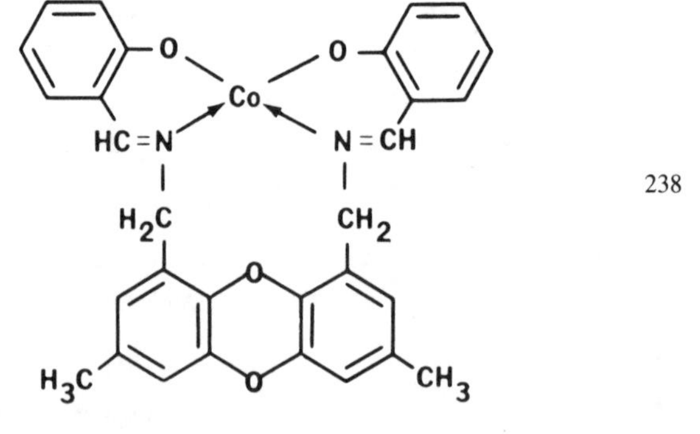

238

Cellulose

Complex	Ref.
Co(ox)$_3^{3-}$	218
Co(gly)$_3$	239
Co(ala)$_3$	239
cis-Co(ox)$_2$(H$_2$O)$_2^-$	239
Cod(en)$_n$(*l*-pn)$_{3-n}^{3+}$ where n = 1, 2, 3	240

Table 15 (continued)
LIST OF COMPLEXES RESOLVED INTO ENANTIOMERS ON CHIRAL CARBOHYDRATES

Complex	Ref.
Cellulose	
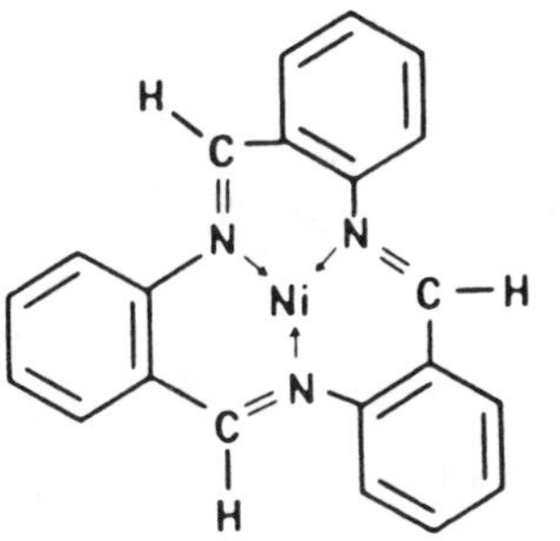	224
CM-Cellulose	
cis-$Co(ida)_2^-$	241
$Co(tmdata)^-$	241
$Co(en)(edda)^+$ and *N*-methyl derivatives	241
$(H_2NCH_2CH_2S)_6Co_3{}^{3+}$	242
$Rh(phen)_2(Cl)_2^+$	243
P-Cellulose	
$Co(en)_3^{3+}$	244
$Co(dien)_2^{3+}$	244
DEAE-Cellulose	
$Co(ida)_2^-$	244
$Co(edta)^-$	244

complexes, whereas the anion-exchanging diethylaminoethyl cellulose should be combined with negatively charged complex ions. A more complete list of resolutions on ion-exchanging cellulose sorbents can be found in the review[273] by Yoshikawa and Yamazaki.

A series of synthetic chiral packings are presented in Table 16. The simplest one was prepared by adsorbing D-lactose on alumina. Optically active tartrate anions also serve as efficient chiral resolving agents with respect to octahedral complexes when adsorbed on the surface of alumina or immobilized electrostatically on anion-exchange resins. Similarly, chiral octahedral complexes themselves can be packed into a chromatographic column in the form of small crystals, as well as in the adsorbed state on particles of montmorillonite clay or beads of ion-exchange resins. An interesting attempt to prepare a silica-bonded chiral phase of the type of λ-$[Co(en)_3]^{3+}$ for outer-sphere LEC by impregnating a (2-aminoethyl)aminopropyl-syliated packing with λ(+)$[Co(en)_2(NO_3)_2]Br$ is described.[268] Both racemic complexes and organic ligands can be resolved with the aid of the above chiral sorbents based on chiral kinetically inert complexes (Table 16).

However, a more promising and rapidly developing approach[214] is the ion-exchange chromatography of racemic complexes in the presence of optically active additives to the eluent, in particular sodium *d*-tartrate or potassium antimony *d*-tartrate, $K_2[Sb_2(d\text{-tart})_2]$. Yoshikawa and Yamasaki, the authors of this "method of chiral eluent," were able to list in their review

Table 16
LIST OF COMPLEXES AND LIGANDS RESOLVED INTO ENANTIOMERS ON CHIRAL PACKINGS

Alumina + *d*-tartrate
 $Co(acac)_3$, $Cr(acac)_3$[215,245]
Alumina + D-lactose
 $Cr(acac)_3$;[245] $Co(acac)_3$;[246] Al(acac), and $Fe(acac)_3$[247,248]
Carboxylic cation exchanger + (+) $[Co(en)_3]^{3+}$
 $Co(edta)^-$;[249] aspartic and mandelic acid[250]
Anion exchanger IRA 400 + *d*-tartrate
 $Co(en)_3^{3+}$, $Co(pn)_3^{3+}$, $Co(en)_2(ox)^+$,
 $Co(en)_2(H_2O)_2^{3+}$, $Co(en)_2(bipy)^{3+}$,
 $[(en)_2Co(OH)_2]_3Co^{6+}$, $Cr(en)_3^{3+}$, $Cr(pn)_3^{3+}$
 $Rh(en)_3^{3+}$, $Rh(pn)_3^{3+}$, $Co(en)_3(phen)^{3+}$[251]
$(+)K[Co(gly)_2(NO_2)_2]$
 $Co(acac)_3$, $Cr(acac)_3$[252]
$(+)K[Co(L\text{-}ala)_2(NO_2)_2]$
 $Co(acac)_3$, $Cr(acac)_3$[252]
$(+)[Co(L\text{-}arg)_2(NO_2)_2]ce$
 $Co(acac)_3$, $Cr(acac)_3$[252]
$(-)[Co(L\text{-}arg)_2(NO_2)_2] \cdot (-)[Co(gly)_2(NO_2)_2]$
 $Co(acac)_3$[252]
$(-)[Co(L\text{-}arg)_2(NO_2)_2] \cdot (+)[Co(gly)_2(NO_2)_2]$
 $Co(acac)_3$[252]
Montmorillonite $^+\Delta[Ni(phen)_3]^{2+}$
 $Co(acac)_3$;[253-255] Cr(acac);[254] $Rh(acac)_3$;[254] $Ru(acac)_3$;[256] $Co(acac)_2$ (gly), Co(acac) $(gly)_2$[255]
 $Co(acac)_2(AA)$ where AA = val, leu, thr, met, phe, gly, ala, ser, pro, trp[257]
Montmorillonite + $\lambda[Ru(phen)_3]^{2+}$
 $Co(acac)_3$, $Cr(acac_3)$, $Rh(acac)_3$;[258] 2,3-dihydro-2-methyl-5,6-diphenylpyrazine;[259] proline, hydroxyproline, 5-methyl-2-pyrolidinone, γ-valerolactone, 2-piperidinecarboxylic acid, 4-thiazolidinecarboxylic acid;[260] $Co(dimethylglyoximato)_3$, $As(catecholato)_3^-$, $Fe(bathophenanthroline\text{-}sulfonic\ acid)_3^{4-}$ and 15 others[261] 2,2′-dihydroxy-2,2′-dimethyl-, and 2,2′-diamino-1,1′-binaphthyl;[262,263] *trans*-2,3-diphenyl-5,6,7,8,9,10-hexahydroquinoline, benzoin, hydrobenzoin, 1,2-diphenylethanol[263]
SP-Sephadex® + $\Delta[Ni(phen)_3]^{2+}$
 $Co(acac)_3$[264]
L(+)-tartrate bonded to Sephadex®
 $Co(en)_3^{3+}$[265]
D(−)-tartrate bonded to Sephadex® in combination with L(+)-tartrate-containing chiral eluent
 $Co(en)_3^{3+}$;[216] $Co(tn)_3^{3+}$[266]
L-hydroxy acids bonded to Sephadex®, Sephadex®-O-CH(R)-COOH, where R = CH_3, $CH(CH_3)_2$, $CH(OH)CH_3$, or CH_2COOH
 $Co(en)_3^{3+}$, $Co(en)_2(gly)^{2+}$, $Co(en)_2(ox)^+$, *cis*-$Co(en)_2(NO_2)_2^+$[267]
$\lambda[Co(en)_3]^{3+}$ bonded onto silica gel
 $Co(ox)_3^{3-}$, $Cr(ox)_3^{3-}$, tartaric acid[268]
Anion exchanger QAE-Sephadex® + $[Sb_2(d\text{-}tart)_2]^{2-}$
 -$Co(\beta\text{-}ala)_3$[269]
Anion exchanger TSK-Gel® 2000 + $[Sb_2(d\text{-}tart)_2]^2$ or *d*-tart
 fac-$Co(\beta\text{-}ala)_n(gly)_{3-n}$, fac-$Co(\beta\text{-}ala)_n(D\text{- or }L\text{-}ala)_{3-n}$[270]
Chiralpack®-OT, poly(triphenylmethyl-methacrylate) coated on silica gel
 $Co(acac)_3$, $Cr(acac)_3$;[271] $Al(acac)_3$[272]

paper[273] more than 100 successful resolutions of octahedral complexes on ion-exchanging dextran gels, mainly sulfopropylated Sephadex®. These sorbents have a high swelling ability and a relatively low concentration of ionic groups and therefore do not retain multiply-charged complex ions as long as do conventional polystyrene-based ion exchangers. The chiral matrix of Sephadex® gels itself plays but a minimal role in the process of chiral recognition of octahedral complexes. The enantiomeric recognition is mainly based upon the difference in the binding of (+)-tartrate ions in the outer coordination sphere of λ and

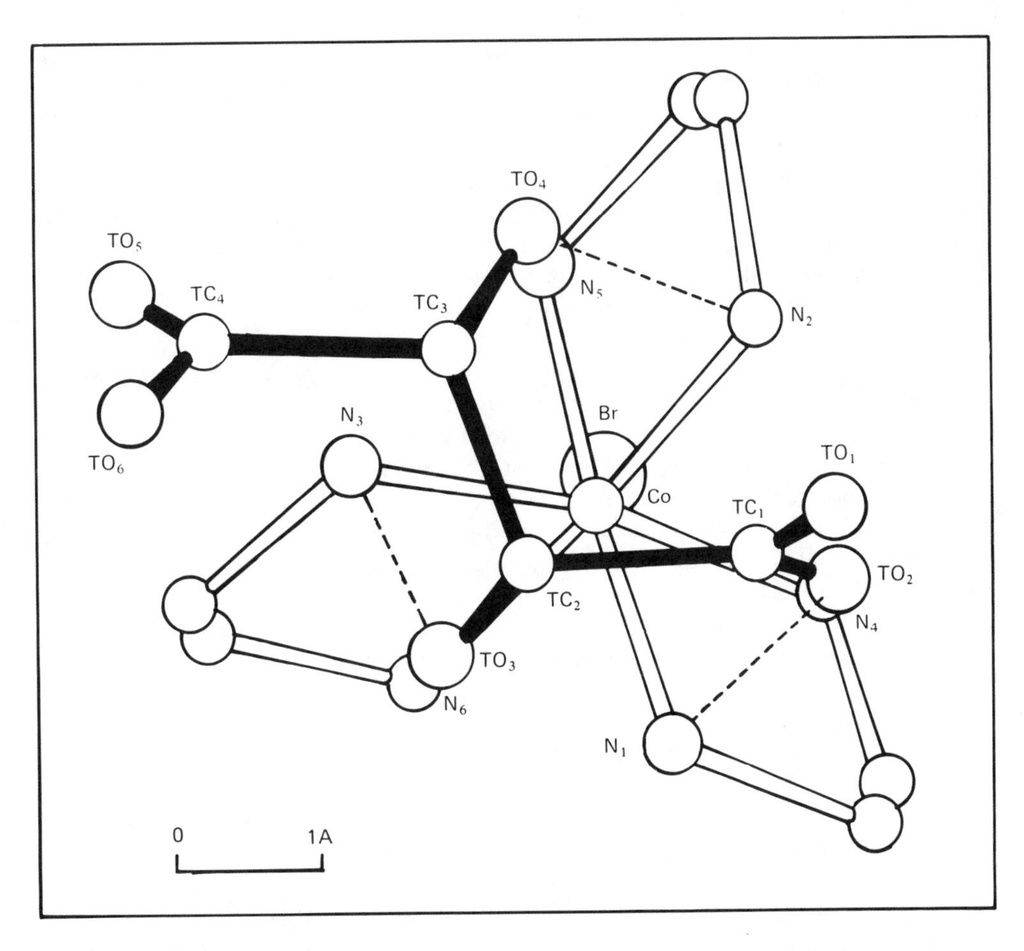

FIGURE 22. The ion pair structure between λ-$[Co(en)_3]^{3+}$ and *d*-$tart^{2-}$, viewed along with three fold axis of the complex cation. (From Yoneda, H., *J. Liquid Chromatogr.*, 2, 1157, 1979. With permission.)

Δ complexes. Thus, the association constant of (+)-tartrate with λ(+)$[Co(en)_3]^{3+}$ was found[274] to be 20% larger (25°C, I = 0.1) than for (+)-tartrate with Δ(−)$[Co(en)_3]^{3+}$. In light of this fact, one can easily understand that with (+)-tartrate ions in the mobile phase, the λ-complex would elute first from the negatively charged cation exchanger. On the contrary, λ-complexes were observed to be retained longer where the (+)-tartrate ions were chemically bonded (via ester or ether links) onto the Sephadex® matrix.[216,266] Especially effective are systems combining enantioselectivity contributions from (−)-tartrate groups in the stationary phase with those from (+)-tartrate ions in the chiral eluent.[216,265]

In an important series of papers, Yoneda and co-workers[275-279] examined the mechanisms of the enantioselective association between tartrate anions and *tris*-ethylenediamine-cobalt cations. The origin of the discrimination was found to lie in the face-to-face structure of the ion pair. As shown in Figure 22, the four carbon atoms of *d*-tartrate make a plane which is perpendicular to the threefold axis of the complex. Three oxygen atoms (one carbonyl atom and two hydroxyl oxygens) are projecting toward the complex and face the triangular facet formed by three NH_2 groups of the complex. Three hydrogen bonds can be supposed to exist in the intimate ion pair with the λ-complex, while *d*-tart can not form such a favorable ion pair with the Δ(−)$[Co(en)_3]^{3+}$ enantiomer.

If this idea is valid, *d*-tart should resolve not only all $[M(en)_3]^{3+}$ complexes where M(III) is cobalt, chromium, or rhodium, but also any complex having three NH_2 groups in a triangular facet of their octahedral coordination sphere. In fact, complete resolution by *d*-tart-containing eluents is reported for a series of complexes (see Table 17) $[Co(en)_2(AA)]^{2+}$ and fac-$[Co(AA)_3]^{\circ}$, where AA is an amino acid anion; for instance glycine, serine, or β-alanine.

Stable divalent anions of antimony *d*-tartrate $[Sb_2(d\text{-tart})_2]^{2-}$ (abbreviated hereafter as *d*-antart) were found to be still better resolving agents with respect to octahedral complexes than *d*-tartrate itself.[279] This agent has resolved more types of complexes than *d*-tartrate, indicating that the mode of association of *d*-antart is different from that of *d*-tartrate and that three neighboring NH_2 groups in the complex are not necessarily required. According to Yoneda et al.,[275-279] *d*-antart is presumed to approach the *tris*-chelate from the direction other than along the threefold axis. As is seen in Figure 23, λ and Δ configurations of the complex are characterized by the mode of coordination of its three ligands, and at the same time, by the shape of the opening between three chelate rings. In the Δ-configuration, all openings between three chelate rings have an L-shape, whereas the λ-complex possesses six openings of a reverse J-shape. The chiral resolving counter ion, $[Sb_2(d\text{-tart})_2]^{2-}$, easily fits the L-shaped channels and does not fit the J-shaped channels. As shown in Figure 24, *d*-tartrate ions bridging two Sb atoms in the resolving agent are the parts that can fit only the L-shaped channels.

The above model of enantioselective ion-pair formation explains well a series of experimental observations; for instance, the dependence of resolution selectivities on the number of L- and J-shaped channels in complexes with polydentate ligands such as $[Co(dien)_2]^{3+}$ or $[Co(en)(trien)]^{3+}$,[279] as well as the influence of polarity of complexes of the *cis*-$[Co(en)_2(X)_2]^{+}$-type on their association and resolution with *d*-antart.[286] Typical examples are shown in Figure 25.

More recently, Yoneda and co-workers proposed mechanisms of enantioselective association between octahedral complex anions of the type of $[Co(edta)]^{-}$, $[Co(ox)_2(en)]^{-}$, and $[Co(mal)_2(en)]^{-}$ and chiral Δ-$[Co(en)_3]^{3+}$ cations[290,291] and alkaloid molecules (cinchonine, cinchonidine, quinine, quinidine,[292] as well as the perspectives of mutual chromatographic resolutions of the above compounds using outer-sphere ligand exchange in a reversed-phase[292] and ion-exchange[290,291] mode. The mechanism of chiral recognition of octahedral complexes via binding of various chiral resolving agents in the outer coordination sphere of the former is discussed in detail in a recent extensive review by Yoneda.[293]

Based on theoretical analyses of two equilibria, ion exchange and ion association, Yamasaki and Yoneda derived general equations for the retention and resolution of enantiomeric complex cations on cation exchange[294] and anion exchange[295] columns which are percolated with a chiral eluent containing bivalent anions, e.g., *d*-tart^{2-} or *d*-antart^{2-}. Since the retention of cations on the cation exchanger always decreases and the resolution selectivity increases (up to a certain limiting value) with the rise in concentration of the chiral eluent, the optimal eluent concentration for achieving maximum resolution in a short period of time appears to depend on the association constants of two diastereomeric ion pairs, β_λ and β_Δ, and the enantioselectivity of the system, $\alpha = \beta_\lambda/\beta_\Delta$. With the anion exchange column, the retention of complex cations was shown to pass through a maximum at the eluent concentration $[X] = 1/\beta$. Thereafter, the enantioselectivity of the system gradually decreases from the value $\alpha = \beta_\lambda/\beta_\Delta$ at $[X] \rightarrow 0$ to unity at $[X] \rightarrow \infty$.

Before closing the section on ligand exchange in the outer coordination sphere of chiral complexes, we would like to note that a great amount of important information has been collected in this field since successful resolutions of stable racemic complexes on starch and cellulose were first ascribed to ligand exchange.[6,8] However, little has been done thus far to achieve high efficiency of columns operating in this mode, though ligand exchange in outer coordination spheres of stable complexes is expected to be a fast process.[296] In the near future, we may expect substantial progress in developing reversed-phase HPLC systems

Table 17
RESOLUTION OF STABLE COMPLEXES INTO ENANTIOMERS USING CHIRAL ELUENTS (SEE ALSO REFERENCE 273)

Sorbent	Eluent	Racemate
Dowex® 50 × 8	*d*-antart	$Co(Gly)_2(NH_3)_2^{+}$ [280]
SP-Sephadex®	*d*-tart or *d*-antart	$Co(dien)(medien)^{3+}$, $Co(dien)_2^{3+}$, $Co(medien)_2^{3+}$ [281]
SP-Sephadex®	*d*-tart or *d*-antart	$Co(en)_3^{3+}$, $Co(en)_2(NH)_3)_2^{3+}$ [282]
Dowex® 50W × 2	*d*-tart or *d*-antart	$Co(en)_2(X)_2$ where $(X)_2$ = gly, β-ala, acac, ox, mal, $(CN)_2$, $(N_3)_2$, $(NO_2)_2$ or $(NCS)_2$ [283]
SP-Sephadex®	*d*-tart	$Co(en)_2(AA)^{+}$ where AA = L- or D-phe, L- or D-leu[284]
CM-Sephadex®	*d*-tart	fac-Co(β-ala)$_3$ [285]
TSK-Gel® 211	*d*-tart	fac-Co(gly)$_2$(β-ala), Co(gly)(β-ala)$_2$, fac-Co(D- or L-ser)$_n$(β-ala)$_{3-n}$ [277]
SP-Sephadex®	*d*-antart	$Co(en)_3^{3+}$; *cis*-α- and *cis*-β-Co(trien)(en)$^{3+}$; u-fac-, s-fac-, and mer-Co(dien)$_2^{3+}$ [278,279]
SP-Sephadex®	*d*-antart	*cis*-$Co(en)_2(N_3)_2^{+}$, *cis*-α-Co(trien)$(N_3)_2^{+}$ [286]
SP-Sephadex®	*d*-antart	*cis,cis,cis*-Co(gly)$_2$(NH$_3$)$_2^{+}$, *cis(O),trans(N), cis*(NH$_3$)-Co(gly)$_2$(NH$_3$)$_2^{+}$, *cis(O)*-Co(gly)$_2$(en)$^{+}$ [287]
SP-Sephadex®	*d*-antart	mer-*trans(S)*-, *trans(S)*-, *trans*(N)-, and *cis, cis,cis*-Co(actp)$_2^{+}$ [288]
Dowex® 50W × 8	*d*-antart	mer-(*N*)-Co(mal)(en)(py)(H$_2$O)$^{+}$, mer(N)-*cis*(NH$_3$)-Co(mal)(NH$_3$)$_2$(py)(H$_2$O)$^{+}$ [289]
IEX-220	$Co(en)_3^{3+}$, $Co(sep)^{3+}$, or Co(1-chxn)$_3^{3+}$	$Co(edta)^{-}$, $Co(tdta)^{-}$, $Co(ox)_2(en)^{-}$, and $Co(mal)_2(en)^{-}$ [290,291]
RP-Silica	Quinine, quinidine, and cinchonidine	$Co(edta)^{-}$, *cis*-$Co(ida)^{2-}$, $Co(cydta)^{-}$, $Co(ox)_2(en)^{-}$, and *cis*-α-Co(ox)(edda)$^{-}$ [292]

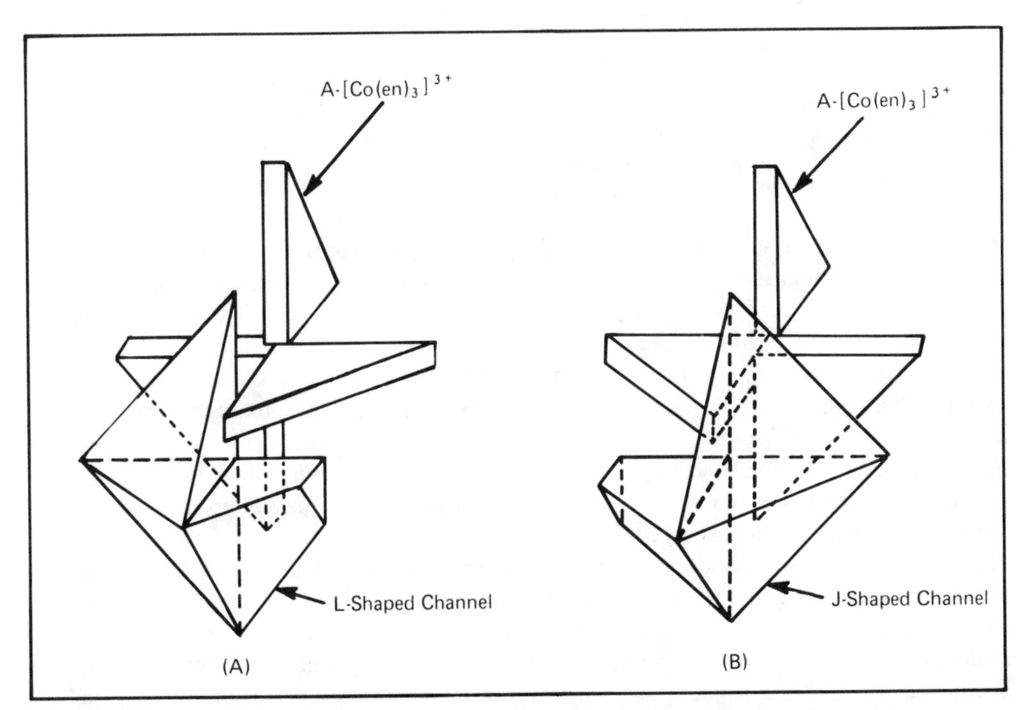

FIGURE 23. A model proposed for the association of $[Sb_2(d\text{-tart})_2]^{2-}$ ion with λ-tris(chelate)cobalt(III) complex having L-shaped channels between the chelate rings (A), and the enantiomeric J-shaped ion pair (B). (From Nakazawa, H. and Yoneda, H., *J. Chromatogr.*, 160, 89, 1978. With permission.)

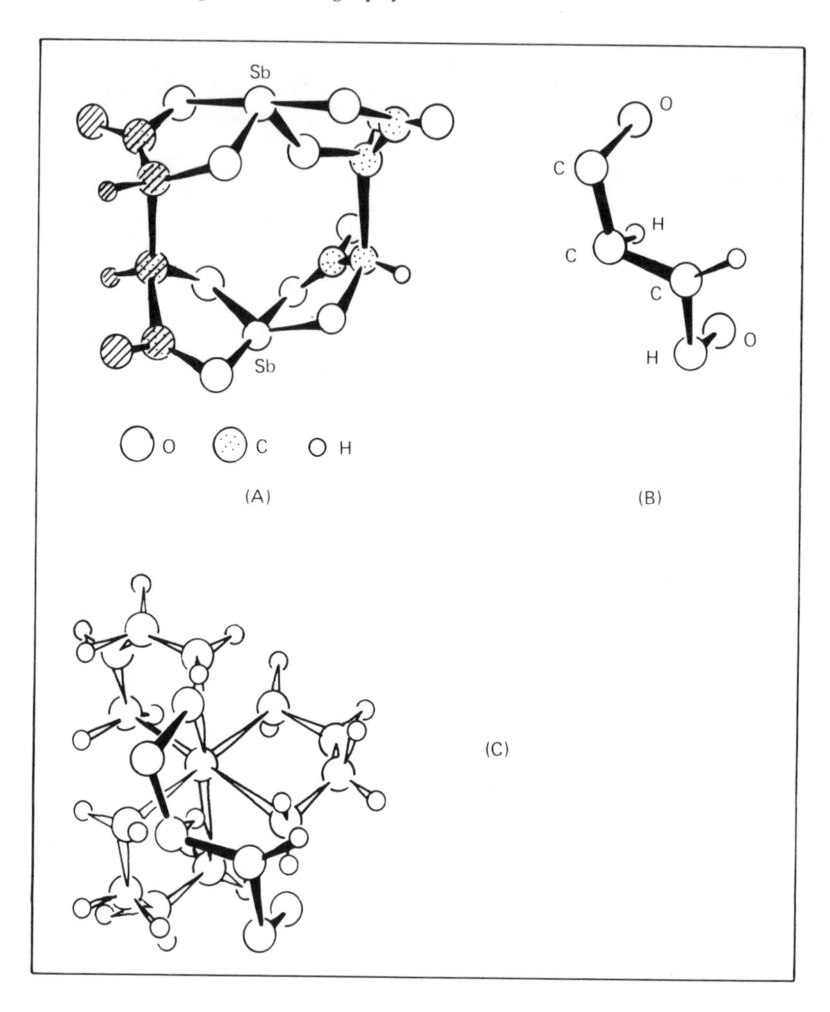

FIGURE 24. The overall structure of the $[Sb_2(d\text{-tart})_2]^{2-}$ ion (A), of nonbonded atoms of the *d*-tart ligands as viewed from the center of the ion (B), and of the adduct with λ-$[Co(en)_3]^{3+}$ (C). (From Nakazawa, H. and Yoneda, H., *J. Chromatogr.*, 160, 89, 1978. With permission.)

that are similar to those involving stable complexes and alkaloids.[292] And indeed, in his recent review,[293] Yoneda mentions complete resolutions of cobalt-*tris*-ethylenediamine and cobalt-*tris*-(1,3-propanediamine) complexes as well as some other complex cations on reversed-phase packings with chiral antimony-*d*-tartrate eluents.

Finally, it is worth mentioning that column liquid chromatography is by no means the only possible technique for applying enantioselectivity in complex formation to practical resolutions of optical isomers. As early as 1970, Yoneda and co-workers succeeded in resolving racemic *tris*(ethylenediamine) cobalt(III) by thin-layer chromatography,[297] as well as paper electrophoresis[298] in the presence of *d*-tartrate and aluminum(III) ions in the eluent. Similar complexes of chromium(III) and rhodium(III) were also resolved.[299,300] It can be suggested that Al^{3+} ions associate with *d*-tartrate to form complex chiral resolving agents which are more efficient in recognizing configuration of the $[M(en)_3]^{3+}$ cations than are *d*-tart^{2-} anions themselves. Octahedral complex anions, $[Co(ox)_3]^{3-}$ and $[Cr(ox)_3]^{3-}$, display marked enantioselectivity effects on binding alkaloid cations (brucine, cinchonine, quinine) in their outer coordination sphere, which permits enantiomeric resolution of complexes by paper electrophoresis in concentrated alkaloid solutions.[301] Enantioselective ion association is likewise responsible for the electrophoretic resolution of cationic cobalt(III) diamine

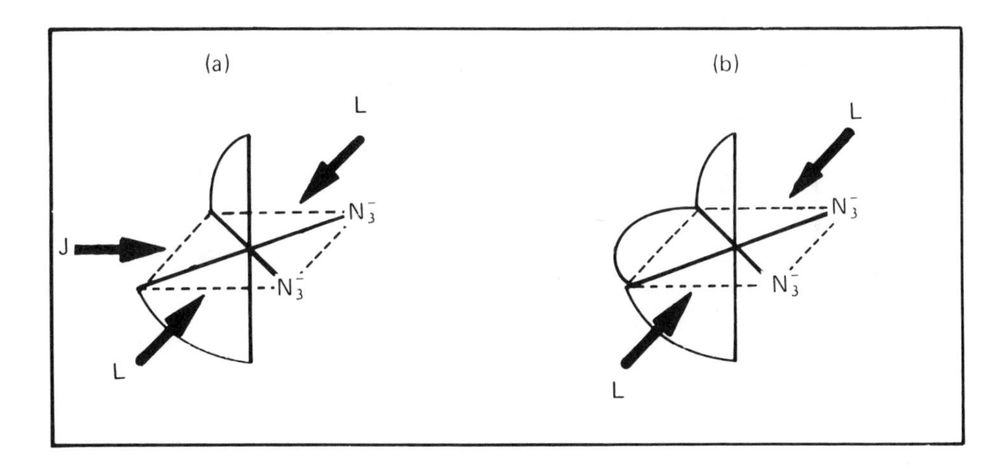

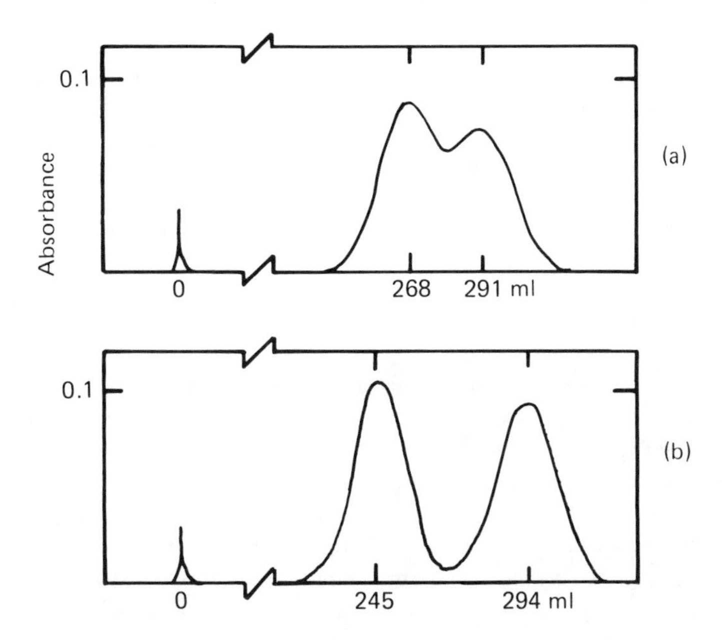

FIGURE 25. Preferential approach directions of $[Sb_2(d\text{-tart})_2]^{2-}$ to *cis*-$[Co(N_3)_2(en)_2]^+$ (a) and *cis*-α-$[Co(N_3)_2(trien)]^+$ (b), and typical chromatograms of these complexes on SP-Sephadex® C-25 packing in 0.075 *M* $K_2Sb_2(d\text{-tart})_2$ aqueous solution: column, 98 × 1.1-cm glass bore, flow rate 0.5 to 0.7 mℓ/min. (From Yamazaki, S. and Yoneda, H., *J. Chromatogr.*, 177, 227, 1979. With permission.)

complexes in the presence of antimony *d*-tartrate or arsenic sodium *d*-tartrate in the buffer solution.[302,303]

VII. SEPARATION OF ENANTIOMERS BY GAS CHROMATOGRAPHY

Though the main principle of LEC, retardation of solutes in the form of mixed ligand sorption complexes, is valid both for liquid chromatography in all its versions and for gas chromatography, the latter technique cannot employ the same coordination systems. In liquid chromatography, solvent molecules or specially added components of the mobile phase actively compete with the solute species for vacant positions in the coordination sphere of the complexing metal ion. A real ligand exchange thus proceeds in the moving chromatographic zone of each solute. On the contrary, nitrogen, helium, or other inert gases used in

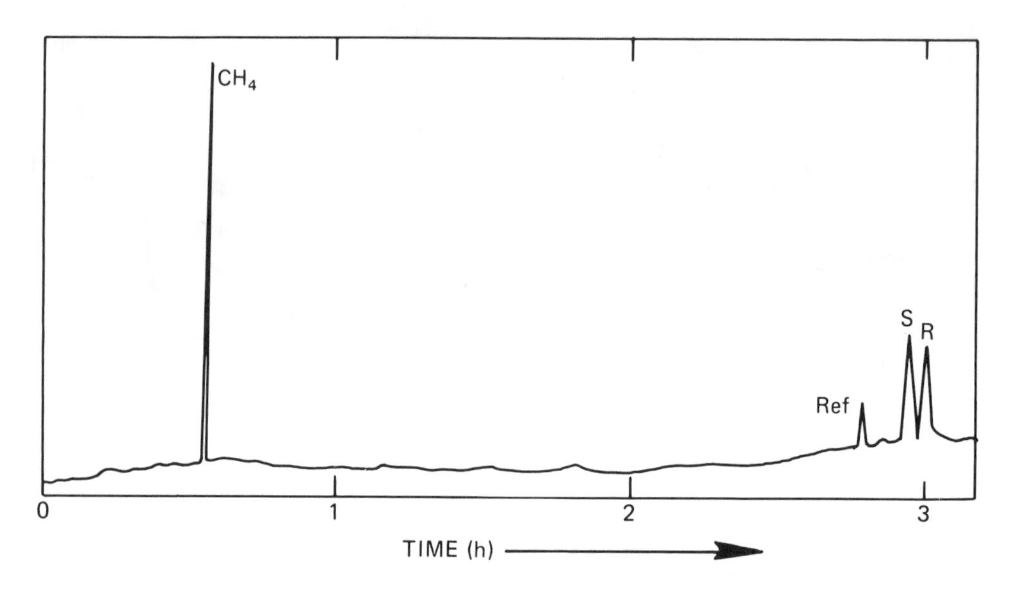

FIGURE 26. Resolution of R, *S*-3-methylcyclopentene on dicarbonyl-Rh(I)-trifluoroacetyl-(1R)-camphorate dissolved in squalane (0.15 *M*). Column, 200 m × 0.5 mm; temperature, 22°C; α = 1.025. (From Schurig, V., *Angew. Chem.*, 89, 113, 1977. With permission.)

gas chromatography as mobile phases are not able to displace ligands from their sorption complexes. Therefore, only extremely labile complexes that easily dissociate at moderate temperatures can be employed in gas chromatographic systems. The central metal ion extends and then reduces its coordination number, and some reversible sterical rearrangements of its coordination sphere occur on binding and releasing the mobile ligand. Sterically overloaded complexes of rhodium(I) and manganese(II) proved especially useful in such reactions. Another possibility, of course, is to bind the mobile ligand in the outer coordination sphere of stable complexes.

In 1977, Schurig and Gil-Av[304-306] for the first time resolved into enantiomers 3-methylcyclopentene using a capillary gas chromatographic column which contained a chiral rhodium(I) complex dissolved in squalane (Figure 26). The resolution was ascribed to formation of labile mixed-ligand complexes comprising the chiral resolving agent, dicarbonyl-Rh(I)-trifluoroacetyl-(1R)-camphorate, and the π-bonded olefin:

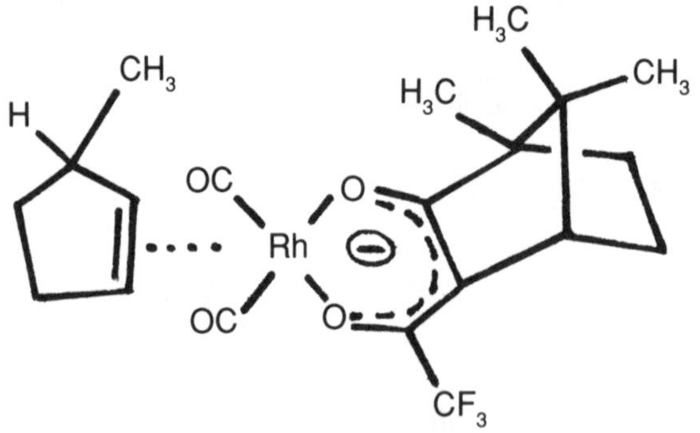

Under the same experimental conditions, 3-ethylcyclopentene was also observed to resolve into enantiomers,[305,307] but 3-methyl-1-pentene, 1,4- and 1,5-dimethylcyclopentene, and 3- and 4-methylcyclohexene were not.[307]

Asymmetrical epoxydes, first of all 1,2-epoxypropane and *trans*-2,3-epoxybutane, were found to be readily resolved on europium(III)-*tris*(3-trifluoroacetyl-1R-camphorate)[308] and nickel(II)-*bis*(3-perfluoroacyl-1R-camphorate):[307,309]

1s   2s, 3s   R = CF_3

α = 1.24   α = 1.43 C_3F_7

The above represented *S*-enantiomers of oxiranes are retained longer on the chiral nickel complex than the corresponding R-enantiomers. This empirical stereochemical quadrant rule also holds for analogous heterocycles with *S* and NH atoms instead of oxygen[310] (a remarkable exception to the quadrant rule was later found for *trans*-2,3-dimethylthiirane).

The above method of analyzing the enantiomeric composition of oxiranes and assigning their absolute configuration made it possible to follow the enzyme-catalyzed aerobic epoxidation of prochiral olefins to chiral oxirans by liver microsomes of rat and the subsequent hydration of oxiranes to 1,2-diols. The testing was performed on the nanogram-level scale by head-space analysis of incubation mixtures without interruption of the reaction.[311] The highest enantiospecificity was observed in the epoxidation of olefins with terminal double bonds and with the smallest degree of substitution of the double bond. The (*S*)-enantiomer was always formed preferably. However, the (*S*)-isomers were enantioselectively hydrolyzed to diols by the epoxy hydrolase at a higher rate than were the (R)-isomers.

Tricyclic acetal lineatin, 3,3,7-trimethyl-2,9-dioxatricyclo[3·3·2·0]nonane, contains four asymmetric bridge-head carbon atoms interlinked in a tricyclic structure (**A**). It has been possible to resolve this compound into enantiomeric and diastereomeric pairs using copper(II)-*bis*(3-heptafluorobutyryl-1R-camphorate).[312,313] Similarly, chiral spiroketal (**B**), 2-ethyl-1,6-dioxaspiro[4·4]nonane, which is the aggregation pheromone of the forest pest *Pityogenes chalcographus* was quantitatively resolved into four individual peaks on a capillary column[314,315] coated with nickel(II)-*bis*(6-heptafluorobutyryl-R-pulegonate) (**C**). The latter optically active stationary phase combines the advantage of considerable enantiomeric differentiation with short retention times.

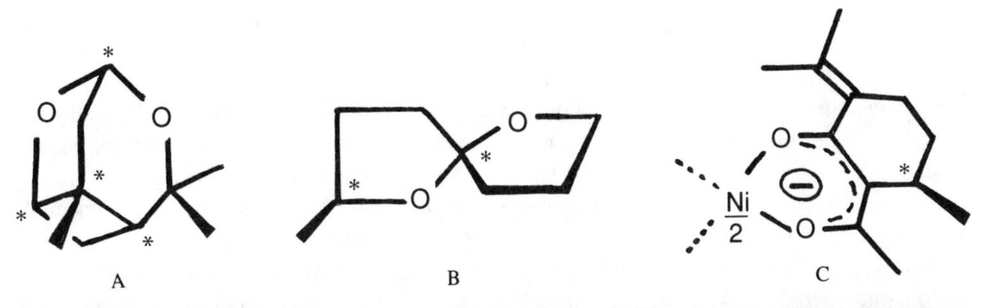

However, the preferred resolving complex appears to be[316] manganese(II)-*bis*(3-perfluoroacyl-1R-camphorate). In regard to cyclic ethers, it may show a 50-fold weaker interaction

Table 18
SELECTIVITY, α, OF ENANTIOMER SEPARATION FOR RACEMIC CYCLIC ETHERS ON FOUR DIFFERENT 3-PERFLUOROACYL-1R-CAMPHORATE-*BIS*-CHELATES OF MANGANESE(II), NICKEL(II), AND COBALT(II) IN SQUALANE AT 60°C[316,320]

Solute	Mn^{II} PFB	Mn^{II} TFA	Mn^{II} HFB	Mn^{II} PFO	Ni^{II} HFB	Co^{II} HFB
Methyloxirane	NR	1.02	1.08	1.06	1.19	1.18
Ethyloxirane	1.03	1.04	1.15	1.11	1.08	1.17
Isopropyloxirane	1.03	1.04	1.17	1.17	NR	1.13
sec.-Butyloxirane (erythro)	1.02	—	1.07	—	—	—
sec.-Butyloxirane (threo)	1.02	—	1.19	—	—	—
tert.-Butyloxirane	1.05	1.04	1.15	1.16	1.11	1.11
2-Ethyl-2-methyloxirane	NR	NR	1.02	1.01	1.05	NR
trans-2,3-Dimethyloxirane	1.05	1.06	1.30	1.25	1.32	1.24
cis-2-Ethyl-3-methyloxirane	1.02	1.03	1.04	1.02	—	—
Trimethyloxirane	1.02	1.03	1.20	1.17	1.30	1.12
2-Methyloxetane	1.13	1.02	1.11	1.11	—[a]	1.11
2-Methyltetrahydrofuran	1.09	1.01	1.04	1.03	1.06	1.01
3-Methyltetrahydrofuran	1.06	1.01	1.02	NR	—	—
trans-2,5-Dimethyltetrahydrofuran	1.01	NR	1.02	NR	1.18	NR
2-Methyltetrahydropyran	1.02	NR	NR	NR	1.05	NR
2-Ethyltetrahydropyran	1.01	NR	NR	NR	—	—
2-Isopropyltetrahdropyran	NR	NR	NR	NR	—	—

Note: PFB = pentafluorobenzoyl, TFA = trifluoroacetyl, HFB = heptafluorobutyryl, PFO = perfluorooctanoyl, NR = not resolved.

[a] Exceedingly long retention time.

and, consequently, much shorter retention as compared with corresponding nickel complexes, while displaying at the same time enantioselectivity of equal order of magnitude (Table 18). Many oxygen-containing three- to six-membered heterocycles (oxirane, oxetane, tetrahydrofurane, and tetrahydropyrane) can be resolved into enantiomeric pairs using this chiral complex. A typical chromatogram is shown in Figure 27. Worth mentioning is quantitation of four components in a mixture of exo- and endo-brevicomin,[317] which are produced in different proportions by males and females of several pine beetles.

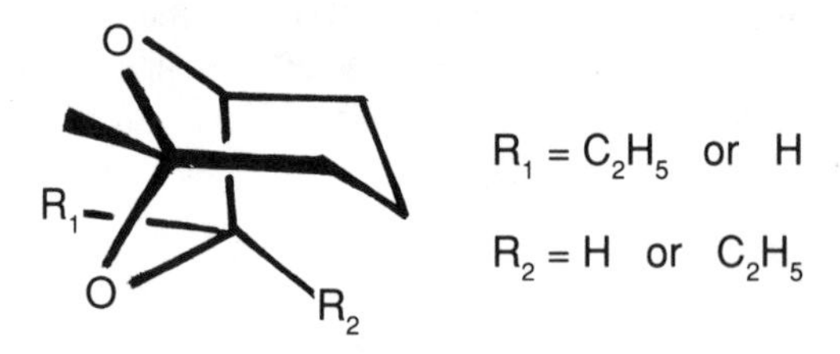

Response of insects to racemic brevicomin varies from rejection to attraction depending on the pheromone release rate and population. This observation underlines the fact that related insects often achieve species selectivity of chemical communication through differential productions and/or response to enantiomers. Enantiomeric analysis in such cases is especially important, as it also is in stereodifferentiated organic synthesis.[318,319]

Noteworthy, though not well understood, is the extremely strong coordination interaction of four-membered oxetanes with the metal chelates, with the overall coordination ability of methyl-substituted cyclic ethers increasing in the order:

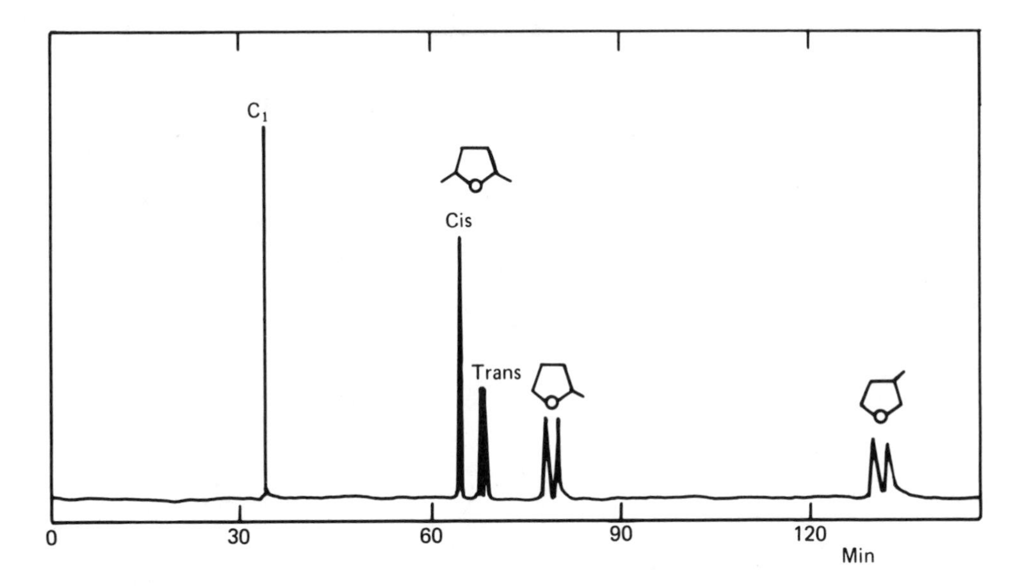

FIGURE 27. Separation of diatereomeric and enantiomeric alkyl substituted tetrahydroufuranes; 0.05 *M* manganese(II)-*bis*-(3-heptafluorobutyryl-1R-camphorate) in squalane. Column, stainless steel, 160 m × 0.4 mm; temperature, 60°C. (From Schurig, V. and Weber, R., *J. Chromatogr.*, 217, 51, 1981. With permission.)

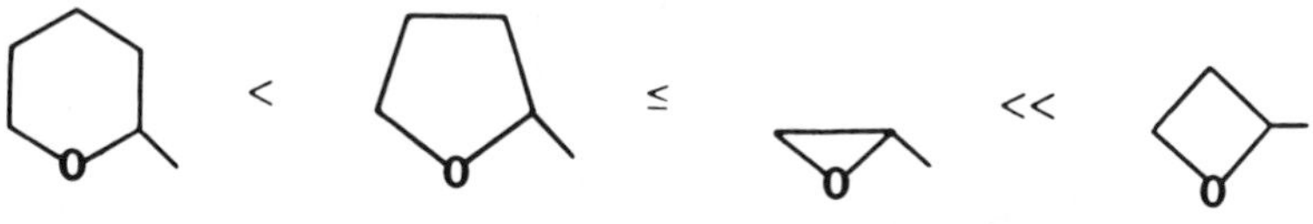

For a given cyclic ether, retention increases dramatically in the order manganese(II) $<$ cobalt(II) $\ll$ nickel(II). Hence, the strongly coordinating nickel chelates are recommended for enantiomer resolutions of weak racemic donor molecules, tetrahydropyranes, and tetrahydrofuranes. Increasing alkyl substitution at the heterocycle carbon atoms lead to a decrease of interaction with metal chelates, but, unexpectedly, the efficiency of chiral recognition is improved. Curiously, the increase of branching in the alkyl group does not necessarily enhance the enantiomeric resolution.[320]

Enantiomer resolution by ligand-exchange gas chromatography is not only limited to simple cyclic ethers. The scope of the method has been successfully extended on resolution of a chlorinated oxirane (i.e., epichlorohydrin, **A**), an underivatized cyclic ketone (i.e., *trans*-2,5-dimethylcyclopentanone, **B**), an underivatized aliphatic alcohol (i.e., *tert*-butylmethylcarbinol, **C**), a mixture of menthone, isomenthone and menthol (Figure 28), and aziridine (i.e., 1-chloro-2,2-dimethylaziridine, **D**).

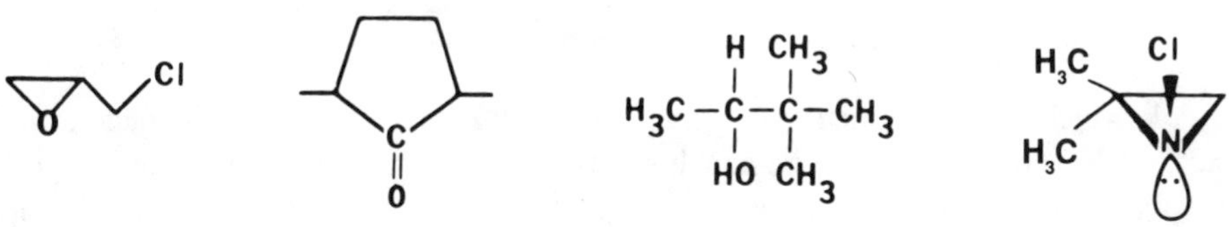

A B C D

These compounds have been quantitatively resolved for the first time on nickel(II)-*bis*(3-heptafluorobutyryl-1R-camphorate).[320] The latter racemate, 1-chloro-2,2-dimethylaziridine, is especially interesting in that the nitrogen atom constitutes the sole chiral center in the

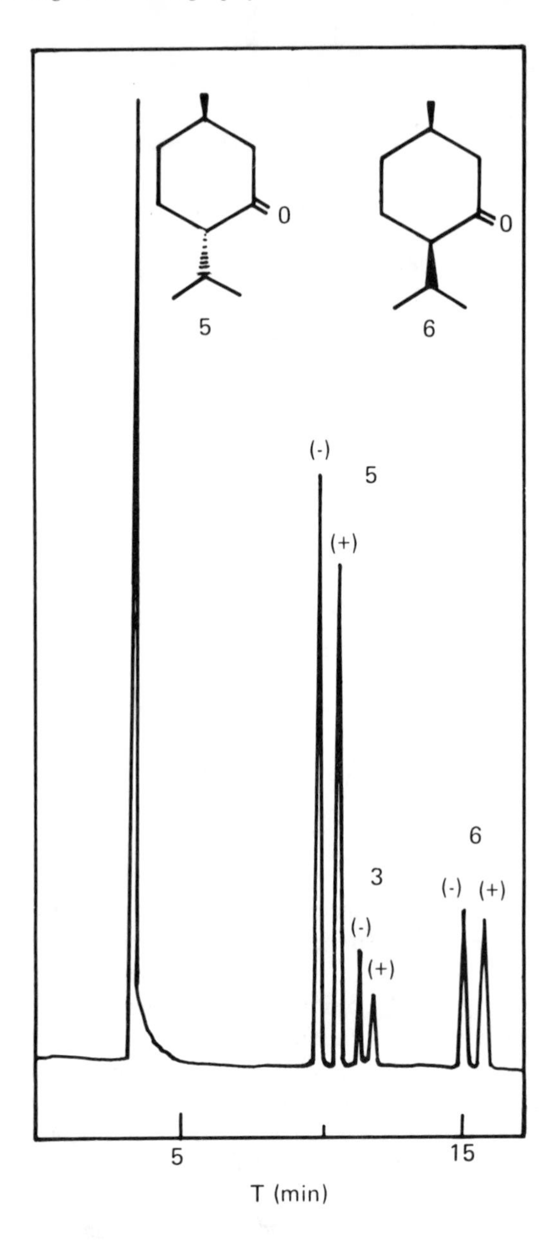

FIGURE 28. Enantiomeric resolution of menthone 5, menthol 3, and isomenthone 6 on nickel(II)*bis*(3-heptafluorobutyryl-1R-camphorate) (0.08 *M* in the methyl silicone fluid OV-101). Column, Duran® glass capillary 37 m × 0.75 mm; temperature, 133°C; carrier gas, 1.0 bar N_2. (From Schurig, V., *Angew. Chem. Int. Ed. Engl.*, 22, 772, 1983. With permission.)

molecule which, in this particular constrained-ring structure, is sufficiently stable to inversion.[320,321] Nevertheless, as can be concluded from Figure 29, partial inversion of the configuration of the chiral nitrogen atom takes place during the process of chromatography, provided 3-trifluoroacetyl-1R-camphorate are the chiral ligands in the resolving nickel complex. Obviously, this complex catalyzes racemization of the 1-chloro-2,2-dimethylaziridine under given experimental conditions. The inversion process manifests itself in that the chromatographic elution curve between the terminal enantiomeric peaks does not approach the zero base line but forms a plateau caused by molecules that have inverted their config-

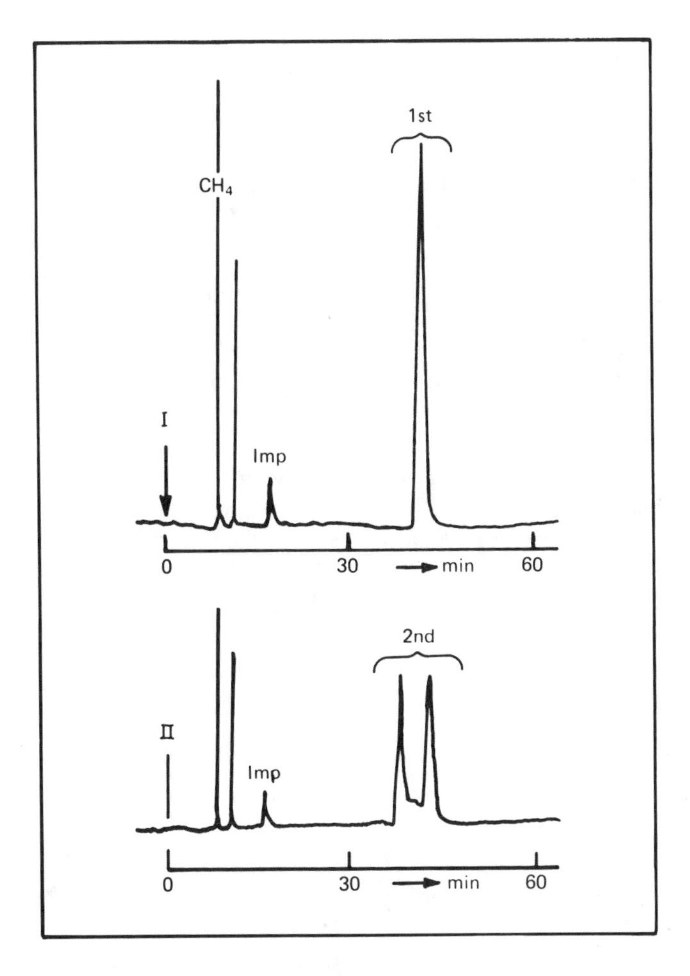

FIGURE 29. Coalescence phenomena for 1-chloro-2,3-dimethyl-aziridine on nickel(II)-*bis*-(3-trifluoroacetyl-camphorate) (0.13 *M* in squalane) at 60°C. Column, 100 m × 0.5 mm nickel capillary; carrier gas, nitrogen at a flow rate of 2.4 mℓ/min. Peak coalescence (first kind) on a racemic stationary phase (top); Peak coalescence (second kind) during resolution on the optically active (1R) stationary phase (bottom). (From Schurig, V. and Burkle, W., *J. Am. Chem. Soc.*, 104, 7573, 1982. With permission.)

uration during resolution and that now travel with the speed of the antipode (Figure 29,II). It is evident that the shape of the plateau should depend on the rate of inversion, and indeed, kinetic parameters of this process are accessible by peak-shape analysis.[322] This type of dynamic peak coalescence (of the ''second'' kind) has nothing in common with the trivial, ''first'' kind of peak coalescence (Figure 29,I) which is observed when the optically active stationary phase is replaced by the racemic one. In this case, the two enantiomeric fractions coalesce to one sharp peak positioned in the middle between two enantiomeric peaks (not shown). (By the way, a chromatographic test on a racemic stationary phase is the method of choice to distinguish a true enantiomeric resolution from the separation of diastereomers with an accidental peak ratio of 1:1.)

The following racemic thiiranes and thiethanes

have been quantitatively resolved by complexation gas chromatography on optically active nickel(II)-*bis*(3-heptafluorobutyryl-1R-camphorate) and the absolute configurations of individual enantiomers have been assigned.[320] Interestingly, the elution order of *trans*-2,3-dimethylthiirane enantiomers was found to be opposite to that of enantiomers of *trans*-2,3-dimethyloxirane. Besides this, alkylated thiiranes seem to resolve less readily than the corresponding epoxides.

More recently,[323] it has been suggested to resolve monoalkylsubstituted 1,2-diols in the form of *n*-butylboronates and 1,2-, 2,3-, 1,3-, and 1,4-diols in the form of acetals, acetonides, or ketals. There is a consistent relationship between the absolute configuration of the diols and the order of elution. Thus, the last eluting 1,2-diol-*n*-boronate enantiomer has configuration R on nickel(II) chelates obtained from 1R-camphor, for instance:

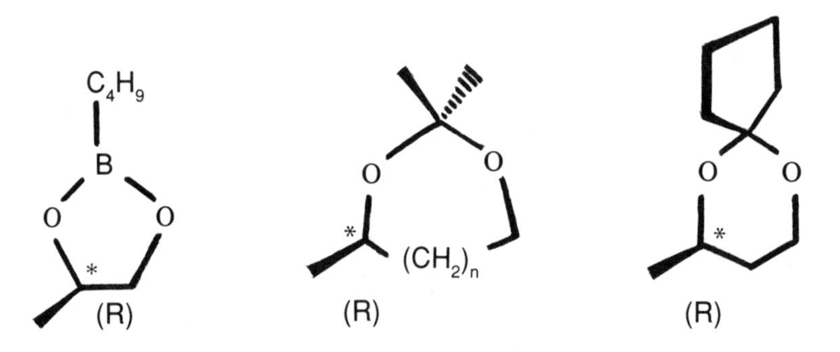

Great progress in the derivatization free enantiomer separation of chiral alcohols and ketones, which are highly polar compounds, has been achieved after special methods of coating glass and fused silica capillary columns with resolving metal chelates had been developed. The above polar compounds strongly interact with metal surfaces of nickel or stainless-steel capillary columns, which causes long retention times and poor column efficiency. On the other hand, the surface silanol groups of glass columns would block the coorination sphere of the chiral metal chelates. Therefore, the preparation procedure of glass and fused silica open tubular columns[324,325] includes acid leaching, hydrothermal treatment, and deactivation of silanol groups by thermal polymerization of a dynamically coated Carbowax® 20 M film. This last step makes the silica surface totally inert and amenable to smooth coating with appropriate resolving metal chelate solution in a silicon liquid OV-101 with coating efficiency higher than 80%.

The use of glass and fused silica capillary columns coated with manganese(II)*bis*[3-heptafluorobutyryl-(1R)-camphorate] or nickel(II)*bis*[3-heptafluorobutyryl-(1R)-camphorate] in polysilicone OV-101 permitted Schurig et al.[325-327] to screen enantiomeric composition and absolute configuration of terpinene-4-ol, the aggregation pheromone of *Polygraphus poligraphus,* and directly separate structural, configurational, and optical isomers of a series of *sec*-methylpentanols and *sec*-methylhexanols as well as corresponding chiral aliphatic ketones.

A new series of chiral resolving complexes for the ligand exchange gas chromatography of various hydroxy group-containing compounds have been introduced by the Japanese team of Oi and co-workers.[328-331] These are binuclear copper(II) complexes of *N*-salicyliden-(R)-2-amino-1,1-*bis*-(5-*tert*-butyl-2-octyloxyphenyl)-propan-1-ol (**A**), *N*-salicyliden-(*S*)-2-amino-1,1-*bis*-(5-*tert*-butyl-2-heptyloxyphenyl)-3-phenyl-propan-1-ol (**B**), and *N*-salicyliden-(*S*)-2-amino-1,1-diphenylpropan-1-ol (**C**).

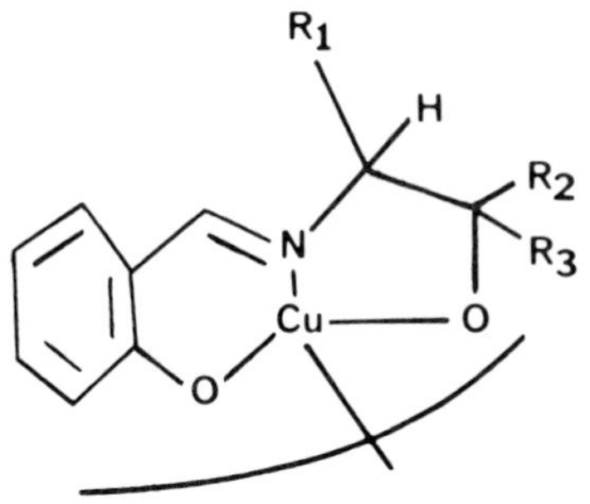

(A) $R_1 = CH_3$

$R_2 =$ tert C_4H_9–(phenylene)–O–n–C_8H_{17}

(B) $R_1 = CH_2-C_6H_5$

$R_2 =$ tert C_4H_9–(phenylene)–O–n–C_7H_{15}

(C) $R_1 = CH_3$

$R_2 = C_6H_5$

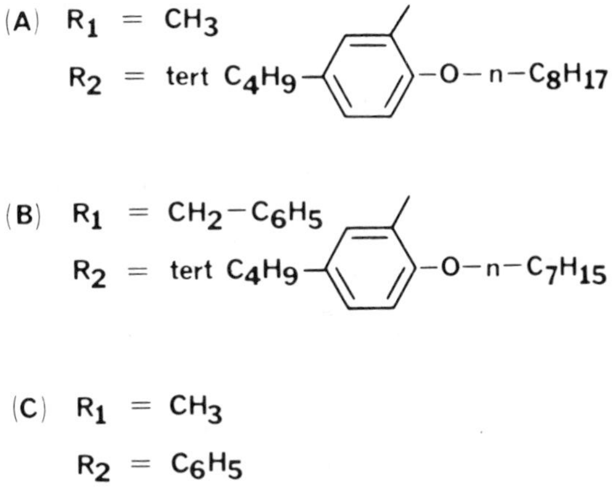

Glass capillary columns (40 m × 0.25 mm) coated with the 1:1 mixtures of silicone OV-101 and the above chiral copper complexes proved able to resolve at 60 to 120°C racemates of some α-hydroxypropionic and α-hydroxybutyric acid esters,[328,329] alcohols, amine, amino alcohols, and amino esters.[330,331] Though on this first stage of development the peak tailing was observed to be quite severe, enantioselectivity values, α, were relatively high (Table 19) permitting occasional separations to be performed on packed columns,[308,328,330] which opens perspectives for preparative applications of the ligand-exchange gas chromatography.

Ligand exchange, in its gas chromatographic version, represents herewith a new approach to gas chromatographic resolution of optical isomers, which cannot be replaced by any other technique at the moment (see References 9 and 332).

Table 19
GAS CHROMATOGRAPHIC RESOLUTION OF RACEMATES ON COPPER(II) CHELATES WITH CHIRAL SCHIFF BASES, A, B, AND C, IN SILICONE OV-101[328,330]

Compound	Stat. phase	Temp. (°C)	Ret. time (min.)		α
Lactic acid esters $CH_3CH(OH)COOR$					
R = CH_3	A	70	3.6	4.2	1.17
C_2H_5	A	70	6.3	7.1	1.13
iso-C_3H_7	A	70	7.3	8.5	1.16
	B	100	2.4	3.2	1.33
iso-C_4H_9	A	70	21.9	24.8	1.13
	B	110	4.3	5.8	1.35
α-Hydroxybutyric acid iso-C_3H_7 ester	'A	70	15.7	16.8	1.07
α-Amino acid isopropyl esters:					
Alanine	A	70	96.0	111.6	1.16
	C	80	25.9	28.5	1.10
Aminoisobutyric acid	A	70	74.6	91.6	1.23
Valine	A	70	91.6	105.2	1.15
Tetrahydrofurfuryl alcohol	A[a]	60	3.58	3.84	1.07
α-Phenylpropargyl alcohol	A[a]	80	23.06	25.59	1.11
2-Ethylpiperidine	A[a]	60	37.39	43.45	1.16
	C[a]	60	60.00	78.4	1.31
1-Phenylethylamine	A[a]	80	35.80	37.3	1.04
	C[a]	120	10.40	11.0	1.06
Dimethylaminopropanol-2	A[a]	20	2.38	5.94	2.50
	C[a]	60	2.63	4.88	1.86
Dibutylaminopropanol-2	A[a]	60	42.80	46.6	1.10

[a] Glass capillary 20 m × 0.25 mm, helium at a flow rate 1.2 mℓ/min; in all other cases column 40 m × 0.25 mm, flow rate 0.6 mℓ/min.

REFERENCES

1. **Davankov, V. A.,** Synthesis and Investigation of Dissymmetric Ion-Exchange Resins, Institute of Organo-Element Compounds, Ph.D. thesis, Moscow, 1966.
2. **Rogozhin, S. V., Davankov, V. A., Korshak, V. V., Vesa, V., and Belchich, L. A.,** Dissymmetric ion exchange resins based on α-amino acids and their derivatives, *Izv. Akad. Nauk S.S.S.R. Ser. Khim.*, p. 502, 1971.
3. **Davankov, V. A., Rogozhin, S. V., Semechkin, A. V., and Sachkova, T. P.,** Ligand exchange chromatography of racemates. I. Influence of the degree of saturation of the asymmetric resin by metal ions on ligand exchange, *J. Chromatogr.*, 82, 359, 1973.
4. **Eliel, E. L.,** *Stereochemistry of Carbon Compounds,* McGraw-Hill, New York, 1962, 51.
5. **Beitler, U. and Feibush, B.,** Interaction between asymmetric solutes and solvents — diamides derived from L-valine as stationary phases in gas-liquid partition chromatography, *J. Chromatogr.*, 123, 149, 1976.
6. **Rogozhin, S. V. and Davankov, V. A.,** Chromatographic resolution of racemates on dissymmetric sorbents, *Usp. Khim.*, 37, 1327, 1968.
7. **Buss, D. R. and Vermeulen, T.,** Optical isomer separation, *Ind. Eng. Chem.*, 60, 12, 1968.
8. **Rogozhin, S. V., Davankov, V. A., and Piesliakas, I. I.,** Recent advances in the field of chromatographic resolution of racemates, in *Chemistry and Technology of High Molecular Compounds,* Vol. 4, Sladkov, A. M., Ed., VINITI, Moscow, 1973, 45.
9. **Lochmüller, C. H. and Souter, R. W.,** Chromatographic resolution of enantiomers (selective review), *J. Chromatogr.*, 113, 283, 1975.

10. **Krull, I. S.,** The liquid chromatographic resolution of enantiomers, in *Advances in Chromatography,* Vol.16, Giddings, J. C., Grushka, E., Cazes, J., and Brown, P. R., Eds., Marcel Dekker, New York, 1978, 175.
11. **Audebert, R.,** Direct resolution of enantiomers in column liquid chromatography, *J. Liq. Chromatogr.,* 2, 1063, 1979.
12. **Blaschke, G.,** Chromatographische Racemattrennung, *Angew, Chem.,* 92, 14, 1980; *Angew. Chem. Int. Ed. Engl.,* 19, 13, 1980.
13. **Lindner, W.,** Trennung von Enantiomeren mittels moderner Flüssigkeits-Chromatographie, *Chimia,* 35, 294, 1981.
14. **Allenmark, S.,** Recent advances in methods of direct optical resolution, *J. Biochem, Biophys. Methods,* 9, 1, 1984.
15. **Davankov, V. A.,** Resolution of racemates by ligand exchange chromatography, in *Advances in Chromatography,* Vol. 18, Giddings, J. C., Grushka, E., Cazes, J., and Brown, P. R., Eds., Marcel Dekker, New York, 1980, 139.
16. **Davankov, V. A., Kurganov, A. A., and Bochkov, A. S.,** Resolution of racemates by high-performance liquid chromatography, in *Advances in Chromatography,* Vol. 22, Giddings, J. C., Grushka, E., Cazes, J., and Brown, P. R., Eds., Marcel Dekker, New York, 1983, 71.
17. **Davankov, V. A.,** Review of ligand exchange chromatography, in *Handbook of HPLC for the Separation of Amino Acids, Peptides, and Proteins,* Vol. 1, Hancock, W. S., Ed., CRC Press, Boca Raton, Fla., 1984, 393.
18. **Caude, M. H., Jardy, A. P., and Rosset, R. H.,** Separations where the ligand is bound to a normal-phase column, in *Handbook of HPLC for the Separation of Amino Acids, Peptides, and Proteins,* Vol. 1, Hancock, W. S., Ed., CRC Press, Boca Raton, Fla., 1984, 411.
19. **Sugden, K.,** Separations where the ligand is bound to a reversed phase system, in *Handbook of HPLC for the Separation of Amino Acids, Peptides, and Proteins,* Vol. 1, Hancock, W. S., Ed., CRC Press, Boca Raton, Fla., 1984, 423.
20. **Gil-Av, E. and Weinstein, S.,** Resolution of α-amino acids and dns-α-amino acids by HPLC with mobile phases containing a chiral ligand, in *Handbook of HPLC for the Separation of Amino Acids, Peptides, and Proteins,* Vol. 1, Hancock, W. S., Ed., CRC Press, Boca Raton, Fla., 1984, 429.
21. **Davankov, V. A.,** Chiral chelating resins in chromatography of optical isomers, *Pure Appl. Chem.,* 54, 2159, 1982.
22. **Rogozhin, S. V. and Davankov, V. A.,** On the properties of halogenmethylated copolymers of styrene with divinylbenzene, *Vysokomol. Soedin.,* 9A, 1286, 1967.
23. **Rogozhin, S. V., Korshak, V. V., Davankov, V. A., and Maslova, L. A.,** Halogenmethylation of copolymers of styrene with divinylbenzene, *Vysokomol. Soedin.,* 8, 1275, 1966.
24. **Davankov, V. A., Rogozhin, S. V., Korshak, V. V., and Tsyurupa, M. P.,** Reactivity of halogen-methylated copolymers of styrene with divinylbenzene, *Isv. Akad. Nauk S.S.S.R. Ser. Khim.,* p. 1612, 1967.
25. **Rogozhin, S. V., Davankov, V. A., and Korshak, V. V.,** Iodomethylation of aromatic compounds, *Isv. Akad. Nauk S.S.S.R. Ser. Khim.,* p. 1498, 1966.
26. **Korshak, V. V., Rogozhin, S. V., and Davankov, V. A.,** Halogen exchange in chloromethylated styrene-type copolymers, *Isv. Akad. Nauk S.S.S.R. Ser. Khim.,* p. 1912, 1965.
27. **Korshak, V. V., Rogozhin, S. V., and Davankov, V. A.,** Synthesis of bromo- and iodemethylated copolymers of styrene with divinylbenzene, *Vysokomol. Soedin;* 8, 1686, 1966; *Polym. Sci. U.S.S.R.,* 8, 1860, 1966.
28. **Rogozhin, S. V., Davankov, V. A., and Korshak, V. V.,** Synthesis and properties of ion exchangers based on halogenmethylated copolymers of styrene with divinylbenzene and C-derivatives of α-amino acids, *Vysokomol. Soedin.,* 10A, 1283, 1968.
29. **Rogozhin, S. V., Davankov, V. A., Vyrbanov, S. G., and Korshak, V. V.,** Synthesis and properties of ion exchangers based on α-amino acids and halogenmethylated copolymers of styrene with divinylbenzene, *Vysokomol. Soedin.,* 10A, 1277, 1968.
30. **Davankov, V. A., Rogozhin, S. V., and Piesliakas, I. I.,** Synthesis and properties of ion exchangers based on neutral and hydroxyl containing α-amino acids and chloromethylated macronet isoporous styrene copolymers, *Vysokomol. Soedin.,* 14B, 276, 1972.
31. **Davankov, V. A., Rogozhin, S. V., and Tsyurupa, M. P.,** A Method of Obtaining Macronet Copolymers of Styrene, U.S.A. Patent 3,729,457, 1970; Appl. U.S.S.R., 1969; *Chem. Abstr.,* 75, 6841v, 1971.
32. **Davankov, V. A., Rogozhin, S. V., and Tsyurupa, M. P.,** Influence of polymeric matrix structure on performance of ion exchange resins, in *Ion Exchange and Solvent Extraction,* Vol. 7, Marinsky, J. A. and Marcus, Y., Eds., Marcel Dekker, New York, 1977, 29.
33. **Davankov, V. A. and Tsyurupa, M. P.,** Macronet isoporous styrene copolymers — unusual structure and properties, *Angew. Makromol. Chem.,* 91, 127, 1980.

34. **Rosenberg, G. I., Shabaeva, A. S., Moryakov, V. S., Musin, T. G., Tsyurupa, M. P., and Davankov, V. A.,** Sorption properties of hypercross-linked polystyrene sorbents, *React. Polym.*, 1, 175, 1983.
35. **Davankov, V. A., Rogozhin, S. V., and Tsyurupa, M. P.,** New approach to obtaining evenly cross-linked macronet polystyrene structures, *Vysokomol. Soedin.*, 15B, 463, 1973.
36. **Davankov, V. A., Rogozhin, S. V., and Tsyurupa, M. P.,** Macronet isoporous gels through cross-linking of dissolved polystyrene, *J. Polym. Sci. Polym. Symp.*, 47, 95, 1974.
37. **Tsyurupa, M. P., Davankov, V. A., and Rogozhin, S. V.,** Macronet isoporous ion exchange resins, *J. Polym. Sci. Polym. Symp.*, 47, 189, 1974.
38. **Davankov, V. A., Rogozhin, S. V., and Tsyurupa, M. P.,** Uber Faktoren, die das Quellvermögen von vernetzten Polymeren bestimmen. I, *Angew. Makromol. Chem.*, 32, 145, 1973.
39. **Davankov, V. A., Tsyurupa, M. P., and Rogozhin, S. V.,** On factors determining the swelling ability of cross-linked polymers. II, *Angew. Makromol. Chem.*, 53, 19, 1976.
40. **Davankov, V. A.,** Ligand Exchange Chromatography of Racemates, D.Sc. thesis, Institute of Organo Element Compounds, Moscow, 1975.
41. **Davankov, V. A., Rogozhin, S. V., Piesliakas, I. I., and Vesa, V. S.,** Use of α-amino acid *t*-butyl esters in the synthesis of dissymmetric complex forming ion exchangers, *Vysokomol. Soedin.*, 15B, 115, 1973.
42. **Rogozhin, S. V., Yamskov, I. A., Davankov, V. A., Kolesova, T. F., and Voevodin, V. M.,** Ligand chromatography of racemates on dissymmetric sorbents with L-aspartic and L-glutamic acids, *Vysokomol. Soedin.*, 17A, 564, 1975.
43. **Rogozhin, S. V., Yamskov, I. A., and Davankov, V. A.,** Resolution of racemates on the dissymmetric sorbent with L-α,γ-diaminobutyric acid, *Vysokomol. Soedin.*, 17B, 107, 1975.
44. **Rogozhin, S. V., Yamskov, I. A., and Davankov, V. A.,** Synthesis of dissymmetric sorbents by reacting copper complexes of diamino carboxylic acids with poly-*p*-vinylbenzyl-dimethylsulfonium chloride, *Vysokomol. Soedin.*, 16B, 849, 1974.
45. **Rogozhin, S. V., Davankov, V. A., and Yamskov, I. A.,** Synthesis of chelate forming ion exchangers by interaction of α-amino acids with poly-*p*-vinylbenzyldimethylsulfonium chloride, *Vysokomol. Soedin.*, 15B, 216, 1973.
46. **Rogozhin, S. V., Davankov, V. A., and Yamskov, I. A.,** Synthesis and properties of the dissymetric complex forming sorbent with L-histidine, *Isv. Akad. Nauk S.S.S.R. Ser. Khim.*, p. 2325, 1971.
47. **Davankov, V. A., Rogozhin, S. V., Yamskov, I. A., and Kabanov, V. P.,** Dissymmetric complex forming sorbent with D-methionine and its use in chromatography of racemates, *Isv. Akad. Nauk S.S.S.R. Ser. Khim.*, p. 2327, 1971.
48. **Roberts, C. W. and Haigh, D. H.,** Partial amino acid resolutions on a new resolving resin, *J. Org. Chem.*, 27, 3375, 1962.
49. **Rogozhin, S. V., Davankov, V. A., Yamskov, I. A., and Kabanov, V. P.,** Study on the reaction of L-cysteic acid with a chloromethylated styrene copolymer. Chromatography of racemates on the sorbent obtained, *Vysokomol. Soedin.*, 14B, 472, 1972.
50. **Yamskov, I. A., Berezin, B. B., Tikhonov, V. E., Belchich, L. A., and Davankov, V. A.,** Ligand exchange chromatography of amino acid enantiomers on dissymmetric sorbents with groups of L-phenyalanine, *Bioorg. Khim.*, 4, 1170, 1978.
51. **Yamskov, I. A., Berezin, B. B., and Davankov, V. A.,** Ligand exchange chromatography of amino acid enantiomers on asymmetric sorbents with polydentate sulfur containing groupings, *Makromol. Chem.*, 179, 2121, 1978.
52. **Belov, Yu. P., Rogozhin, S. V., and Davankov, V. A.,** Dissymmetric sorbents based on (−)-α-amino benzylphosphonic acid, *Isv. Akad, Nauk S.S.S.R. Ser. Khim.*, p. 2320, 1973.
53. **Belov, Yu. P., Davankov, V. A., and Rogozhin, S. V.,** Sorbents with groups of optically active α-amino ethylphosphonic acids for the ligand exchange chromatography of racemates, *Isv. Akad. Nauk S.S.S.R. Ser. Khim.*, p. 1856, 1977.
54. **Rogozhin, S. V., Davankov, V. A., and Belov, Yu. P.,** Optically active α-amino benzylphosphonic acid diethyl ester, *Isv. Akad. Nauk S.S.S.R. Ser. Khim.*, p. 955, 1973.
55. **Belov, Yu. P., Davankov, V. A., and Rogozhin, S. V.,** Optically active α-amino ethylphosphonic acids and their ethyl esters, *Isv. Akad. Nauk S.S.S.R. Ser. Khim.*, p. 1596, 1976.
56. **Rogozhin, S. V., Davankov V. A., Yamskov, I. A., and Kabanov, V. P.,** Chromatography of racemates on the dissymetric complex-forming sorbents with D-methionine-*dl*-sulfoxide, *J. Gen. Chem. U.S.S.R.*, 42, 1614, 1972.
57. **Berezin, B. B., Yamskov, I. A., and Davankov, V. A.,** Ligand exchange chromatography of amino acid racemates on polystyrene sorbents containing L-methionine-*d*-sulfoxide or L-methionine-*l*-sulfoxide groups, *J. Chromatogr.*, 261, 301, 1983.
58. **Davankov, V. A. and Zolotarev, Yu. A.,** Ligand exchange chromatography of racemates. VI. Separation of optical isomers of amino acids on polystyrene resins containing L-proline or L-azetidine carboxylic acid, *J. Chromatogr.*, 155, 295, 1978.

59. **Rogozhin, S. V., Yamskov, I. A., Pushkin, A. S., Belchich, L. A., Zhuchkova, L. Ya., and Davankov, V. A.,** Dissymmetric sorbents with *N*-carboxymethyl-L-valine and *N*-carboxymethyl-L-aspartic acid, *Isv. Akad. Nauk S.S.S.R. Ser. Khim.*, p. 2378, 1976.
60. **Angelici, R. J., Snyder, R. V., and Meck, R. B.,** Amino acids resolved with metal complexes, *Chem. Eng. News*, 49(15), 34, 1971.
61. **Snyder, R. V., Angelici, R. J., and Meck, R. B.,** Partial resolution of amino acids by column chromatography on a polystyrene resin containing an optically active copper(II) complex, *J. Am. Chem. Soc.*, 94, 2660, 1972.
62. **Petit, M. A. and Josefonvicz, J.,** Synthesis of copper (II) complexes of asymmetric resins prepared by attachment of α amino acids to cross-linked polystyrene, *J. Appl. Polym. Sci.*, 21, 2589, 1977.
63. **Jozefonvicz, J., Petit, M. A., and Szubarga, A.,** Preparative resolution of DL-proline by liquid chromatography on a polystyrene resin containing the L-proline copper(II) complex, *J. Chromatogr.*, 147, 177, 1978.
64. **Spassky, N., Reix, M., Quette, J., Quette, M., Sepulchre, M., and Blanchard, J.,** Dedoublement des acides α-amines par hydrolyse stereoselective de leurs esters catalysee par des complexes chiraux de metaux de transition immobilises sur des supports macromoleculaires, *C. R. Acad. Sci. Ser. C*, 287, 589, 1978.
65. **Tsuchida, E., Nishikawa, H., and Terada, E.,** The resolution of α-amino acid by chiral polymer copper complex, *Eur. Polym. J.*, 12, 611, 1976.
66. **Tsuchida, E. and Nishikawa, H.,** Optical resolution of racemic mixtures, *Jpn. Kokai*, 77, 85101; *Chem, Abstr.*, 88, 23398s, 1978.
67. **Vesa, V. S.,** Resolution of racemic amino acids on a dissymmetric polysulfonamide sorbent, *Zh. Obshch. Khim.*, 42, 2780, 1972.
68. **Vesa, V. S. and Moroshchikas, R. K.,** Dissymmetric ion exchange resins based on α-amino acids and chlorosulfonated copolymers of styrene with divinylbenzene, *Tr. Akad. Nauk Lit. SSR Ser. B*, 2(69), 93, 1972; *Chem. Abstr.*, 77, 115177v, 1972.
69. **Hatano, M., Murakami, T., and Kitagawa, S.,** Optical resolutions of DL-amino acids, *Jpn. Kokai*, 76, 29438, 1976; *Chem. Abstr.* 85, 143513k, 1976.
70. **Hatano, M., Murakami, T., and Kitagawa, S.,** Tetraamines stereospecifically coordinated with metal ions, *Jpn. Kokai*, 76, 34148, 1976; *Chem. Abstr.*, 85, 108541w, 1976.
71. **Kurganov, A. A., Zhuchkova, L. Ya., and Davankov, V. A.,** Ligand exchange chromatography of racemates. X. Asymmetric sorbents based on mono- and diamines and resolution of racemates by ligand exchange chromatography, *Makromol. Chem.*, 180, 2101, 1979.
72. **Kurganov, A. A., Zhuchkova, L. Ya., and Davankov, V. A.,** Synthesis of isomeric *N*-benzyl derivatives of 1,2-propane-diamine, *Liebig's Ann. Chem.*, p. 786, 1980.
73. **Bernauer, K.,** Separating Racemates Using Column Chromatography on Optically Active Metal Complexes, Swiss Patent 509239, 1971; *Chem. Abstr.*, 76, 60602b, 1972.
74. **Lefebvre, B., Audebert, R., and Quivoron, C.,** Direct resolution of amino acid enantiomers by high pressure liquid chromatography, *Isr. J. Chem.*, 15, 69, 1977.
75. **Lefebvre, B., Audebert, R., and Quivoron, C.,** Use of new chiral hydrophilic gels for the direct resolution of α-amino acids by high pressure liquid chromatography, *J. Liq. Chromatogr.*, 1, 761, 1978.
76. **Yamskov, I. A., Berezin, B. B., Davankov, V. A., Zolotarev, Yu. A., Dostovalov, I. N., and Myasoedov, N. F.,** Ligand exchange chromatography of racemates. XVII. Ligand exchange chromatography of amino acid racemates on "Separon" gels containing L-proline and L-hydroxyproline groupings, *J. Chromatogr.*, 217, 539, 1981.
77. **Watanabe, N., Ohzeki, H., and Niki, E.,** Enantiomeric resolution of amino acids by high performance ligand exchange chromatography using a chemically modified hydrophilic porous polymer gel, *J. Chromatogr.*, 216, 406, 1981.
78. **Piesliakas, I. I., Rogozhin, S. V., and Davankov, V. A.,** On the optical stability of dissymmetric sorbents containing residues of α-amino acids, *Zh. Obshch. Khim.*, 44, 468, 1974.
79. **Rogozhin, S. V. and Davankov, V. A.,** A Chromatographic Method of resolving Racemates of Optically Active Compounds, German Patent 1,932,190, 1970; Appl. U.S.S.R., 1968; *Chem. Abstr.*, 72, 90875c, 1970.
80. **Rogozhin, S. V. and Davankov, V. A.,** Ligand chromatography on dissymmetric complex forming sorbents as a new principle of resolving racemates, *Dokl. Akad. Nauk S.S.S.R.*, 192, 1288, 1970.
81. **Davankov, V. A., Rogozhin, S. V., Piesliakas, I. I., Yamskov, I. A., and Semechkin, A. V.,** Ligand chromatography — a new principle of separating optical isomers, in *Ion Exchange and Chromatography*, Part 1, Voronesh, 62, 1971.
82. **Davankov, V. A. and Rogozhin, S. V.,** A new principle of resolving racemates of α-amino acids, *Seventh Int. Symp. on Chem. of Natural Compounds*, Riga, A-32, 1970.
83. **Rogozhin, S. V. and Davankov, V. A.,** Ligand chromatography on assymmetric complex-forming sorbents as a new method for resolution of racemates, *Chem. Commun.*, p. 490, 1971.

84. **Davankov, V. A., Rogozhin, S. V., Piesliakas, I. I., Semechkin, A. V., and Sachkova, T. P.,** Resolution of racemates by ligand chromatography on sorbents based on L-proline and L-hydroxyproline, *Dokl. Akad. Nauk S.S.S.R.*, 201, 854, 1971.
85. **Davankov, V. A., Rogozhin, S. V., and Semechkin, A. V.,** Ligand exchange chromatography of racemates. III. Resolution of α-amino acids, *J. Chromatogr.*, 91, 493, 1974.
86. **Davankov, V. A. and Zolotarev, Yu. A.,** Ligand exchange chromatography of racemates. V. Separation of optical isomers of amino acids on a polystyrene resin containing L-hydroxyproline, *J. Chromatogr.*, 155, 285, 1978.
87. **Davankov, V. A. and Zolotarev, Yu. A.,** Ligand exchange chromatography of racemates. VII. Separation of optical isomers of amino acids on a polystyrene resin containing L-allo-hydroxyproline as the fixed ligand, *J. Chromatogr.*, 155, 303, 1978.
88. **Davankov, V. A., Zolotarev, Yu. A., and Tevlin, A. V.,** Ligand exchange chromatography of racemates. VIII. A quantitative resolution of racemates of amino acids on a polystyrene resin with L-hydroxyproline groups for the analysis of the enantiomeric composition of amino acids, *Bioorg. Khim.*, 4, 1164, 1978.
89. **Davankov, V. A., Zolotarev, Yu. A., and Kurganov, A. A.,** Ligand exchange chromatography of racemates. XI. Complete resolution of some chelating racemic compounds and nature of sorption enantioselectivity, *J. Liquid Chromatogr.*, 2, 1191, 1979.
90. **Piesliakas, I. I., Rogozhin, S. V., and Davankov, V. A.,** Resolution of racemates by ligand exchange chromatography on sorbents based on D-alanine, L-valine, and L-isoleucine, *Isv. Akad. Nauk S.S.S.R. Ser. Khim.*, p. 174, 1974.
91. **Piesliakas, I. I., Rogozhin, S. V., and Davankov, V. A.,** Resolution of racemates by ligand exchange chromatography on sorbents based on L-serine, L-threonine and L-tyrosine, *Isv. Akad. Nauk S.S.S.R. Ser. Khim.*, p. 1872, 1974.
92. **Yamskov, I. A., Rogozhin, S. V., and Davankov, V. A.,** Ligand exchange chromatography of racemates on dissymmeteric sorbents with trifunctional amino acid groups (histidine, methionine, methionine sulfoxide), *Bioorg. Khim.*, 3, 200, 1977.
93. **Davankov, V. A. and Kurganov, A. A.,** Ligand exchange chromatography of racemates. XII. High performance liquid chromatography of α-amino acids on a polystyrene resin with fixed ligands of the type (R)-N^1,N^1-dibenzyl-1,2-propanediamine, *Chromatographia,* 13, 339, 1980.
94. **Shirokov, V. A., Tsyryapkin, V. A., Nedospasova, L. V., Kurganov, A. A., and Davankov, V. A.,** Enantiomeric analysis of proline, valine, phenylalanine, tyrosine, tryptophan, and their esters by ligand exchange chromatography, *Bioorgan. Khim.*, 9, 878, 1983.
95. **Zolotarev, Yu. A., Myasoedov, N. F., Penkina, V. I., Petrenik, O. V., and Davankov, V. A.,** Ligand exchange chromatography of racemates. XIV. Micropreparative resolution of LD-leucine, *J. Chromatogr.*, 207, 63, 1981.
96. **Myasoedov, N. F., Kuznetsova, O. B., Petrenik, O. V., Davankov, V. A., and Zoloatrev, Yu. A.,** Resolution of tritium labelled amino acid racemates by ligand exchange chromatography. I. Method of obtaining L- and D-[^{3}H]valine using a polystyrene resin with L-hydroxyproline groupings, *J. Labeled Compds. Radiopharm.*, 17, 439, 1979.
97. **Zolotarev, Yu. A., Myasoedov, N. F., Penkina, V. I., Dostovalov, I. N., Petrenik, O. V., and Davankov, V. A.,** Resolution of tritium labelled amino acid racemates by ligand exchange chromatography. II. L-hydroxy-proline- and L-phenylalanine-modified resins for the resolution of common α-amino acids, *J. Chromatogr.*, 207, 231, 1981.
98. **Davankov, V. A., Rogozhin, S. V., and Kurganov, A. A.,** The first example of stereoselectivity in copper complexes with bidentate amino acids, *Isv. Akad. Nauk S.S.S.R. Ser. Khim.*, p. 204, 1971.
99. **Davankov, V. A., Rogozhin, S. V., and Kurganov, A. A.,** Stereoselectivity in copper-*bis*-*N*-benzylproline complexes, *Zh. Neorg. Khim.*, 17, 2163, 1972.
100. **Senyukova, G. A., Kurganov, A. A., Nikitaev, A. T., Davankov, V. A., and Zamaraev, K. I.,** EPR-study of the electronic state of the copper atom in the complexes with L- and D-forms of *N*-benzylproline; EPR-study of thermodynamic properties and structure of adducts of methanol with *bis*(*N*-benzylprolinato)copper, *Koordin. Khim.*, 1, 396 and 400, 1975.
101. **Kurganov, A. A., Davankov, V. A., and Zhuchkova, L. Ya.,** Spectrophotometric investigation of the influence of the inner- and outer-sphere coordination of methanol on the enantioselectivity effects in copper(II) complexes with *N*-benzylproline, *Koordin. Khim.*, 3, 988, 1977.
102. **Alexandrov, G. G., Struchkov, Yu. T., Kurganov, A. A., Rogozhin, S. V., and Davankov, V. A.,** Crystal and molecular structure of *bis*(*N*-benzyl-prolinato)-copper(II) complexes, *Zh. Strukt. Khim.* 13, 671, 1972; *Chem. Commun.*, 1328, 1972.
103. **Kurganov, A. A., Zhuchkova, L. Ya., and Davankov, V. A.,** Stereoselectivity in *bis*(α-amino acid) copper(II) complexes. VII. Thermodynamics of *N*-benzylproline coordination to copper(II), *J. Inorg. Nucl. Chem.*, 40, 1081, 1978.

104. **Davankov, V. A. and Mitchell, P. R.,** Stereoselectivity in *bis*(α-amino acid) copper(II) complexes. I. Stability constants from circular dichroism and electronic spectra, *J. Chem. Soc. Dalton Trans.*, p. 1012, 1972.
105. **Muller, D., Jozefonvicz, J., and Petit, M.,** Mise en evidence d'effets stereoselectifs par determination des constantes de formation des complexes cuivriques mixtes associant le derive *N*-benzyl L-proline et l'histidine, la proline ou la phenylalanine, *C. R. Acad. Sci. Ser. C*, 288, 45, 1979.
106. **Jozefonvicz, J., Muller, D., and Petit, M.,** Stereoselectivity in the ternary complexes copper(II)-*N*-benzyl-L-proline-D- or L-α-amino acids, *J. Chem. Soc. Dalton Trans.*, p. 76, 1980.
107. **Muller, D., Petit, M., Szubarga, A., and Jozefonvicz, J.,** Effet stereoselectif du derive *N*-benzyl L-proline dans la formation de complexes metalliques mixtes associant l'histidine, *C. R. Acad. Sci. Ser. C*, 285, 531, 1977.
108. **Davankov, V. A., Rogozhin, S. V., Struchkov, Yu. T., Alexandrov, G. G., and Kurganov, A. A.,** Stereoselectivity in *bis*-(α-amino acid) copper(II) complexes. IV. Structure of Cu(II)-*N*-benzylvaline complexes and the nature of enantioselectivity effects, *J. Inorg. Nucl. Chem.*, 38, 631, 1976.
109. **Leach, B. E. and Angelici, R. J.,** Stereoselective interaction of optically active amino acids and esters with (L-valine-*N*-monoacetato)-copper(II), *J. Am. Chem. Soc.*, 91, 6296, 1969.
110. **Snyder, R. V. and Angelici, R. J.,** Stereoselectivity of *N*-carboxymethyl-amino acid complexes of copper(II) toward optically active amino acids, *J. Inorg. Nucl. Chem.*, 35, 523, 1973.
111. **Muller, D., Jozefonvicz, J., and Petit, M.,** Stereoselective binding of D- or L-α-amino acids by copper(II) complexes of *N*-benzenesulfonyl-L-α-phenylalanine, *J. Inorg. Nucl. Chem.*, 42, 1665, 1980.
112. **Kurganov, A. A., Davankov, V. A., Zhuchkova, L. Ya., and Ponomareva, T. M.,** Copper(II) complexes with optically active diamines. I. Synthesis and properties of copper(II) complexes with *N*-benzyl- and *N*-methyl derivatives of 1,2-diaminopropane, *Inorg. Chim. Acta*, 39, 237, 1980.
113. **Kurganov, A. A., Davankov, V. A., Zhuchkova, L. Ya., and Ponomareva, T. M.,** Copper(II) complexes with optically active diamines. II. The effect of solvent, temperature, and alkyl subtituents at nitrogen atoms on circular dichroism spectra of copper(II) equally paired and mixed ligand complexes with (R)-1,2-diaminopropane, *Inorg. Chim. Acta*, 39, 243, 1980.
114. **Kurganov, A. A., Ponomareva, T. M., and Davankov, V. A.,** Copper(II) complexes with optically active diamines. III. The complexes of 2-(amino-methyl)-pyrrolidine and its benzyl derivatives: synthesis, properties, and circular dichroism spectra, *Inorg. Chim. Acta*, 45, L23, 1980.
115. **Kurganov, A. A., Ponomareva, T. M., and Davankov, V. A.,** Copper(II) complexes with optically active diamines. IV. Between ligand interactions in mixed ligand copper(II) complexes containing (*S*)-2-(amino-methyl)-pyrrolidine, ethylenediamine, 1,2-diaminopropane, and their *N*-benzyl derivatives, *Inorg. Chim. Acta*, 68, 51, 1983.
116. **Kurganov, A. A., Ponomareva, T. M., and Davankov, V. A.,** Copper(II) complexes with optically active diamines. V. Enantioselective effects in equally paired and mixed ligand copper(II) complexes with diamines, *Inorg. Chim. Acta*, 86, 145, 1984.
117. **Davankov, V. A., Rogozhin, S. V., Semechkin, A. V., Baranov, V. A., and Sannikova, G. S.,** Ligand exchange chromatography of racemates. II. Influence of temperature and concentration of eluent on ligand exchange chromatography, *J. Chromatogr.*, 93, 363, 1974.
118. **Muller, D., Jozefonvich, J., and Petit, M.,** Importance in ligand exchange chromatography of stereoselectivity in the ternary complexes copper(II)-*N*-substituted L-proline-D/L-α-amino acids, *J. Inorg. Nucl. Chem.*, 42, 1083, 1980.
119. **Boue, J., Audebert, R., and Quivoron, C.,** Direct resolution of α-amino acid enantiomers by ligand exchange — stereoselection mechanism on silica packings coated with a chiral polymer, *J. Chromatogr.*, 204, 185, 1981.
120. **Myasoedov, N. F., Penkina, V. I., Petrenik, O. V., and Zolotarev, Yu. A.,** Chromatographic study on the resolution of racemates of tritium labeled amino acids, in First All Union Conference on Isolation of Cyclotron and Fission Isotopes, Tashkent, Uzbeck S.S.R., 1980, Abstr., 53.
121. **Szczepaniak, W. and Ciszewska, W.,** Resolution of some racemic D,L amino acids on an ion exchanger containing iminodi(methanephosphonic) groups in amino copper form, *Chromatographia*, 15, 38, 1982; **Bayer, E.,** *Chromatographia*, 18, 220, 1984; and **Szczepaniak, W.,** *Chromatographia*, 18, 221, 1984.
122. **Bochkov, A. S., Zolotarev, Yu. A., Belov, Yu. P., and Davankov, V. A.,** Ligand exchange chromatographic separation of amino acid enantiomers on silica gel with fixed chiral ligands, Paper B 3.23, Proc. Second Danube Symp. Progress in Chromatography, Carlsbad, Czechoslovakia, 1979.
123. **Gübitz, G., Jellenz, W., Löfler, G., and Santi, W.,** Chemically bonded chiral stationary phases for the separation of racemates by high performance liquid chromatography, *J. High Resolut. Chromatogr. Chromatogr. Commun.*, 2, 145, 1979.
124. **Foucault, A., Caude, M., and Oliveros, L.,** Ligand exchange chromatography of enantiomeric amino acids on copper loaded chiral bonded silica gel and of amino acids on copper(II) modified silica gel, *J. Chromatogr.*, 185, 345, 1979.

125. **Gübitz, G., Jellenz, W., and Santi, W.,** Resolution of the optical isomers of underivatisized amino acids on chemically bonded chiral phases by ligand exchange chromatography, *J. Liquid Chromatogr.*, 4, 701, 1981.
126. **Gübitz, G., Jellenz, W., and Santi, W.,** Separation of the optical isomers of amino acids by ligand exchange chromatography using chemically bonded chiral phases, *J. Chromatogr.*, 203, 377, 1981.
127. **Gübitz, G., Juffman, F., and Jellenz, W.,** Direct separation of amino acid enantiomers by high performance ligand exchange chromatography on chemically bonded chiral phases, *Chromatographia*, 16, 103, 1983.
128. **Gübitz, G. and Mihellyes, S.,** Direct separation of 2-hydroxy acid enantiomers by high performance liquid chromatography on chemically bonded chiral phases, *Chromatographia*, 19, 257, 1984.
129. **Lindner, W.,** HPLC-Enantiomerentrennung an gebundenen chiralen Phasen, *Naturwissenchaften*, 67, 354, 1980.
130. **Watanabe, N.,** Enantiomeric resolution of amino acids by high performance ligand exchange chromatography using histidine bonded silica gel, *J. Chromatogr.*, 260, 75, 1983.
131. **Engelhardt, H. and Kromidas, S.,** HPLC an chiralen chemisch gebundenen Phasen auf Kieselgel, *Naturwissenchaften*, 67, 353, 1980.
132. **Engelhardt, H. and Ahr, G.,** Properties of chemically bonded phases, *Chromatographia*, 14, 227, 1981.
133. **Sugden, K., Hunter, C., and Lloyd-Jones, G.,** Ligand exchange chromatography. I. Resolution of L- and D-proline on a copper(II) proline complex bond to microparticulate silica gel, *J. Chromatogr.*, 192, 228, 1980.
134. **Roumeliotis, P., Unger, K. K., Kurganov, A. A., and Davankov, V. A.,** Chiral silica packings with L-proline or L-hydroxyproline bonded via alkyl or alkylbenzyl chains for the separation of the enantiomers of α-amino acids by HPLC, *Angew. Chem. Int. Ed. Engl.*, 21, 930, 1982.
135. **Roumeliotis, P., Unger, K. K., Kurganov, A. A., and Davankov, V. A.,** High performance ligand exchange chromatography of α-amino acid enantiomers — studies on monomerically bonded 3-(L-prolyl)- and 3-(L-hydroxy-prolyl) propyl silicas, *J. Chromatogr.*, 255, 51, 1983.
136. **Roumeliotis, P., Kurganov, A. A., and Davankov, V. A.,** Effect of the hydrophobic spacer in bonded $[Cu(L\text{-hydroxyprolyl})alkyl]^+$ silicas on retention and enantioselectivity of α-amino acids in high performance liquid chromatography, *J. Chromatogr.*, 266, 439, 1983.
137. **Roumeliotis, P., Kurganov, A. A., and Davankov, V. A.,** High performance ligand exchange chromatography of α-amino acid enantiomers on bonded chiral silica phases of the *N*-(*p*-ethylenebenzyl)-L-hydroxyproline type, in press.
138. **Davankov, V. A. and Rogozhin, S. V.,** Ligand chromatography as a novel method for the investigation of mixed complexes: stereoselective effects in α-amino acid copper(II) complexes, *Dokl. Akad. Nauk S.S.S.R.*, 193, 94, 1970; *J. Chromatogr.*, 60, 280, 1971.
139. **Davankov, V. A., Kurganov, A. A., and Rogozhin, S. V.,** Enantioselectivity effects in coordination compounds, *Russ. Chem. Rev.*, 43, 764, 1974.
140. **Davankov, V. A. and Kurganov, A. A.,** Chiral monomeric versus polymeric ligands bonded to silica surface in the ligand exchange chromatography of optical isomers, in Symp. on Liquid Chromatography in the Biomedical Sciences, Abstr., Ronneby, Sweden, June 18 to 21, 1984.
141. **Feibush, B., Cohen, M. J., and Karger, B. L.,** The role of bonded phase composition on the ligand exchange chromatography of dansyl-D,L-amino acids, *J. Chromatogr.*, 282, 3, 1983.
142. **Gelber, L. R., Karger, B. L., Neumeyer, J. L., and Feibush, B.,** Ligand exchange chromatography of amino alcohols. Use of Schiff bases in enantiomer resolution, *J. Am. Chem. Soc.*, 106, 7729, 1984.
143. **Kurganov, A. A., Tevlin, A. B., and Davankov, V. A.,** High performance ligand exchange chromatography of enantiomers — studies on polystyrene type chiral phases bonded to microparticulate silicas, *J. Chromatogr.*, 261, 223, 1983.
144. **Kicinski, H. G. and Kettrup, A.,** Synthesis and characterization of L(+)-diacetyltartaric acid silica gel for high-performance liquid chromatography, *Fresenius Z. Anal. Chem.*, 316, 39, 1983.
145. **Kicinski, H. G. and Kettrup, A.,** Determination of enantiomeric catecholamines by ligand-exchange chromatography using chemically modified L(+)-tartaric acid silica gel, *Fresenius Z. Anal. Chem.*, 320, 51, 1985.
146. **Bernauer, K.,** Separation of racemic mixtures by chromatography on an optically active support, Swiss Patent 490292, 1970; *Chem. Abstr.*, 74, 64395t, 1971.
147. **Humbel, F., Vonderschmitt, D., and Bernauer, K.,** Effets stereoselectifs dans les reactions des complexes metalliques. I. Stereoselectivité de la reaction de l'anion D-*N*-(hydroxy-2-ethyl)-propylenediaminetriacetatoferrate(III) avec quelques derivés d'acides aminés, *Helv. Chim. Acta*, 53, 1983, 1970.
148. **Bernauer, K., Jeanneret, M. F., and Vonderschmitt, D.,** Stereoselective effects in reactions of metal complexes. II. Selectivity in the complex formation of nickel(II)- , copper(II)- , and zinc(II)-(D)-propylenediaminetetraacetate with racemic 1-phenylethyl-amine, *Helv. Chim. Acta*, 54, 297, 1971.
149. **Davankov, V. A., Bochkov, A. S., Kurganov, A. A., Roumeliotis, P., and Unger, K. K.,** Separation of unmodified α-amino acid enantiomers by reverse phase HPLC, *Chromatographia*, 13, 677, 1980.

150. **Davankov, V. A., Bochkov, A. S., and Belov, Yu. P.,** Ligand exchange chromatography of racemates. XV. Resolution of α-amino acids on reversed phase silica gels coated with *N*-decyl-L-histidine, *J. Chromatogr.*, 218, 547, 1981.
151. **Klemisch, W., von Hodenberg, A., and Vollmer, K. O.,** Resolution of the enantiomers of *m*- and *p*-hydroxymandelic acid by high performance liquid chromatography, *High Resolut. Chromatogr. Chromatogr. Commun.*, 4, 535, 1981.
152. **Lam-Thang, H., Fermandjian, S., and Fromageot, P.,** High performance liquid chromatography and magnetic circular dichroism — study of the "palladium(II)-thioether peptide" complexes, *J. Chromatogr.*, 235, 139, 1982.
153. **Krasutskii, P. A., Rodionov, V. N., Tiochonov, V. P., and Yurchenko, A. G.,** Silver *d*-camphor-10-sulphonate — a new chiral reagent for enantiomeric resolution of olephines, *Theor. Exp. Khim.*, 20, 58, 1984.
154. **Guenther, K., Martens, J., and Schickedanz, M.,** Thin layer chromatographic enantiomer separation by using ligand exchange, *Angew. Chem.*, 96, 514, 1984; *Angew. Chem. Int. Ed. Engl.*, 23, 506, 1984.
155. **Günther, K., Schickedanz, M., and Martens, J.,** Thin-layer chromatographic enantiomeric resolution, *Naturwissenschaften*, 72, 149, 1985.
156. **Weinstein, S.,** Resolution of optical isomers by thin layer chromatography, *Tetrahedron Lett.*, 25, 985, 1984.
157. **Grinberg, N. and Weinstein, S.,** Enantiomeric separation of DNS-amino acids by reversed phase thin-layer chromatography, *J. Chromatogr.*, 303, 251, 1984.
158. **LePage, J. N., Lindner, W., Davies, G., Seitz, D. E., and Karger, B. L.,** Resolution of the optical isomers of dansyl-amino acids by reversed phase liquid chromatography with optically active metal chelate additives, *Anal. Chem.*, 51, 433, 1979.
159. **Hare, P. E. and Gil-Av, E.,** Separation of D- and L-amino acids by liquid chromatography — use of chiral eluants, *Science*, 204, 1226, 1979.
160. **Sousa, L. R., Hoffman, D. H., Kaplan, L., and Cram, D. J.,** Total optical resolution of amino esters by designed host/guest relationship in molecular complexation, *J. Am. Chem. Soc.*, 96, 7100, 1974.
161. **Gil-Av, E., Tishbee, A., and Hare, P. E.,** Resolution of underivatized amino acids by reversed phase chromatography, *J. Am. Chem. Soc.*, 102, 5115, 1980.
162. **Oelrich, E., Preusch, H., and Wilhelm, E.,** Separation of enantiomers by high performance liquid chromatography using chiral eluents, *J. High Resolut. Chromatogr. Chromatogr. Commun.*, 3, 269, 1980.
163. **Gelber, L. R. and Neumeyer, J. L.,** Determination of the enantiomeric purity of levo-DOPA, methyl-DOPA, and tryptophan by use of chiral mobile phase high performance liquid chromatography, *J. Chromatogr.*, 257, 317, 1983.
164. **Wernicke, R.,** Separation of underivatized amino acid enantiomers by means of a chiral solvent-generated phase, *J. Chromatogr. Sci.*, 23, 39, 1985.
165. **Weinstein, S., Engel, M. H., and Hare, P. E.,** The enantiomeric analysis of a mixture of all common protein amino acids by high performance liquid chromatography using a new chiral mobile phase, *Anal. Biochem.*, 121, 370, 1982.
166. **Weinstein, S.,** Enantiomeric analysis of the common protein amino acids by liquid chromatography, *Trends Anal. Chem.*, 3, 16, 1984.
167. **Weinstein, S.,** Resolution of D- and L-amino acids by HPLC with copper complexes of *N,N*-dialkyl-α-amino acids as novel chiral additives — a structure selectivity study, *Angew. Chem. Suppl.*, p. 425, 1982; *Angew. Chem. Int. Ed. Engl.*, 21, 218, 1982.
168. **Weinstein, S. and Grinberg, N.,** Enantiomer separation of underivatized α-methyl-α-amino acids by high performance liquid chromatography, *J. Chromatogr.*, 318, 117, 1985.
169. **Gilon, C., Leshem, R., Tapuhi, Y., and Grushka, E.,** Reversed phase chromatographic resolution of amino acid enantiomers with metal aspartame eluents, *J. Am. Chem. Soc.*, 101, 7612, 1979.
170. **Gilon, C., Leshem, R., and Grushka, E.,** Determination of enantiomers of amino acids by reversed phase high performance liquid chromatography, *Anal. Chem.*, 52, 1206, 1980.
171. **Gilon, C., Leshem, R., and Grushka, E.,** Structure resolution relationship. I. The effect of the alkylamide side chain of aspartyl derivatives on the resolution of amino acid enantiomers, *J. Chromatogr.*, 203, 365, 1981.
172. **Gilon, C., Leshem, R., and Grushka, E.,** Determination of amino acid enantiomers by reversed phase HPLC: the effect of alkylamide chain of aspartylamides on resolution, in *Peptides*, 16th Proc. Eur. Pept. Symp., Copenhagen, 1980, Brunfeldt, B., Ed., 1981, 700; as cited in *Chem. Abstr.*, 97, 169563, 1982.
173. **Grushka, E., Leshem, R., and Gilon, C.,** Retention behavior of amino acid enantiomers in reversed phase liquid chromatography, *J. Chromatogr.*, 255, 41, 1983.
174. **Nimura, N., Suzuki, T., Kasahara, Y., and Kinoshita, T.,** Reversed phase liquid chromatographic resolution of amino acid enantiomers by mixed chelate complexation, *Anal. Chem.*, 53, 1380, 1981.

175. **Nimura, N., Toyama, A., and Kinoshita, T.,** Optical resolution of DL-proline by reversed phase high performance liquid chromatography using *N*-(*p*-toluenesulphonyl)-L-phenylalanine-copper(II) as a chiral additive, *J. Chromatogr.*, 234, 482, 1982.
176. **Nimura, N., Toyama, A., Kasahara, Y., and Kinoshita, T.,** Reversed phase liquid chromatographic resolution of underivatized DL-amino acids using chiral eluents, *J. Chromatogr.*, 239, 671, 1982.
177. **Nimura, N., Toyama, A., and Kinoshita, T.,** Optical resolution of amino acid enantiomers by high performance liquid chromatography, *J. Chromatogr.*, 316, 547, 1984.
178. **Kurganov, A. A. and Davankov, V. A.,** Ligand exchange chromatography of racemates. XVI. Microbore column chromatography of amino acid racemates using *N,N,N′,N′*-tetramethyl-(R)-propanediamine-1,2-copper(II) complexes as chiral additives to the eluent, *J. Chromatogr.*, 218, 559, 1981.
179. **Lindner, W., LePage, J. N., Davies, G., Seitz, D. E., and Karger, B. L.,** Reversed phase separation of optical isomers of Dns-amino acids and peptides using chiral metal chelate additives, *J. Chromatogr.*, 185, 323, 1979.
180. **Lam, S. K. and Chow, F. K.,** Resolution of DL-dansyl amino acids by HPLC with a copper(II)-L-proline eluent, *J. Liq. Chromatogr.*, 3, 1579, 1980.
181. **Lam, S., Chow, F., and Karmen, A.,** Reversed phase high performance liquid chromatographic resolution of D- and L-DNS-amino acids by mixed chelate complexation, *J. Chromatogr.*, 199, 295, 1980.
182. **Lam, S. K.,** Improved enantiomeric resolution of DL-dns-amino acids, *J. Chromatogr.*, 234, 485, 1982.
183. **Lam, S.,** Stereoselective analysis of D and L dansyl-amino acids as the mixed chelate copper(II) complexes by HPLC, *J. Chromatogr. Sci.*, 22, 416, 1984.
184. **Lam, S., Azumaya, H., and Karmen, A.,** High performance liquid chromatography of amino acids in urine and cerebrospinal fluid, *J. Chromatogr.*, 302, 21, 1984.
185. **Tapuhi, Y., Miller, N., and Karger, B. L.,** Practical consideration in the chiral separation of Dns-amino acids by reversed phase liquid chromatography using metal chelate additives, *J. Chromatogr.*, 205, 325, 1981.
186. **Weinstein, S. and Weiner, S.,** Enantiomeric analysis of a mixture of the common protein amino acids as their Dns derivatives. Single-analysis reversed-phase high-performance liquid chromatographic procedure using a chiral mobile phase additive, *J. Chromatogr.*, 303, 244, 1984.
187. **Marchelli, R., Dossena, A., Casnati, G., Dallavalle, F., and Weinstein, S.,** Chiral copper(II) complexes for the enantioselective resolution of DL-dansyl-amino acids, *Angew. Chem. Int. Ed. Engl.*, 24, 336, 1985.
188. **Lam, S. and Karmen, A.,** Resolution of optical isomers of Dns-amino acids by high performance liquid chromatography with L-histidine and its derivatives in the mobile phase, *J. Chromatogr.*, 239, 451, 1982.
189. **Lam, S. and Karmen, A.,** Stereoselective D- and L-amino acid analysis by high performance liquid chromatography, *J. Chromatogr.*, 289, 339, 1984.
190. **Lebl, M., Hrbas, P., Skopkova, J., Slaninova, J., Machova, A., Barth, T., and Jost, K.,** Synthesis and properties of oxytocin analogues with high and selective natriuretic activity, *Coll. Czech. Chem. Commun.*, 47, 2540, 1982.
191. **Gundlach, G., Sattler, E. L., and Wagenbach, U.,** Trennung von racemischen Alkylaminosäuren durch HPLC mit chiraler mobiler Phase, *Fresenius Z. Anal. Chem.*, 311, 684, 1982.
192. **DuBois, G. E. and Stephenson, R. A.,** Dehydrochalcone sweetners — synthesis, sensory evaluation, and chiral eluent chromatography of the D and L antipodes of a potentially sweet, sucrose-like homoserine-dehydrochalcone conjugate, *J. Agric. Food Chem.*, 30, 676, 1982.
193. **Forsman, U.,** Enantiomeric resolution of an optically active guanine derivative by high-performance liquid chromatography with phenylalanine-Cu(II) in the mobile phase, *J. Chromatogr.*, 303, 217, 1984.
194. **Benecke, I.,** Resolution of underivatized 2-hydroxy acids by high performance liquid chromatography, *J. Chromatogr.*, 291, 155, 1984.
195. **Davankov, V. A., Rogozhin, S. V., and Kurganov, A. A.,** The presence of two forms in solutions of copper *bis*-complexes with bidentate α-amino acids, *Isv. Akad. Nauk S.S.S.R. Ser. Khim.*, p. 486, 1972.
196. **Rogozhin, S. V., Davankov, V. A., Kurganov, A. A., and Timofeeva, G. I.,** On the structure of three forms of copper(II) *bis*-complexes with bidentate α-amino acid ligands, *Zh. Neorg. Khim.*, 19, 3294, 1974.
197. **Gillard, R. D.,** Stereoselectivity and reactivity in complexes of amino acids and peptides, *Inorg. Chim. Acta Rev.*, 1, 69, 1967.
198. **Kurganov, A. A. and Davankov, V. A.,** The rule of average environment and the axial coordination in copper(II) complexes, *Inorg. Nucl. Chem. Lett.*, 12, 743, 1976.
199. **Lafuma, F., Boue, J., Audebert, R., and Quivoron, C.,** Copper(II) and nickel(II) stereocomplexes involved in the separation of α-amino acid enantiomers by ligand exchange chromatography — stability constants and structural investigations, *Inorg. Chim. Acta*, 66, 167, 1982.
200. **Takeuchi, T., Horikava, R., and Tanimura, T.,** Enantioselective solvent extraction of neutral D,L-amino acids in two-phase system containing *N*-*n*-alkyl-L-proline derivatives and copper ions, *Anal. Chem.*, 56, 1152, 1984.
201. **Takeuchi, T., Horikawa, R., and Tanimura, T.,** Complete resolution of D,L-isoleucine by droplet counter current chromatography, *J. Chromatogr.*, 284, 285, 1984.

202. **Davankov, V. A. and Kurganov, A. A.,** The role of the achiral sorbent matrix in chiral recognition of amino acid enantiomers in ligand exchange chromatography, *Chromatographia,* 17, 686, 1983.
203. **Petit-Ramel, M. M. and Paris, M. R.,** Etude polarimetrique des complexes metalliques des amino acides. II. Les complex mixtes du cuivre avec deux amino acides, *Bull. Soc. Chim. Fr.,* p. 2791, 1968.
204. **Al-Ani, N. and Olin, A.,** Stereoselective effects in the formation of mixed copper(II) complexes with proline and some other α-aminoacids, *Chem. Scr.,* 23, 165, 1984.
205. **Davankov, V. A., Rogozhin, S. V., Kurganov, A. A., and Zhuchkova, L. Ya.,** Stereoselectivity in *bis*(α-amino acid) copper(II) complexes. II. Spectrophotometric and spectropolarimetric investigation of the solvent effect, *J. Inorg. Nucl. Chem.,* 37, 369, 1975.
206. **Zolotarev, Yu. A. and Myasoedov, N. F.,** Effect of skeletal structure of sorbents containing L-hydroxyproline groups on enantioselectivity in ligand exchange chromatography of amino acid racemates, *J. Chromatogr.,* 264, 377, 1983.
207. **Kurganov, A. A., Ponomareva, T. M., and Davankov, V. A.,** Study into mechanism of chiral recognition of amino acid enantiomers in ligand exchange chromatography using microbore column technique, Sixth Int. Symp. on Capillary Chromatography, Riva del Garda, Italy, May 14 to 16, 1985, 871.
208. **Zolotarev, Yu. A., Kurganov, A. A., and Davankov, V. A.,** Potentiometric determination of the dissociation constants of an asymmetric sorbent containing L-proline, and the stability constants of its copper(II) complexes, *Talanta,* 25, 493, 1978.
209. **Zolotarev, Yu. A., Kurganov, A. A., Semechkin, A. V., and Davankov, V. A.,** Determination of stability constants of fixed site complexes of copper(II) ions and of sorbed copper-L-proline complexes with an asymmetric resin containing L-proline groups, *Talanta,* 25, 499, 1978.
210. **Hering, R.,** *Chelatbildende Ionenaustauscher,* Akademie Verlag, Berlin, 1967, 77.
211. **Davankov, V. A., Rogozhin, S. V., Semechkin, A. V., and Sachkova, T. P.,** Resolution of racemates by means of ligand chromatography — influence of the degree of saturation of the sorbent by metal ions on the distribution of mobile ligands, *Zh. Physic. Khim.,* 47, 1254, 1973.
212. **Krebs, H. and Rasche, R.,** Über chromatographische Spaltung von Racematen. I. Optisch aktive Kobaltkomplexe von Dithiosäuren, *Z. Anorg. Allg. Chem.,* 276, 236, 1954.
213. **Krebs, H., Rasche, R., Wagner, J. A., and Diewald, J.,** Über Spaltung von racemischen Komplexsalzen auf chromatographischem Wege, *Angew. Chem.,* 66, 329, 1954.
214. **Yoshikawa, Y. and Yamasaki, K.,** Complete chromatographic resolution of cobalt(III) complexes on ion exchange Sephadex, *Inorg. Nucl. Chem. Lett.,* 6, 523, 1970.
215. **Piper, T. S.,** Partial chromatographic resolution, rotatory dispersion, and absolute configuration of octahedral complexes containing three identical bidentate ligands, *J. Am. Chem. Soc.,* 83, 3908, 1961.
216. **Fujita, M., Yoshikawa, Y., and Yamatera, H.,** Highly efficient chromatographic resolution of $[Co(en)_3]^{3+}$ ion with a column of TA(ES)-Sephadex containing D-tartrate groups, *Chem. Lett.,* 11, 473, 1975.
217. **Krebs, H. and Rasche, R.,** Über ein chromatographisches Verfahren zur optischen Aktivierung von Racematen, *Naturwissenschaften,* 41, 63, 1954.
218. **Krebs, H., Diewald, J., Arlitt, H., and Wagner, J. A.,** Über die chromatographische Spaltung von Racematen. II. Versuche zur Aktivierung von oktaederförmig gebauten Komplexen, *Z. Anorg. Allg. Chem.,* 287, 98, 1956.
219. **Douglas, B. E. and Yamada, S.,** Configurational and vicinal contributions to the optical activity of the isomers of tris(alaninato) cobalt(III), *Inorg. Chem.,* 4, 1561, 1965.
220. **Gillard, R. D., Harrison, P. M., and McKenzie, E. D.,** Optically active coordination compounds. IX. Complexes of dipeptides with cobalt(III), *J. Chem. Soc. A,* p. 618, 1967.
221. **Jursik, F., Wollmanova, D., and Hayek, B.,** The synthesis and partial resolution of *tris*(α-aminoisobutyrato) cobalt(III), *Collection,* 38, 3627, 1973.
222. **Legg, J. I., Cooke, D. W., and Douglas, B. E.,** Circular dichroism of trans-*N,N'*-ethylenediaminediacetic acid cobalt(III) complexes, *Inorg. Chem.,* 6, 700, 1967.
223. **Warner, L. G., Rose, N. J., and Busch, D. H.,** Stereochemistry of a macrocyclic complex — Chelate ring conformations and unusual isomers, *J. Am. Chem. Soc.,* 89, 703, 1967.
224. **Taylor, L. T. and Bush, D. H.,** Chromatographic resolution of antipodes of a helical complex of nickel(II), *J. Am. Chem. Soc.,* 89, 5372, 1967.
225. **Moeller, T. and Gulyas, E.,** The partial resolution of certain inner complexes by means of a chromatographic technique, *J. Inorg. Nucl. Chem.,* 5, 245, 1958.
226. **Collman, J. P., Blair, R. P., Marshall, R. L., and Slade, A. L.,** The chemistry of metal chelate rings. IV. Electrophilic substitution of optically active tris-acetylacetonates, *Inorg. Chem.,* 2, 576, 1963.
227. **Fay, R. C., Girgis, A. Y., and Klabunde, U.,** Stereochemical rearrangements of metal tris-β-diketonates. I. Partial resolution and racemization of some tris(acetylacetonates), *J. Am. Chem. Soc.,* 92, 7056, 1970.
228. **Irving, H. M. N. H., Simpson, R. B., and Smith, I. S.,** The examination by viscosimetry of possible stereospecific interaction between cromium(III) acetylacetonate and an optically active ester in benzene solution, *J. Inorg. Nucl. Chem.,* 32, 2275, 1970.

229. **Collman, J. P., Blair, R. P., Slade, A. L., and Marshall, R. L.,** Optical stability of metal acetylacetonates — racemization during crystallisation, *Chem, Ind.(London)*, 141, 1962.
230. **Collman, J. P.,** Complete resolution of cobalt(III)acetylacetonate, *Diss. Abstr. B*, 28, 4482, 1968.
231. **Moeller, T., Gulias, E., and Marshall, R. H.,** Observations on the rare earths. LXVIII. Partial resolution of yttrium and gadolinium acetylacetonates by means of a chromatographic technique, *J. Inorg. Nucl. Chem.*, 9, 82, 1959.
232. **Fay, R. C. and Piper, T. S.,** Coordination compounds of trivalent metals with unsymmetrical 1,3-diketones. III. Mechanism of stereochemical rearrangements, *Inorg. Chem.*, 3, 348, 1964.
233. **Gordon, J. G. and Holm, R. H.,** Intramolecular rearrangement reactions of tris-chelate complexes. I. General theory and the kinetics and probable mechanism of the isomerization and racemization of tris(5-methylhexane-2,4-dionato) cobalt(III), *J. Am. Chem, Soc.*, 92, 5319, 1970.
234. **Hseu, T. M., Martin, D. F., and Moeller, T.,** Partial resolution of some copper(II) and nickel(II) β-ketoimine compounds by means of a chromatographic technique, *Inorg. Chem.*, 2, 587, 1963.
235. **Ramaiah, K., Anderson, F. E., and Martin, D. F.,** Preparation and resolution of a five coordinate complex — bis-acetylacetonepropylenediiminooxovanadium(IV), *Inorg. Chem.*, 3, 296, 1964.
236. **Ramaiah, K. and Martin, D. F.,** Five coordinate compounds. II. Studies on β-diketo and β-ketoimino complexes of oxovanadium(IV) ion, *J. Inorg. Nucl. Chem.*, 27, 1663, 1965.
237. **Markovic, V. G. and Schweitzer, G. K.,** The partial separations of some neutral β-diketone complexes into optically active isomers, *J. Inorg. Nucl. Chem.*, 33, 3197, 1971.
238. **Okawa, H. and Yoshino, T.,** Macro chelate rings. II. Synthesis and properties of 2,7-dimethyl-4,5-bis(salicylideneaminomethyl)xanthone as a quadridentate chelating agent and its metal complexes, *Bull. Chem. Soc. Jpn.*, 43, 805, 1970.
239. **Buchar, E. and Suchy, K.,** Chromatography of certain inorganic isomers and complex compounds, *Mag. Kem. Foly.*, 64, 45, 1958; *Chem. Abstr.*, 52, 11518f, 1958.
240. **Dwyer, F. P., McDermott, T. E., and Sargeson, A. M.,** Stereospecific influences in metal complexes containing optically active ligands. VII. The isolation of the D- and L-isomers of *bis*-[(−)-propylenediamine]-ethylenediamine cobalt(III) and *bis*(ethylenediamine)-[(−)-propylenediamine)] cobalt(III) ions, *J. Am. Chem. Soc.*, 85, 2913, 1963.
241. **Legg, J. I. and Douglas, B. E.,** Partial resolution of cobalt(III) chelate complexes by ion exchange cellulose chromatography, *Inorg. Chem.*, 7, 1452, 1968.
242. **Brubaker, G. R., Legg, J. I., and Douglas, B. E.,** Total resolution of the hexakis(2-aminoethanethio)tricobalt(III) cation by ion exchange chromatography, *J. Am. Chem. Soc.*, 88, 3446, 1966.
243. **Schwab, D. E. and Rund, J. V.,** Resolution of the optical isomers — Geometry of (di-chloro-*bis*-phenanthroline)rhodium chloride, *J. Inorg. Nucl. Chem.*, 32, 3949, 1970.
244. **Yoshikawa, Y. and Yamasaki, K.,** Chromatographic resolution of cobalt(III) complexes on ion exchange cellulose, *Inorg. Nucl. Chem. Lett.*, 4, 697, 1968.
245. **Norden, B.,** Circular dichroism spectrum and absolute configuration of tris-(acetylacetonato)chromium(III), *Inorg. Nucl. Chem. Lett.*, 11, 387, 1975.
246. **Jonas, I. and Norden, B.,** Circular dichroism and absolute configuration of tris-(acetylacetonato)cobalt(III), *Inorg. Nucl. Chem. Lett.*, 12, 43, 1976.
247. **Jonas, I. and Norden, B.,** Optical resolution by chromatography at low temperature, *Nature (London)*, 258, 597, 1975.
248. **Norden, B. and Jonas, I.,** Optical resolution of tris-(acetylacetonato)aluminium(III) by low temperature chromatography, *Inorg. Nucl. Chem. Lett.*, 12, 33, 1976.
249. **Yoshino, Y., Sugiyama, H., Nogaito, S., and Kinoshita, H.,** Resolution of racemic complexes on ion exchange resins. *Sci. Pap. Coll. Gen. Educ. Univ. Tokyo*, 16, 57, 1966.
250. **Gaal, J. and Inszedy, J.,** Chromatographic separation of optical isomers by means of outer sphere complex formation reactions, *Talanta*, 23, 78, 1976.
251. **Gillard, R. D. and Mitchell, P. R.,** An easy novel chromatographic resolution of complex cations, *Transition Met. Chem.*, 1, 223, 1976.
252. **Celap, M. B., Hodzic, I. M., and Janjic, T. J.,** Resolution of neutral complexes of transition metals by stereoselective adsorption on optically active complexes. I. Partial resolution of tris(acetylacetonato)cobalt(III) and tris(acetylacetonato)-chromium(III) on cobalt(III) complexes, *J. Chromatogr.*, 198, 172, 1980.
253. **Yamagishi, A., Ohnishi, R., and Soma, M.,** Chromatographic resolution of tris(acetylacetonate)cobalt(III) on a Δ-tris-(1,10-phenanthroline)-nickel(II) montmorillonite column, *Chem. Lett.*, 85, 1982.
254. **Yamagishi, A.,** Clay column chromatography — partial resolution of metal(III) tris(acetylacetonate) on a Δ-nickel(II) tris(1,10-phenanthroline)-montmorillonite column, *Inorg. Chem.*, 21, 3393, 1982.
255. **Yamagishi, A. and Ohnishi, R.,** Clay column chromatography for optical resolution: initial resolution of *bis*(acetylacetonato)(glycinato)cobalt(III) and (acetylacetonato)*bis*(glycinato)cobalt(III) on a Δ-tris(1,10-phenanthroline)nickel(II)-montmorillonite column, *Inorg. Chem.*, 21, 4233, 1982.
256. **Yamagishi, A.,** Partial resolution of tris(acetylacetonato)-ruthenium(III) on a liquid chromatography column of Δ-tris-(1,10-phenanthroline)nickel(II) montmorillonite, *Chem. Commun.*, p. 1168, 1981.

257. **Yamagishi, A.,** Chirality recognition of a clay surface modified by an optically active metal chelate, *J. Chem. Soc. Dalton Trans.*, p. 679, 1983.
258. **Yamagishi, A. and Ohnishi, R.,** Clay column chromatography for optical resolution — partial resolution of tris(acetylacetonato)metal(III) on a λ-tris(1,10-phenanthroline)ruthenium(II) montmorillonite column, *J. Chromatogr.*, 245, 213, 1982.
259. **Yamagishi, A.,** Clay as a medium for optical resolution — resolution of 2,3-dihydro-2-methyl-5,6-diphenylpyrazine on a λ-$[Ru(phen)_3]^{2+}$ (phen = 1,10-phenanthroline) montmorillonite column, *Chem. Commun.*, p. 9, 1983.
260. **Yamagishi, A. and Ohnishi, R.,** Chromatographic separation of enantiomers of cyclic organic compounds on a λ-tris(1,10-phenanthroline)ruthenium(II) montmorillonite column, *Angew. Chem.*, 95, 158, 1983.
261. **Yamagishi, A.,** Clay column chromatography for optical resolution of tris(chelated) and *bis*(chelated) complexes on a λ-Ru-$(1,10\text{-phenanthroline})_3^{2+}$ montmorillonite column, *J. Chromatogr.*, 262, 41, 1983.
262. **Yamagishi, A.,** Chromatographic resolution of enantiomers having aromatic groups by an optically active clay-chelate adduct, *J. Am. Chem. Soc.*, 107, 732, 1985.
263. **Yamagishi, A.,** Clay column chromatography for optical resolution. Resolution of aromatic compounds on a λ-Ru$(1,10\text{-phenanthroline})_3^{2+}$-montmorillonite column, *J. Chromatogr.*, 319, 299, 1985.
264. **Yamamoto, M., Iwamoto, E., Kozasa, A., Takemoto, K., Yamamoto, Y., and Tatehara, A.,** Partial resolution of $Co(acac)_3$ by salting-in chromatography on SP-Sephadex in Δ-$[Ni(phen)_3]^{2+}$ form, *Inorg. Nucl. Chem. Lett.*, 16, 71, 1980.
265. **Fujita, M., Yoshikawa, Y., and Yamatera, H.,** Preparation of optically active cation exchangers with L-tartrate groups and its application to the resolution of $[Co(en)_3]^{3+}$ ion, *Chem. Lett.*, p. 1515, 1974.
266. **Fujita, M., Yoshikawa, Y., and Yamatera, H.,** Enantiomers of tris(trimethylenediamine)cobalt(III) ion of 100% optical purity — preparation and circular dichroism study, *Chem. Commun.*, p. 941, 1975.
267. **Fujita, M., Sakano, M., Yoshikawa, Y., and Yamatera, H.,** Preparation of Sephadex derivatives with optically active groups and column chromatographic application to the resolution of some cobalt(III) complexes, *Bull. Chem. Soc. Jpn.*, 54, 3211, 1981.
268. **Carunchio, V., Messina, A., Sinibaldi, M., and Corradini, D.,** Outer sphere ligand exchange chromatography on bonded chiral silica gel, *J. Liq. Chromatogr.*, 5, 819, 1982.
269. **Yamazaki, S. and Yoneda, H.,** Chromatographic study of optical resolution. IV. Complete resolution of the neutral complex, meridional isomer of tris(β-alaninato)cobalt(III), *Inorg. Nucl. Chem. Lett.*, 15, 195, 1979.
270. **Yukimoto, T. and Yoneda, H.,** Chromatographic study of optical resolution. VI. Separation of isomers of facial tris-(aminoacidato) mixed ligand chelates with *d*-tartrate and antimony *d*-tartrate adsorbed on an anion exchange resin, *J. Chromatogr.*, 210, 477, 1981.
271. **Okamoto, Y., Honda, S., Yashima, E., and Yuki, H.,** Chromatographic resolution. V. Complete chromatographic resolution of tris(acetylacetonato)-cobalt(III) and -chromium(III) on an optically active poly(triphenylmethylmethacrylate) column, *Chem Lett.*, p. 1221, 1983.
272. **Okamoto, Y., Yashima, E., and Hatada, K.,** Chromatographic resolution of tris(acetylacetonato)aluminum on an optically active poly(triphenylmethylmethacrylate) column, *Chem. Commun.*, p. 1051, 1984.
273. **Yoshikawa, Y. and Yamasaki, K.,** Chromatographic resolution of metal complexes on Sephadex ion exchangers, *Coord. Chem. Rev.*, 28, 205, 1979.
274. **Ogino, K. and Saito, U.,** Association involving optically active ions. I. Association constants of tris(ethylenediamine) cobalt(III) and tartrate ions, *Bull. Chem. Soc. Jpn.*, 40, 826, 1967.
275. **Yoneda, H.,** Stereochemical aspects of optical resolution of octahedral metal chelates by liquid chromatography, *J. Liq. Chromatogr.*, 2, 1157, 1979.
276. **Yoneda, H., Yamazaki, S., and Yukimoto, T.,** Optical resolution of facial and meridional tris(aminoacidato) cobalt(III) chelates by *d*-tartrate and antimony *d*-tartrate, in *Stereochemistry of Optically Active Transition Metal Compounds, American Chemical Society Symp. Ser. No. 119,* Douglas, B. E. and Saito, Y., Eds., 1980, 315.
277. **Yamazaki, S., Yukimoto, T., and Yoneda, H.,** Chromatographic study of optical resolution. III. Separation of isomers of facial tris(amino-acidato) cobalt(III) complexes with *d*-tartrate and antimony *d*-tartrate solutions, *J. Chromatogr.*, 175, 317, 1979.
278. **Sakaguchi, U., Tsuge, A., and Yoneda, H.,** Chiral recognition in solution. V. Circular dichroism study of the chiral interaction between *bis*(tartrato)diantimonate(III) ions and trigonal hexanitrido cobalt(3^+) Co $(N)_6$ complexes in solution, *Inorg. Chem.*, 22, 3745, 1983.
279. **Nakazawa, H. and Yoneda, H.,** Chromatographic study of optical resolution. II. Separation of optically active cobalt(III) complexes using potassium antimony *d*-tartrate as eluent, *J. Chromatogr.*, 160, 89, 1978.
280. **Kobayashi, K. and Shibata, M.,** Ion exchange chromatographic studies of the $[Co(N)_4(O)_2]^+$ type complexes, *Bull. Chem. Soc. Jpn.*, 48, 2561, 1975.
281. **Searle, G. H.,** The role of ion association in the chromatographic separation of isomeric cationic cobalt(III) amine complexes on cation exchange resins, particularly SP-Sephadex, *Aust. J. Chem.*, 30, 2625, 1977.

282. **Miyoshi, K., Oh, C. E., Nakazawa, H., and Yoneda, H.,** Induced circular dichroism spectra of some achiral and racemic cobalt(III) amine complexes in aqueous (R,R)-tartrate and (R,R)-tartrateantimonate(III) solution, *Bull. Chem. Soc. Jpn.*, 51, 2946, 1978.
283. **Nakazawa, H., Oh, C. E., Miyoshi, K., and Yoneda, H.,** Induced circular dichroism spectra of some racemic *cis-bis-*(ethylenediamine) cobalt(III) complexes in aqueous (R,R)-tartrate and (R,R)-tartrateantimonate(III) solution and their optical resolution by ion exchange chromatography, *Bull. Chem. Soc. Jpn.*, 53, 273, 1980.
284. **Taura, T., Tamada, H., and Yoneda, H.,** Stereoselectivity in ion pair formation. IV. Stereochemical consideration about specific ion association between optically active (amino acidato)*bis*(ethylenediamine) cobalt (III) and tartrate ions, *Inorg. Chem.*, 17, 3127, 1978.
285. **Yoneda, H. and Yoshizawa, T.,** Chromatographic study of optical resolution. I. Complete resolution of the neutral complex, facial isomer of tris(β-alaninato) cobalt(III), *Chem. Lett.*, 707, 1976.
286. **Yamazaki, S. and Yoneda, H.,** Chromatographic study of optical resolution. V. Resolution of *cis*-diaziodobis(ethylenediamine) and *cis*-α-diazido(triethylenetetramine) cobalt(III) complexes by antimony *d*-tartrate, *J. Chromatogr.*, 177, 227, 1979.
287. **Nakazawa, H., Sakaguchi, U., and Yoneda, H.,** Chromatographic study of optical resolution. VII. Directional ion association model for the optical resolution of cis-$[Co(O)_2(N)_4]^+$ type of complexes by the antimony *d*-tartrate ion, *J. Chromatogr.*, 213, 323, 1981.
288. **Yamanari, K., Hidaka, J., and Shimura, Y.,** Stereochemistry of cobalt(III) complexes with thioethers. II. Geometrical isomers, absorption, and circular dichroism spectra of *bis*(tetradentate-*N,S,O*) complexes, *Bull. Soc. Chem. Jpn.*, 59, 2643, 1977.
289. **Fujinami, S., Tsuji, K., Minegishi, K., and Shibata, M.,** Synthesis and circular dichroism spectra of mer(*N*)-*cis*-(NH_3)-$[Co(OO)(NH_3)(py)(H_2O]^+$ and mer(*N*)-$[Co(OO)(en)(py)(H_2O)]^+$ complexes, *Bull. Soc. Chem. Jpn.*, 55, 1319, 1982.
290. **Sakaguchi, U., Imamoto, I., Izumoto, S., and Yoneda, H.,** The mode of stereoselective association between complex cation and complex anion, *Bull. Soc. Chem. Jpn.*, 56, 153, 1983.
291. **Sakaguchi, U., Yamamoto, I., Izumota, S., and Yoneda, H.,** Inversion of the retention volume order of enantiomers caused by the concentration of eluent, *Bull. Soc. Chem. Jpn.*, 56, 1407, 1983.
292. **Izumoto, S., Sakaguchi, U., and Yoneda, H.,** Stereochemical aspects of the optical resolution of *cis*(*N*)-$[Co(N)_2(O)_4]^-$ complexes by reversed phase ion pair chromatography with cinchona alkaloid cations as the ion pairing reagents, *Bull. Soc. Chem. Jpn.*, 56, 1646, 1983.
293. **Yoneda, H.,** Mechanism of chromatographic separation of optically active metal complexes, *J. Chromatogr.*, 313, 59, 1984.
294. **Yamazaki, S. and Yoneda, H.,** Chromatographic study of optical resolution. VIII. Theoretical study of the chromatographic behavior of the enantiomers of racemic complex cations on a cation exchange column, *J. Chromatogr.*, 219, 29, 1981.
295. **Yamazaki, S. and Yoneda, H.,** Chromatographic study of optical resolution. IX. Optical resolution of monovalent complex cations on an anion exchange column, *J. Chromatogr.*, 235, 289, 1982.
296. **Hewkin, D. J. and Prince, R. H.,** The mechanism of octahedral complex formation by labile metal ions, *Coord. Chem. Rev.*, 5, 45, 1970.
297. **Yoneda, H. and Baba, T.,** Studies of thin layer chromatography of inorganic salts. V. Resolution of the racemic tris-ethylenediamine-cobalt(III) complex by means of thin layer chromatography on silica gel, *J. Chromatogr.*, 53, 610, 1970.
298. **Yoneda, H. and Miura, T.,** Complete resolution of the racemic tris-ethylenediaminecobalt(III) complex into its optical antipodes by means of electrophoresis, *Bull. Chem. Soc. Jpn.*, 43, 574, 1970.
299. **Yoneda, H. and Miura, T.,** Study of hydration and association of ions in solution. II. Complete resolution of tris(ethylenediamine) cobalt(III), chromium(III), and rhodium(III) complexes by means of paper electrophoresis, *Bull. Chem. Soc. Jpn.*, 45, 2126, 1972.
300. **Corradini, C. and Lederer, M.,** Separation of optical isomers of metal complexes by paper electrophoresis in mixtures of aluminum chloride and tartrate, *J. Chromatogr.*, 157, 455, 1978.
301. **Cardaci, V. and Ossicini, L.,** Paper electrophoretic study of ion pair formations. XIII. Behavior of anionic trisoxalatocobalt(III) and trisoxalatochromium(III) in optically active electrolytes, *J. Chromatogr.*, 198, 76, 1980.
302. **Fanali, S., Cardici, V., and Ossicini, L.,** Paper electrophoretic study of ion pair formation. XIV. Resolution of optically active cobalt(III) complexes, *J. Chromatogr.*, 265, 131, 1983.
303. **Fanali, S., Ossicini, L., and Prosperi, T.,** Paper electrophoretic study of ion pair formation. XV. Resolution of optically active cobalt(III)cyclohexanediamine complexes, *J. Chromatogr.*, 318, 440, 1985.
304. **Schurig, V.,** Enantiomerentrennung eines chiralen Olefins durch Komplexierungschromatographie an einem optisch aktiven Rhodium(I) Komplex, *Angew. Chem.*, 89, 113, 1977.
305. **Schurig, V. and Gil-Av, E.,** Chromatographic resolution of chiral olefins — specific rotation of 3-methylcyclopentene and related compounds, *Isr. J. Chem.*, 15, 96, 1976/77.

306. **Schurig, V.**, Selektivität und Stereochemie der Olefin-Metall-π-Komplexierung, *Chem. Ztg.*, 101, 173, 1977.
307. **Schurig. V.**, Resolution of enantiomers and isotopic compounds by selective complexation gas chromatography on metal complexes, *Chromatographia*, 13, 263, 1980.
308. **Golding, B. T., Sellars, P. J., and Wong, A. K.**, Resolution of racemic epoxides on G. L. C. columns containing optically active lanthanoid complexes, *Chem. Commun.*, 570, 1977.
309 **Schurig, V. and Bürkle, W.**, Quantitative Enantiomerentrennung von *trans*-2,3-Epoxybutan durch Komplexierungschromatographie an einem optisch aktiven Nickel(II)-Komplex, *Angew. Chem.*, 90, 132, 1978.
310. **Schurig, V., Koppenhöfer, B., and Bürkle, W.**, Korrelation der absoluten Konfiguration chiraler Epoxide durch Komplexierungschromatographie; Synthese und Enantiomerenreinheit von (+)- und (−)-1,2-Epoxypropan, *Angew. Chem. Int. Ed. Engl.*, 17, 937, 1978; *Angew. Chem.*, 90, 993, 1978.
311. **Schurig, V. and Wistuba, D.**, Asymmetric microsomal epoxidation of simple prochiral olefines, *Angew. Chem. Int. Ed. Engl.*, 23, 796, 1984.
312. **Weber, R. and Schurig, V.**, Analytical resolution of Lineatin by complexation gas chromatography, *Naturwissenschaften*, 68, 330, 1981.
313. **Shurig, V., Weber, R., Klimetzek, D., Kohnle, U., and Mori, K.**, Enantiomeric composition of "Lineatin" in three sympatric ambrosia beetles, *Naturwissenschaften*, 69, 602, 1982.
314. **Koppenhöfer, B., Hintzer, K., Weber, R., and Schurig, V.**, Quantitative separation of the enantiomeric pairs of the pheromone 2-ethyl-1, 6-dioxaspiro[4·4]nonane by complexation chromatography on an optically active metal complex, *Angew. Chem. Int. Ed. Engl.*, 19, 471, 1980.
315. **Weber, R., Hintzer, K., and Schurig, V.**, Enantiomer resolution of spiroketales by complexation gas chromatography on an optically active metal complex, *Naturwissenschaften*, 67, 453, 1980.
316. **Schurig, V. and Weber, R.**, Manganese(II)-*bis*(3-heptafluorobutyryl-1R-camphorate — a versatile agent for the resolution of racemic cyclic ethers by complexation gas chromatography, *J. Chromatogr.*, 217, 51, 1981.
317. **Schurig, V., Weber, R., Nicholson, G. J., Oehlschlager, A. C., Pierce, Jr., H., Pierce, A. M., Borden, J. H., and Ryker, L. C.**, Enantiomer composition of natural exo- and endo-brevicomin by complexation gas chromatography/selected ion mass spectrometry, *Naturwissenschaften*, 70, 92, 1983.
318. **Schurig, V., Koppenhöfer, B., and Bürkle, W.**, Preparation and determination of configurationally pure *trans*-(2*S*,3*S*)-2,3-epoxybutane, *J. Org. Chem.*, 45, 538, 1980.
319. **Hintzer, K., Koppenhöfer, B., and Schurig, V.**, Access to (S)-2-methyloxetane and the precursor (*S*)-1,3-butane-diol of high enantiomeric purity, *J. Org. Chem.*, 47, 3850, 1982.
320. **Schurig, V. and Bürkle, W.**, Extending the scope of enantiomer resolution by complexation gas chromatography, *J. Am. Chem. Soc.*, 104, 7573, 1982.
321. **Schurig V., Bürkle, W., Zlatkis, A., and Poole, C. F.**, Quantitative resolution of pyrimidal nitrogen invertomers by complexation chromatography, *Naturwissenschaften*, 66, 423, 1979.
322. **Bürkle, W., Karfunkel, H., and Schurig, V.**, Dynamic phenomena during enantiomeric resolution by complexation gas chromatography — a kinetic study of enantiomerisation, *J. Chromatogr.*, 288, 1, 1984.
323. **Schurig, V. and Wistuba, D.**, Analytical enantiomer separation of aliphatic diols as boronates and acetales by complexation gas chromatography, *Tetrahedron Lett.*, 25, 5633, 1984.
324. **Schurig, V.**, Homogeneous coating of glass and fused silica capillary columns with metal coordination compounds for the selective gas chromatographic separation of structural, stereo- and optical-isomers, German Patent DE 3,247,714; *Chem. Abstr.*, 100, 79243w, 1984.
325. **Schurig, V. and Weber, R.**, Use of glass and fused-silica open tubular columns for the separation of structural, configurational, and optical isomers, *J. Chromatogr.*, 289, 321, 1984.
326. **Schurig, V. and Weber, R.**, Derivatization free enantiomer separation of chiral alcohols and ketones by high resolution complexation gas chromatography, *Angew. Chem. Int. Ed. Engl.*, 22, 772, 1983; *Angew Chem. Suppl.*, p. 1130, 1983.
327. **Schurig, V., Leyrer, U., and Weber, R.**, The use of glass and fused silica open tubular columns for the separation of structural, configurational and optical isomers by selective complexation GC, in Proc. 6th Int. Symp. on Capillary Chromatography, Riva del Garda, Italy, May 1985, 18.
328. **Oi, N., Horiba, M., Kitahara, H., Doi, T., Tani, T., and Sakakibara, T.**, Direct separation of α-hydroxycarboxylic acid ester enantiomers by gas chromatography with optically active copper(II) complexes, *J. Chromatogr.*, 202, 305, 1980.
329. **Oi, N., Horiba, M., Kitahara, H., Doi, T., and Tani, T.**, Gas chromatographic separation of some optical isomers on optically active copper complexes, *Bunseki Kagaku*, 29, 156, 1980.
330. **Oi, N., Shiba, K., Tani, T., Kitahara, H., and Doi, T.**, Gas chromatographic separation of some enantiomers on optically active copper(II) complexes, *J. Chromatogr.*, 211, 274, 1981.
331. **Oi, N., Doi, T., Kitahara, H., and Inda, Y.**, Direct separation of some alcohol enantiomers by gas chromatography with optically active stationary phases, *Bunseki Kagaku*, 30, 78, 1981.
332. **König, W. A.**, Separation of enantiomers by capillary gas chromatography with chiral stationary phases, *J. High Resolut. Chromatogr. Chromatogr. Commun.*, 5, 588, 1982.

Chapter 6

NONNITROGEN LIGANDS

H. F. Walton

I. SULFUR LIGANDS

Sulfur in its oxidation state, −2, is an electron donor, a ''soft base'' that combine preferentially with ''soft acids'' or the ions or bound atoms of Cu, Ag, Zn, Cd, and Hg. Thiols, organic compounds carrying the –SH group, have long been known as ''mercaptans'' for their affinity for mercury.

Thus, it was natural to use cation-exchange resins carrying mercury(II) to absorb sulfur compounds from petroleum.[1] Since conventional gel-type resins do not swell in nonpolar media, macroporous resins were used. They absorbed sulfur compounds very effectively, but held them so tightly that they could not be removed from the absorbent. Copper(II), held on the acrylic resin Bio-Rex® 70, absorbs alkyl and aryl sulfides from hydrocarbon mixtures, but not so strongly that they cannot be removed. Vogh and Dooley[2] passed crude oil through a small column of this copper-loaded resin, 260 × 9 mm, then eluted with *n*-pentane. Aliphatic hydrocarbons were washed out first, then aromatic hydrocarbons, then, in a broad band, the sulfur compounds, most of which were disulfides or cyclic sulfur compounds. Capacity factors were measured for some 15 sulfur compounds, and the well-known steric effect was seen; alkyl substituents close to the nitrogen atom hinder the retention of amines; but the elution bands were very broad, as one expected from the physical character of the resin, and analysis of the sulfide fraction by liquid chromatography was then out of the question. The authors were, however, able to separate the aromatic fraction from the sulfide fraction, stripping the latter from the resin by a mixture of pentane (90%) and diethyl ether (10%). Each fraction was then analyzed by gas chromatography.

Meanwhile reports appeared in the biochemical literature of the separation of thiols and dithiols by ''affinity chromatography''. Sluyterman[3] prepared a mercury-loaded polysaccharide absorbent by treating agarose with cyanogen bromide, then reacting the product with *p*-acetoxymercurianiline, $CH_3COOHg \cdot C_6H_4NH_2$; the structure of the resulting mercury-loaded agarose was the following:

$$\text{Agarose–O–}\underset{\underset{\text{NH}}{\|}}{\text{C}}\text{–NH–}C_6H_4\text{–Hg–O–}\underset{\underset{\text{O}}{\|}}{\text{C}}\text{–}CH_3$$

On a column of this material the enzyme papain was resolved into two fractions, one of which was rich in sulfhydryl (thiol, –SH) groups and had to be eluted from the absorbent by a solution of mercuric chloride.

The drawback to this absorbent is that compounds containing thiol groups must be eluted by mercury salts or by other thiols. Elution with 2-mercaptoethanol was used to recover the thiol fractions of flavin-adenine nucleotide and coenzyme A.[4] To improve this situation, a Swedish group[5] used an absorbent containing arsenic, again prepared from agarose through reaction with cyanogen bromide, followed by addition of *p*-aminophenylarsine oxide to give

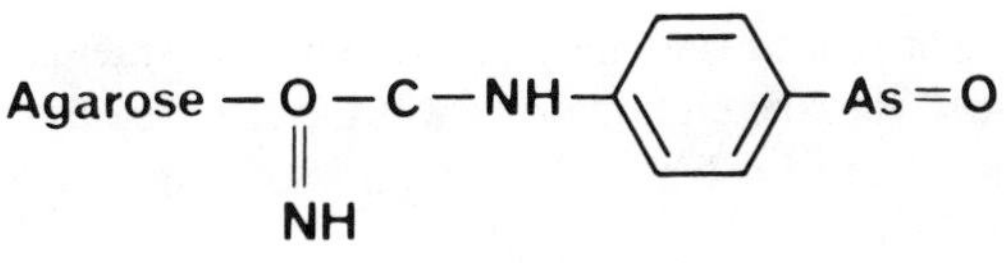

$$\text{Agarose}-\text{O}-\underset{\underset{\text{NH}}{\|}}{\text{C}}-\text{NH}-C_6H_4-\text{As}=\text{O}$$

This compound binds monothiols to give

$$-As\langle S_2 \rangle R'$$

dithiols to give

$$-As(SR)_2$$

Monothiols are displaced by 0.01 *M* sodium hydroxide, dithiols by 0.2 *M* sodium hydroxide.

Chymopapain was separated from papaya latex by chromatography on a common sulfonated polystyrene resin lightly loaded with Hg(II). The eluent was ammonium acetate solution.[6]

High-resolution ligand-exchange chromatography of organic sulfur compounds had to await the development of efficient stationary phases that can be used in nonpolar solvents. Such materials, based on porous silica, have now been made.[7] One of them has 8-mercaptoquinoline functional groups, and others have carbodithioic acid groups, –CSSH. Thus 2-amino-1-cyclopentene-1carbodithioic acid is made to react with propylamino silica to give

$$-Si-C_3H_6-NH-(C_5H_6)-C(=S)SH$$

The spacer group $-NHCH_2CH_2-$ can be interposed between the propyl and aminocyclopentene units, thus:

$$-Si-C_3H_6-NHCH_2CH_2NH-(C_5H_6)-CSH_2$$

Then the absorbents are loaded with Cu(II). A short column of Cu-loaded absorbent, 200 × 4 mm, gave fast, efficient separations of dialkyl sulfides. The eluent was *n*-hexane containing 0.5 to 2. 0% of polar modifier, methanol or acetonitrile; detection was by UV absorbance at 230 nm. Six components were eluted with almost baseline separation in 8 min (see Figure 1). Capacity factors were inversely proportional to the methanol concentration, showing that one methanol molecule displaces one molecule of dialkyl sulfide. The mechanism was truly one of ligand exchange.

The authors list capacity factors for 17 solutes on four stationary phases, with and without Cu. An analysis of the data shows the importance of steric hindrance. The absorbent having the $-NHCH_2CH_2-$ spacer, which may attach to Cu through three of its four coordinate bonds, retains the simpler dialkyl sulfides in the order: di-*tert*-butyl sulfide < di-*n*-butyl < di-*n*-propyl < diethyl < dimethyl < tetrahydrothiophene, the last being retained the most strongly. The first compound is the most sterically hindered and is held the most weakly.

They also tested a silver-loaded stationary phase, with pure methanol as the eluent. Good separations were obtained, but the elution order was the reverse of that with Cu and hexane, pointing to a dominant effect of solvation or ''hydrophobic interaction''.

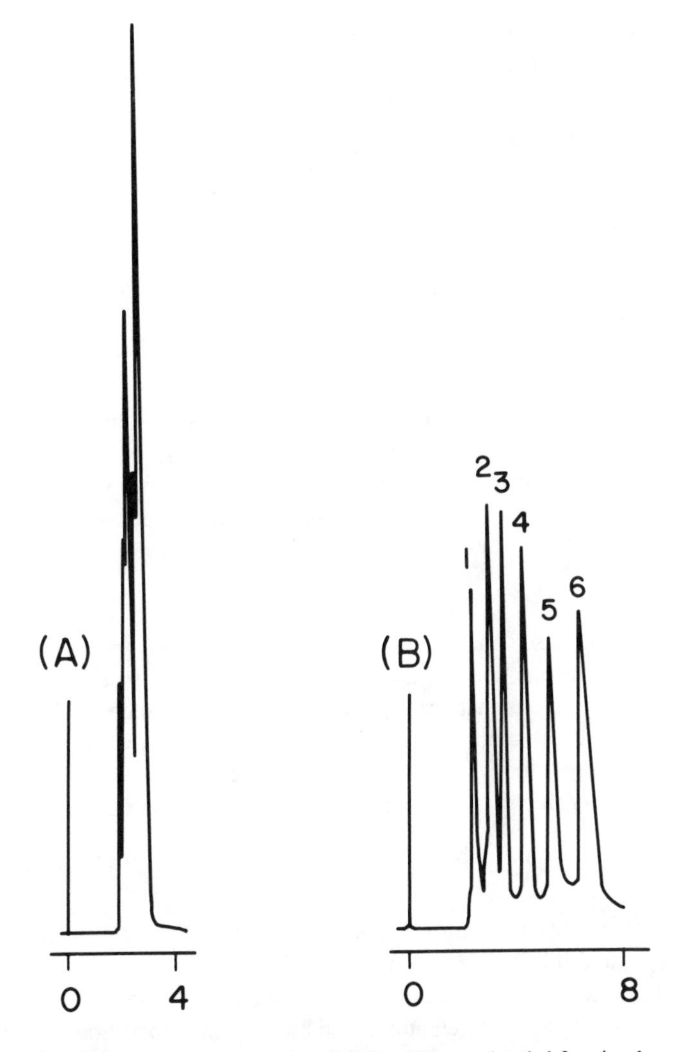

FIGURE 1. Chromatograms of dialkyl sulfides on bonded 2-amino-1-cyclopentene-1-carbodithioic acid. The absorbent is the one with the spacer group, the second cyclopentene formula in the text. (A) is without copper, (B) with copper(II). Peaks in chromatogram (B) are in the order: di-*n*-hexyl sulfide, di-*n*-propyl sulfide, ethyl *n*-propyl sulfide, diethyl sulfide, and tetrahydrothiophene (eluted last). (From Takayanogi, H., Hatano, O., Fujimara, K., and Ando, T., *Anal. Chem.*, 57, 1840, 1985. With permission.)

In Chapter 2 we reviewed the work of Skorokhod and Varavva[8] on metal-thiourea complexes in cation-exchange resins, pointing out that in some cases the complexes were much more stable in the exchanger than in solution. In other words, interactions between the ligand and the exchanger matrix were very important. Even stronger interactions were seen in aluminosilicate exchangers. We are reminded that metal-ligand interactions in ion exchangers are very complex and are affected by factors other than the strength of the metal-ligand bond.

II. OXYGEN LIGANDS

A. Carbohydrates and Polyhydroxy Compounds

In the practice of ligand-exchange chromatography today, the most important oxygen ligands are the carbohydrates. Indeed, of all the applications of ligand-exchange chromatography, after the analysis of amino acids, the analysis of sugars and other carbohydrates is that which is most frequently performed.

Sugars and saccharides can be analyzed by column liquid chromatography in several different ways. A comprehensive review that devotes much space to oligosaccharides was recently published by Honda.[9] We shall present a brief summary that may be useful to the chemist who needs to analyze carbohydrate mixtures, then go on to describe in more detail the techniques that use ligand exchange.

The different combinations of stationary and mobile phases that are used today are the following.

1. Method A

A column of amino-bonded silica (sometimes cyano bonded) with acetonitrile-water mixtures as the mobile phase.[10-15] The proportion of acetonitrile ranges from 50 to 85% by volume. The main mechanism of retention seems to be hydrogen bonding.[12] Raising the water content of the eluent lowers the retention. Good separation factors are obtained; oligosaccharides are retained more strongly, the higher their molecular weight.

2. Method B

A column of strong-base, anion-exchange resin of the low-capacity, surface-functional type, with 0.15 *M* aqueous sodium hydroxide as the eluent. In this medium sugars are deprotonated, forming anions that are separated by anion exchange.[16] They are detected by a pulsed amperometric detector. The method is fast and sensitive; a mixture of common mono- and disaccharides is analyzed in ten min. Oligosaccharides are well resolved, retention being greater the higher the molecular weight. It is likely, however, that degradation occurs in the column.[17]

3. Method C

A column of high-capacity, ion-exchange resin is used. The resin may be of the gel or macroporous type, and it can be a cation or an anion exchanger. The eluent is a mixture of water and ethyl alcohol, the alcohol concentration being 75 to 95%; the higher the alcohol concentration, the stronger is the retention. The mechanism is partition; within the resin the solvent has a higher proportion of water than does the external solvent, and sugars are more soluble in water, because of hydrogen bonding, than they are in alcohol.[18]

Best results are obtained with a strong-base anion-exchange resin in the sulfate form. The effect of the counter-ion in cation-exchange resin partition chromatography was mentioned in Chapter 2, Section C; K^+ gives stronger retention of sugars than Na^+, Na^+ gives stronger retention than Li^+.

High sensitivity (in the low microgram level) can be obtained by post-column derivatization, but compared with newer methods, this one is cumbersome and slow.

4. Method D

A column of strong-base, anion-exchange resin in the borate form, with aqueous borate solutions as eluents.

This method depends on the anion exchange of sugar-borate complexes, and has something in common with ligand exchange. It does not fall under the definition of ligand-exchange chromatography that we gave in Chapter 1 because no metal ions are present; but it does depend on the formation and dissociation of complexes of organic ligands with an inorganic ion. It was one of the very first methods to use ion exchange to separate nonionized organic compounds.[19] Cyclic complexes are formed between borate ions and *cis*-1,2-diols having these structures;[20]

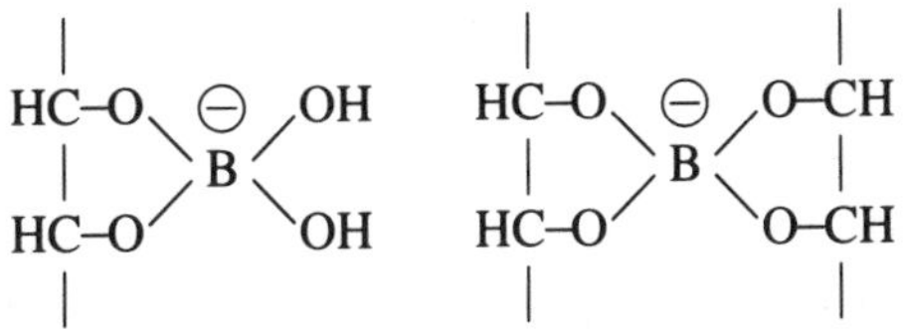

```
  |                     |           |
HC–O   ⊖   OH         HC–O   ⊖   O–CH
  |    \B/               |    \B/    |
HC–O   /  \OH         HC–O   /  \O–CH
  |                     |           |
```

The acids corresponding to these anions are much stronger acids than boric, a fact that makes possible the titration of boric acid after adding mannitol. Neutral complexes of boric acid with 1,2-diols,

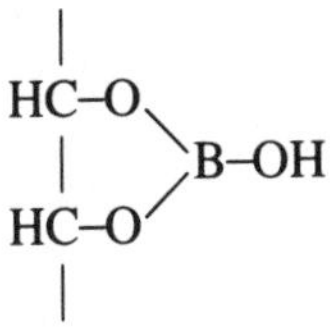

```
  |
HC–O
  |    \B–OH
HC–O   /
  |
```

are unstable, but neutral complexes of 1,3-diols are known and can be extracted from aqueous solutions.[21]

Very little is known about the stabilities of the anionic complexes, particularly with sugars and polyhydric alcohols. Perhaps the complexity of borate ions in aqueous solution has discouraged investigators, or the bewildering number of possibilities for association that exist. The 1:1 complex of ethylene glycol with borate ion has a formation constant of 1.0 $\ell\ mol^{-1}$;[20] the 2:1 complexes of glucose and mannitol with borate were found to have overall formation constants, β_2, of 1.9×10^2 and 1.4×10^5, respectively.[21] In anion-exchange chromatography of borate complexes, the concentration of borate ions is higher inside the exchanger than in the eluent solution, and one expects sugars and polyols to be retained more strongly the more stable are their borate complexes. In general, this is so; Larsson and Samuelson[23] found a parallel between the retention volumes of sugars and their electrophoretic mobilities in borate solutions, but the parallel was not exact. Even though mannitol forms a 2:1 borate complex that is 700 times more stable than the glucose complex (see above) the difference in retentions is quite small. Sorbitol, glucose, and mannitol were eluted in that order from an anion exchanger by a borate buffer.[24] Thus, it seems that the stabilities of the complexes in solution, even if we knew them, would be a poor guide to elution orders.

Typically the eluents are borate solutions, 0.1 to 0.5 M, pH 7 to 9. Raising the borate concentration and pH lowers the retention.[25,29] Elution volumes range widely, and for this reason gradients are employed. In contrast to amino-bonded phases, which deteriorate with repeated use, the resins used in the borate method are robust and not easily fouled. Therefore, the borate method is well adapted to automatic analysis.[9,24-29] It is hard to devise general methods for all types of carbohydrates because no all-purpose detector exists. Refractive index cannot be used with gradient systems, and in any case, its sensitivity is low. Post-column reactions are commonly used with borate eluents, and many of the reagents detect only reducing sugars; thus, the elution conditions must be matched with the detection method.

A typical isocratic procedure designed for reducing mono- and disaccharides uses a column 75 × 6 mm, packed with a surface-functional, strong-base resin of particle diameter 10 μm. The temperature is 50°C (temperatures higher than this cause slow degradation of the resin), flow rate is 40 mℓ/hr; the eluent is 0.4 M borate, pH 8.0.[26] The compounds are eluted in this order, the retention times, in minutes, being in parentheses:

2-deoxyribose (7)
maltose (8.8)
lactose (9.3)
ribose (14.4)
lyxose (14.8)

mannose (18.6)
fructose (25.0)
arabinose (26.3)
galactose (30.3)
xylose (40.1)
sorbose (41.3)
glucose (55.2)

Detection is done by reduction of Cu(II) bicinchoninate to the purple Cu(I) complex, whose absorbance is measured at 570 nm. Microgram quantities of sugars are easily measured.

The same elution sequence is found in gradient systems, which generally give better resolution in less time. One of the older gradient systems[30] uses three buffers, first 0.13 *M* potassium borate of pH 7.5, second 0.25 *M* borate of pH 9.1, third 0.35 *M* borate of pH 9.6; a combination of step and continuous gradients is used. The entire analysis, finishing with glucose, takes 5.5 hr, the longer time being probably due to the use of gel-type resin rather than a surface-functional resin. Orcinol reagent was used for detection, and sucrose and cellobiose came out before maltose. Fructose emerged between arabinose and galactose.

A more recent paper by Honda et al.[29] gives a definitive description of an automated system for carbohydrate analysis that can be run in a step-gradient or continuous gradient mode, or isocratically. In the step gradient mode, three solutions are used, 0.2 *M* borate of pH 7.4, 0.35 *M* borate of pH 7.7, and 0.50 *M* borate of pH 9.2. The continuous gradient gives somewhat better resolution at the cost of greater complexity. The column for the step gradient is 150 × 4 mm, packed with a specially designed surface-functional, strong-base, anion-exchange resin with 11-micron particles (Hitachi® 2633); temperature is 65°C, flow rate 0.5 mℓ/min. Detection is by the reaction of reducing sugars with 2-cyanoacetamide at 100°C and pH 10.5 to give a product that absorbs at 276 nm and is fluorescent, so that low detection limits, about 0.1 nmol or 20 ng of monosaccharide, are possible. Glucose, the last of the common sugars to be eluted, comes out in 90 min. In contrast to the amino bonded phases, the borate-form anion exchanger retains disaccharides more weakly than monosaccharides, and trisaccharides more weakly still.[27]

Other reagents that can be used for detection in borate effluents include ethylenediamine[31] and periodate,[24,32] the latter being able to detect sugar alcohols like mannitol and sorbitol, as well as aldoses and ketoses.

An interesting variant on anion exchange or borate complexes is the use of bonded silica with a borate functional group;[33]

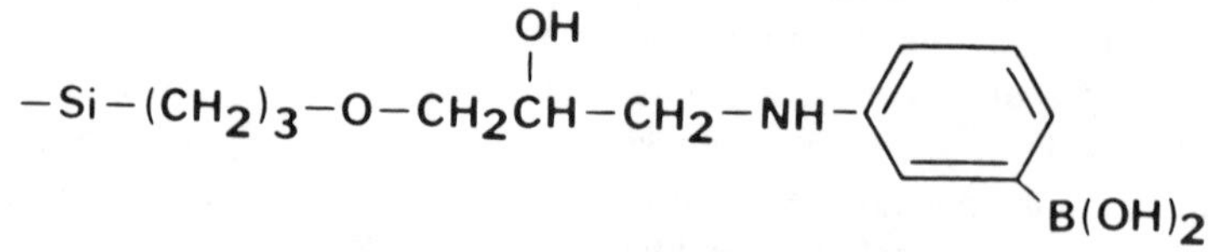

$$-\mathrm{Si}-(\mathrm{CH_2})_3-\mathrm{O}-\mathrm{CH_2}\underset{|}{\overset{\mathrm{OH}}{\mathrm{CH}}}-\mathrm{CH_2}-\mathrm{NH}-\mathrm{C_6H_4}-\mathrm{B(OH)_2}$$

The eluent is 0.01 *M* sodium pyrophosphate; detection is by refractive index. The elution order is quite different from that in anion exchange of borate complexes; glucose is retained very weakly, followed by sucrose and galactose; fructose is retained quite strongly, mannitol and sorbitol even more strongly.

5. *Method E*

A column of cation-exchange resin with a noncomplexing counter-ion.

We have noted (Chapter 2, Section C) that nonionized organic compounds are distributed between water and a swollen ion-exchange resin by at least two mechanisms. One is the nonpolar or van der Waals interaction between the nonpolar carbon chain of the organic molecule and the polystyrene resin "backbone". This attraction is reinforced by π-electron overlap if the organic solute has aromatic character. The other mechanism is hydrogen bonding between polar groups, especially –OH, of the organic solute and water molecules inside and outside the resin. The first interaction draws organic solutes into the resin. The second draws them out, for there is more opportunity for hydrogen bonding in the water outside the resin. The degree of hydration of the inorganic counter-ion of the resin affects these interactions.

Sugar molecules, being rich in –OH groups, tend to stay outside the resin. It is found that the ratio of sugar to water is less inside the resin than out. Comparing the counter-ions Li^+, Na^+, and K^+ in a sulfonated polystyrene cation-exchange resin, it is found experimentally that sugars are absorbed least by the Li^+-form resin, most by the K^+-form resin. The difference is ascribed to the greater hydration of Li^+, which makes less water available for hydrogen bonding with the dissolved sugar.

In 1968, Saunders[34] used a large column of potassium-form, 4% cross-linked cation-exchange resin to separate sucrose, glucose, xylose, and fructose, which were eluted in that order, as well as other sugars. The eluent was pure water. Other workers have used potassium-form resins of small particle size for the same purpose.[35] Recently, Pecina and co-workers[36] have used a column of hydrogen-form resin, 300 × 8 mm, to separate not only carbohydrates and sugar alcohols, but short-chain alcohols, aldehydes, ketones, and carboxylic acids from one another; the eluent was 0.01 *M* sulfuric acid. From the elution volumes it was evident that sugars were eluted beyond the interparticle void volume of the column, but before the total volume of water, counting the water inside the resin beads as well as that between the beads. Sugar molecules were partially excluded from the resin, while less polar compounds, like the butanols, butanal and furfural, were retained to some extent. Hydrophobic interaction with the resin polymer was stronger than hydrogen bonding.

Studying the retention of sugars, sugar alcohols, aliphatic alcohols, and other compounds on a cation-exchange resin in different ionic forms, Walton[37] measured interparticle void volumes by injecting dilute salt solutions into the water eluent, and total water volumes in the column by injecting deuterium oxide. Sugars (sucrose, glucose, fructose, ribose, and others) and polyols (glycerol, pentaerythritol, mannitol, sorbitol) had retention volumes intermediate between the interparticle void volume and the total water volume as measured by D_2O retention. Thus, these compounds were all excluded to some extent from the resin. In every case the retention was greater with K^+-form resin than with Li^+-form resin.

If one adds an organic solvent, like methanol, ethanol, or acetonitrile, to the water eluent, retention of sugars and polyols is greatly increased. Several workers have taken advantage of this effect in chromatography.[38-40] Samuelson based his partition chromatography on solvents rich in ethanol.[18] The rationale for this effect was discussed above.

In cation-exchange resins carrying Li^+, Na^+, K^+, and H^+, there is no evidence that these counter-ions interact directly with the organic solutes; their effects can be interpreted simply by ionic hydration. When the counter-ions are Ca^{2+} or La^{3+}, however, marked differences in the retentions of different solutes are observed, and it becomes clear that direct coordination or complex-ion formation takes place. Ligand exchange is now an important mechanism in chromatographic separations.[37,41,42]

6. Method F and the Techniques that Use Ligand Exchange

A column of cation-exchange resin carrying a counter-ion that forms coordination complexes with sugars.

The first systematic study of the chromatography of sugars and polyols on a cation-exchange resin, and the effect of the inorganic counter-ion on retention, was published in 1975 by Goulding.[41] He needed a fast method of analysis that allowed individual sugar fractions to be recovered in as pure a state as possible for in vivo testing. Elution with borate solutions was not satisfactory; the use of water as the eluent was appealing. Chromatography on a potassium-form polystyrene-type cation-exchange resin had been used successfully by Saunders,[34] and an important review of similar methods using ion-exchange resins had been published in 1974.[43] Goulding used a column 500 × 3 mm, packed with a gel-type cation-exchange resin of particle size 11 μm and run at room temperature with water eluent and refractive index detection. The flow rate was 0.1 mℓ/min. He got useful separations in 20 to 40 min and found, first, that the α- and β-anomers of several sugars were partially separated. Later workers[42] showed that the anomers could be completely separated by using a low temperature (3 to 4°C) and they confirmed Goulding's findings that calcium-form resins gave different elution orders of these isomers than the resins in alkali-metal forms.

Comparing different counter-ions, Goulding showed that the complexing cations, Ag^+, Ca^{2+}, Sr^{2+}, Ba^{2+}, Cd^{2+}, and La^{3+}, gave longer elution times than the alkali-metal cations for the solutes that were the most strongly held; they caused the peaks to be spread out and gave better resolution. He attributed this effect to coordination between the metal ions and –OH groups of sugars and polyols, and showed that the retentions could be correlated with the stereochemistry of these compounds. Those whose configurations have an axial-equatorial-axial sequence of hydroxyl groups favor formation of tridentate complexes:

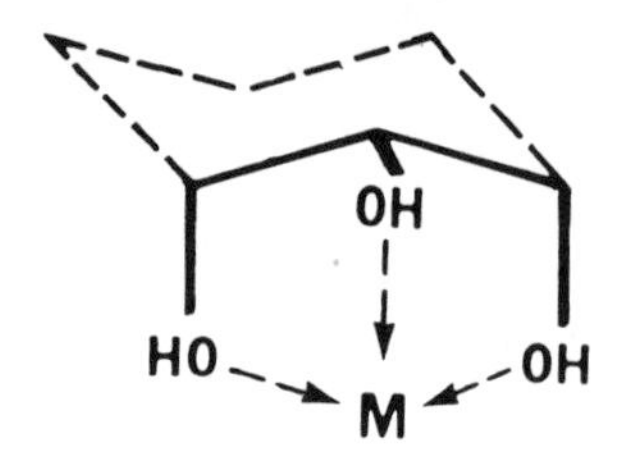

and these are the compounds that are most strongly retained on a calcium-form resin. Thus, ribose is by far the most strongly held of the pentose sugars, talose and allose the most strongly held of the hexoses.

There are a few data in the literature, but only a few, on the association of sugars and polyols with Ca^{2+} and other cations. Angyal[44,45] used NMR to measure the association of Ca^{2+} with D-allose in aqueous solution. Adding $CaCl_2$ to the sugar solution increased the resonance of the α-pyranose form while decreasing that due to β-pyranose and the two furanose forms, and it was deduced that α-D-allopyranose, with the axial-equatorial-axial sequence of hydroxyl groups, formed a complex with Ca^{2+} that had a stability constant 6 $\ell\ mol^{-1}$. Its La^{3+} complex had stability constant 10 $\ell\ mol^{-1}$, while the complexes of Na^+ and Mg^{2+}, if they existed, had stability constants below 0.1 $\ell\ mol^{-1}$. There was no sign of complexes of Ca^{2+} with glucose, mannose, or arabinose.

Using ion-selective electrodes, the lead complexes of ribose, xylose, and arabinose (all pentoses) were found to have stability constants 1.9, 0.4, and 0.4 $\ell\ mol^{-1}$ respectively, while the calcium complex of ribose had stability constant 1.6 $\ell\ mol^{-1}$.[46] The same value for the calcium-ribose complex was found by other workers,[47] who also reported that xylose did not form a complex. The electrophoretic mobilities of sugars in calcium acetate solution, which give a rough guide to complex stabilities, showed that ribose stands out among other sugars, both hexoses and pentoses, indicating the stability of its complex with calcium ions.[48] Ribose also stands out among other sugars in the strength of its retention by a calcium-form ion-exchange resin.[37,42,47]

From the effect of mannitol and sorbitol on the solubility of calcium sulfate it was found that these ligands form complexes with Ca^{2+} having formation constants 0.7 and 1.2 $\ell\ mol^{-1}$, respectively.[37]

These stability constants are small compared to those of metal-amine and metal-amino acid complexes, and the chromatographic retentions they produce are likewise small, but they are enough to give good resolution with efficient columns. The small retentions allow fast chromatography, compared to the anion exchange of borate complexes.

Many studies have been published on the chromatography of sugars and polyols on calcium-loaded cation-exchange resins. As we have observed, this form of chromatography is one of the most popular applications of ligand-exchange chromatography today. The conditions employed may be summarized as follows.

The resin is generally a gel-type resin of particle size 10 μm or less. Fitt[49] reported best results with 4% cross-linking, though cross-linkings of 6 and 8% were almost as good. By contrast, Honda[42] used a highly cross-linked resin. Most authors have been content to use one of the commercially packed columns developed specially for carbohydrate analysis, about whose contents there is little information in the open literature. Column dimensions are typically 300 × 8 mm, though much longer columns have been used. Because capacity factors are small, and many sugars that do not form calcium complexes are eluted soon after the void volume, the column must be large to give sufficient resolution.

The column temperature is generally 80 to 85°C, though lower temperatures (55°C) may be used. Higher temperatures give faster diffusion and narrower bands, but they also give smaller retentions. A reason to use high temperatures is to hasten the conversion between α- and β-anomers, the mutarotation of sugars, so that each sugar will give a single chromatographic peak. Mutarotation may be catalyzed by adding a tertiary amine to the eluent.[50,51]

The hydrolysis, or "inversion", of sucrose and other disaccharides is not wanted. It is acid-catalyzed and will occur if hydrogen ions are allowed to accumulate in the resin column. This condition may be avoided by flushing the column with a calcium salt solution from time to time[52,53] or by adding 10^{-4} *M* calcium acetate to the water eluent.[54]

As a rule, the eluent is pure water, though some workers add up to 25% of methanol or acetonitrile to increase retention.[50,53] Detection is nearly always done by refractive index, though higher sensitivity may be obtained by postcolumn derivatization with 2-cyanoacetamide, followed by fluorescence detection.[55]

Figures 2 and 3 show chromatograms from a commercial column, the first with pure water eluent, the second with acetonitrile:water mixed 1:3 by volume. Figure 4 shows a chromatogram from three commercial columns joined in series.

The elution order is seen from Table 1, where retention volumes on a 6%-cross-linked sulfonated polystyrene resin are compared for four counter-ions. The volumes are normalized to deuterium oxide = 1. The "salt peaks" were obtained by injecting dilute solutions of salts of the cations in question, for example $Ca(NO_3)_2$ for the calcium-loaded column. Assuming salt exclusion by the Donnan equilibrium, these "salt peaks" give the interparticle void volumes. This assumption is not true for $La(NO_3)_3$ because of strong ion association.

All the sugars except ribose are eluted before deuterium oxide, indicating displacement out of the swollen resin into the aqueous mobile phase. Probably the main factor determining the partition of these sugars is hydrogen bonding. Only ribose, mannitol, and sorbitol, of the compounds studied, coordinate with calcium ions to any extent. Taking into account the molal concentration of calcium ions within the resin, one infers that the calcium-polyol complexes have about the same stabilities in the resin that they have in free aqueous solution.[37]

The elution order shown in Table I is almost the same as that reported by other workers.[41,51,56] It is nearly the reverse of the elution order from an amino-bonded phase, a point that has been noted by Baust and others.[57] Oligosaccharides are eluted very early from a calcium-form ion-exchange resin, and the higher the molecular weight, the less they are retained.[58-60] Fitt[49] found a linear relation between the number of glucose units in the oligomer, from n = 1 to n = 6, and the logarithm of the corrected retention volume. On an amino-bonded phase oligomers are retained more strongly, the greater their molecular weight. The same relation holds for partition chromatography on anion- and cation-exchange resins in

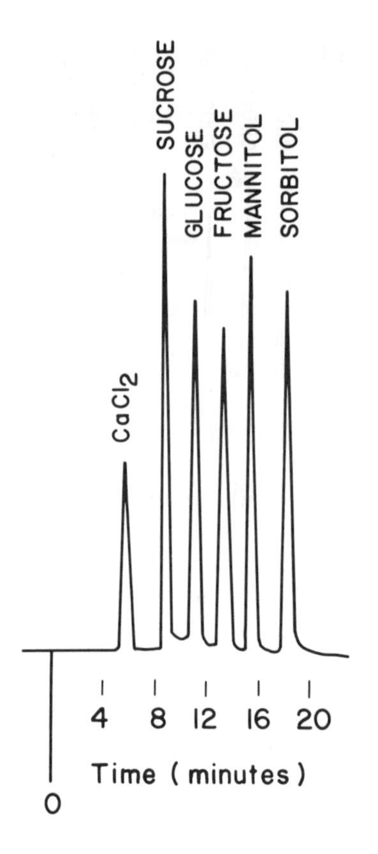

FIGURE 2. Chromatogram of common sugars and sugar alcohols on a calcium-loaded ion-exchange resin. Column, 300 × 6.5 mm; temperature, 85°C, flow rate 0.4 mℓ/min; eluent, water. Quantity of each compound, 100 μg. (From Vidalvalverde, C. and Martin-Villa, C., *J. Liq. Chromatogr.*, 5, 1941, 1982. With permission.)

alcohol-rich solvents; Samuelson[61] found that glucose oligomers were held more strongly the greater their molecular weight, and that log k′ increased linearly with the number of glucose units; in general, the free energy of retention was proportional to the number of hydroxyl groups in the molecule; the mobile phase was 70% ethanol, 30% water. It seems that in the absence of coordination to the metal ions, the dominant force in retention is hydrogen bonding to the water in the stationary phase. For the practical chromatographer who wants to analyze mixtures of oligomers, the Samuelson system of partition chromatography or the use of an amino-bonded phase seem better choices than the use of a calcium-form cation-exchange resin with water as the mobile phase.

Analytical applications of a calcium-loaded resin to simple mixtures of three or four sugars, using water as the eluent, have been described;[62-66] the earlier papers specify 4% cross-linked resins.[59,62] Automated methods are described for saccharides[65] and polyols.[63] The usual detection method is refractive index, and with this method the detection limits are about 1 μg. Precolumns are advised where amino acids, carboxylic acids, or inorganic salts are present.[66,67]

A countercurrent system using calcium-form resin has been devised for the preparative separation of glucose and fructose.[68] A novel stationary phase, a molecular-sieve zeolite, was used by Wortel and van Bekkum[69] to separate glucose, fructose, and mannitol; with sodium as the replaceable cation, glucose and fructose were not separated, but if calcium ions were introduced, these sugars were separated, fructose being held more strongly than

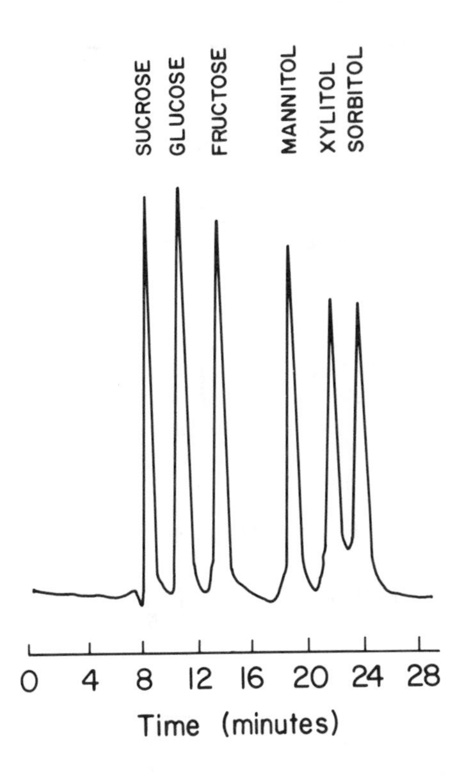

FIGURE 3. Chromatogram of common sugars and sugar alcohols on a calcium-loaded ion-exchange resin. Column as in Figure 2; temperature, 80°C, flow rate 0.5 mℓ/min; eluent, 25% acetonitrile by volume. Note the increased elution volumes, compared to water eluent (Figure 2), and the inclusion of xylitol. (From Vidalvalverde, C., Olmedilla, B., and Martin-Villa, C., *J. Liq. Chromatogr.*, 7, 2003, 1984. With permission.)

glucose, as is the case with calcium-loaded resins. In the inorganic exchanger it is hard to see that hydrogen bonding is important, and it is very likely that coordination of fructose with calcium ions is strong enough to affect the retention. This is Goulding's interpretation.[41]

Though calcium ions are the complexing cations usually used in sugar chromatography, other cations are also effective. We have mentioned the use of lanthanum-loaded resins. Goulding[41] noted this possibility, and his paper gives some retention data, as does Table I. Substituting La^{3+} for Ca^{2+} increases considerably the retention of polyols and increases the separation between them. Petrus et al.[70] separated 14 alditols on a lanthanum-loaded cation-exchange resin with water eluent. Their paper compares La^{3+} with other counter-ions, especially Ca^{2+} and Ba^{2+}. Polyols are retained most strongly and with the largest separation factors on La^{3+}-form resin, but Ba^{2+}-form resin retained some of the aldoses more strongly. In our experience, a drawback to the La^{3+}-resin is the broadness of the elution bands, associated with low swelling and low water content and, therefore, slow diffusion in and out of the resin. Nevertheless, La^{3+} resins do offer high selectivity.

Cation-exchange resins carrying lead ions have been used for chromatography of sugars; it is claimed that Pb^{2+} gives better separations of monomeric sugars (pentoses, hexoses) than does Ca^{2+}.[60] For separating oligosaccharides, Scobell[59] found that a silver-loaded, 4% cross-linked resin gave more efficient separation with much higher plate numbers than did the same resin in the Ca^{2+} form. The Ca-resin gave distinct peaks for oligosaccharides up to D_p = 8 (D_p means degree of polymerization, the number of glucose units) whereas the Ag-resin gave distinct peaks up to D_p = 12 (see Figure 5). Actually, the pure, homoionic silver form was not so efficient; best resolution was obtained with a resin loaded 70% with

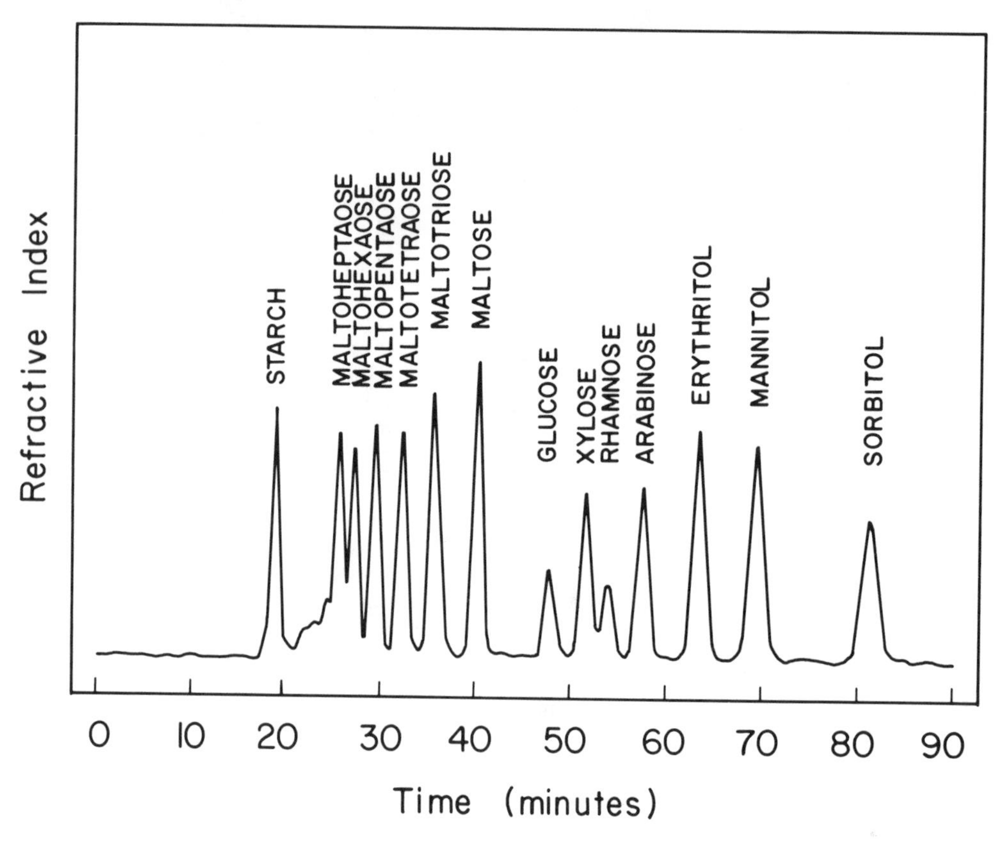

FIGURE 4. Chromatogram of sugars and oligosaccharides on a calcium-loaded ion-exchange resin. Column, 900 × 7.8 mm; temperature, 85°C, flow rate 0.5 mℓ/min; eluent, water. (From Schmidt, J., John, M., and Wandrey, C., *J. Chromatogr.*, 213, 151, 1981. With permission.)

Ag^+, 30% with Ca^{2+} (by equivalents). Figure 5 shows a chromatogram of a hydrolyzed corn syrup on such a resin.

Complex formation in the mobile phase has been used to separate glucose, galactose, and three amino sugars. The stationary phase was amino-bonded silica, the mobile phase an aqueous solution of cadmium sulfate.[71]

Copper(II) ions have naturally been used for ligand-exchange chromatography of sugars and polyhydric alcohols, first in a resin column under gravity flow,[72] then under high-performance conditions[73] with 5-μm spherical porous silica beads impregnated with Cu(II) by the method of Foucault (Chapter 2, References 43 and 44), which is to stir the silica with a solution of copper sulfate in aqueous ammonia, then rinse and dry before packing into the column. For chromatography of sugars, the eluent was acetonitrile-water, 3:1 by volume, 1.5 *M* in ammonia and 2×10^{-5} *M* in Cu(II). Detection was by ultraviolet absorption of the sugar-copper complexes at 254 nm; the limit of detection was one tenth of that obtained by refractive index. Resolution was excellent, and the order of elution was quite different from that on calcium-loaded exchangers. For the sugars tested, it was: rhamnose (first), xylose, fructose, glucose, sucrose, maltose, lactose (last). Retention and peak separation increased with the proportion of acetonitrile. The method seems to have potential for the analysis of disaccharides.

The stabilities of complexes of sugars and polyols with 12 metal ions have been compared by thin-layer chromatography.[74] Copper(II) forms the most stable complexes, with Cr(III) in second place.

Table 1
RETENTION OF SUGARS AND POLYOLS ON 6%-CROSS-LINKED SULFONATED POLYSTYRENE

Counter-ion	Li^+	K^+	Ca^{2+}	La^{3+}
Salt peak	0.385	0.47	0.50	0.72
Sucrose	0.52	0.60	0.57	0.60
Lactose	—	—	0.56	0.60
Glucose	0.58	0.76	0.67	0.65
Xylose	—	0.82	0.77	0.71
Sorbose	—	—	0.80	0.71
Fructose	0.59	0.83	0.82	0.77
Arabinose	—	—	0.90	0.77
Pentaerythritol	0.75	0.73	0.90	1.08
Glycerol	0.77	0.82	0.96	1.10
Mannitol	0.64	0.73	1.10	1.44
Sorbitol	0.665	0.74	1.36	2.60
Ribose	0.72	0.94	1.43	2.05

Note: Deuterium oxide retention = 1.00. Volume retentions of D_2O were; Li^+ 5.81 mℓ, K^+ 5.62 mℓ, Ca^{2+} 5.58 mℓ, La^{3+} 4.95 mℓ. Column temperature, 60°C.

Adapted from Walton, H. F., *J. Chromatogr.*, 332, 203, 1985.

Iron(III) forms complexes with sugars in aqueous solution that have formation constants 100 ℓ mol^{-1} and more,[75] but attempts to separate sugars on a Fe(III)-loaded cation-exchange resin were only marginally successful.[76]

B. Phenols

Phenols form complexes with Fe(III) that are dark red or purple. Their appearance is used as a qualitative test for phenols. Their properties have been studied in detail and their stability constants measured.[77,78]

The primary reaction of phenol with Fe(III) in acidic aqueous solution is the following:

$$C_6H_5OH + Fe(OH_2)_6^{3+} = C_6H_5OFe(OH_2)_5^{2+} + H_3O^+$$

The equilibrium constant for this reaction is $10^{-2.04}$. Combining this value with $pK_a = 9.80$ for the ionization of phenol as an acid, we find that the logarithm of the stability constant of $C_6H_5Fe(OH_2)_5{}^{2+}$ is 7.76. Chloro- and nitrophenols form complexes that for the most part are weaker than those of phenol itself, but they are stronger acids than phenol, so that the equilibrium constants for proton displacement, the reaction written above, are all roughly the same. A quantitative treatment of the association of phenol with Fe(III) must of course take account of the hydrolysis of $Fe(OH_2)_6{}^{3+}$.[78]

It was natural to use the Fe(III)-phenol association to recover phenols selectively from water and other media. Kaolin impregnated with ferric chloride absorbs the more weakly acidic phenols from crude petroleum, along with other compound classes such as carbazoles and amides;[79] the absorbed compounds are recovered by strilling with methylene chloride. Phenols, especially chlorophenols, are recovered from polluted water by sorption on the iminodiacetate chelating resin, Chelex®-100, loaded with Fe(III).[80] It was best to add acid

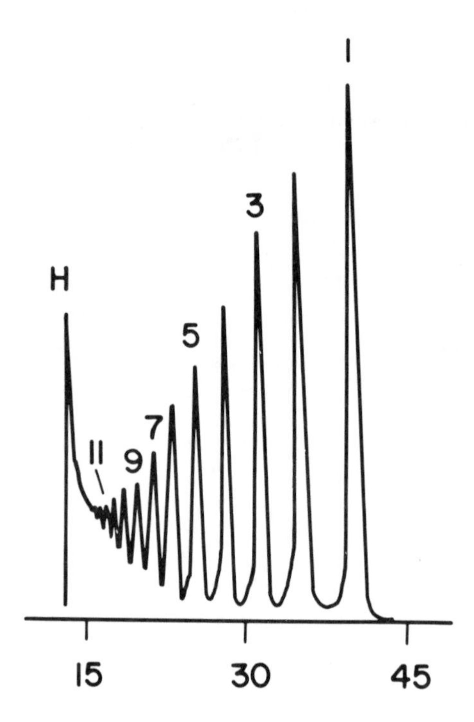

FIGURE 5. Chromatogram of hydrolyzed corn syrup on a 5% cross-linked ion-exchange resin loaded with silver (71 equivalents %) and calcium (29 equivalents %). Column, 300 × 7.8%; particle size, 10 to 15 μm; temperature, 85°C, flow rate 0.25 mℓ/min. Numbers over peaks show degree of polymerization; H means higher, excluded polysaccharides. (From Scobell, H. D. and Probst, K. M., *J. Chromatogr.*, 212, 51, 1981. With permission.)

to the water to bring the pH below 3 before passing it through the column. The phenols were stripped by 1 *M* sodium hydroxide in 20% ethanol. Recoveries were 70 to 80% for phenol and monochlorophenols, and were nearer 100% for the higher chlorinated phenols. In the polluted water their concentrations were only 100 ppb or μg/ℓ.

In later papers, Petronio et al.[81,82] showed that different substituted phenols could be stripped in succession from the iron-loaded resin by using eluents of increasing pH. Phenol was stripped first at pH 8.5, then, above pH 10, nitro-, dichloro-, and pentachlorophenol were desorbed in that order. The differences in sorption strengths were great, and, as we have noted, they cannot be explained by differences in formation constants. Rather they are related to hydrophobic bonding with the polystyrene resin matrix. The same sequence is found in the reversed-phase chromatography of substituted phenols on a C_{18}-silica bonded packing. Nitro groups and chlorine atoms both increase the retention of phenols, as do alkyl substituents.

Chromatographic separations of substituted phenols were accomplished by Maslowska and Pietek, first on Fe(III)-loaded sulfonated polystyrene resin,[83] then on the same resin loaded with cerium(III).[84] In the first case, the eluent was a solution of sodium perchlorate containing perchloric acid; a gradient was used with pH increasing from 3 to 5; the elution order was chloro-, bromo-, nitro-, and methyl (*m*-cresol). With the Ce(III)-loaded resin, the eluent was 0.015 *M* sulfuric acid in 30% acetonitrile. Seven components were separated, starting with unsubstituted phenol and followed by chlorinated phenols. Isomers of chlorophenols were well separated. Column efficiency was poor because a large particle-sized resin was used (200-400 mesh), yet the detection limit was below 1 nmol. A more efficient packing, silica gel bonded with 8-hydroxyquinoline and loaded with Fe(III), was tested by

Shahwan and Jezorek;[85] here the eluent was an acetate buffer in 40% acetonitrile. Again, the substituents Cl, CH_3, and NO_2 increased retention.

Phenols are important as environmental pollutants, and it seems that the possibilities of ligand exchange for concentrating, recovering, and analyzing them are only just being realized. Their economic value is such as to help pay for their removal from polluted water. Chanda et al.[86] report the use of sulfonic and iminodiacetate polystyrene-based cation-exchange resins loaded with iron(III) for removing phenols from industrial waste waters. The absorbed phenols are later stripped and recovered in pure form. Thiocyanate ions are similarly removed and recovered from waste waters from coal coking and gasification.[87] Alumina carrying adsorbed Fe(III) takes up phenols reversibly and may be used to separate phenols by ligand-exchange chromatography; distribution ratios of 17 phenols were measured in six solvent systems.[88]

C. Carboxylic Acids

The anions of carboxylic acids associate with most metal ions to give 1:1, 1:2, and higher complexes whose stability constants are tabulated in reference books.[77] With Cu(II) and Zn(II), for example, acetate ions form complexes with K_1 about 100 and 10, respectively. More stable complexes are formed with hydroxy acids that can form bidentate chelate rings. Thus metal-loaded cation exchangers will absorb carboxylate ions.

In 1974, Bedetti[89] used the iminodiacetate chelating resin Chelex®-100 in the nickel form to absorb a series of hydroxy- and dicarboxylic acids from a dilute ammonia solution. Elution with 10^-5 *M* ammonia brought glycollate, succinate, malonate, and oxalate ions out of the column in that order, which is the order of increasing stabilities of the nickel complexes of these ions. Almost simultaneously, Funasaka and associates[90,91] separated several hydroxybenzoic and hydroxynaphthoic acids on columns of polyacrylic (carboxylic) and sulfonated polystyrene cation exchangers carrying ions of Fe(III), Cu(II), Hg(II), and Rh(IV). They were eluted with ethanol or water. Of the hydroxybenzoic acids, the *ortho*-isomer (salicylic acid) was eluted first, then *meta*-, then *para*-. Previously, these authors had separated aminobenzoic acids and nitrosonaphthols from cation exchangers loaded with Cu(II) and Fe(III), respectively, with dilute ammonia eluents.[92,93] Later, Maslowska and Pietek,[94] compared Al^{3+}, Fe^{3+}, and Ce^{3+} as counter-ions for chromatography of aromatic acids on a sulfonated polystyrene exchanger. The eluent was 0.01 *M* aqueous sulfuric acid. Ferric ion gave the best peak separation. Acids were eluted in 20 min in the order benzoic, cinnamic, 4-hydroxybenzoic, 2,3-dihydroxybenoic, 2,4-dihydroxybenzoic, 4-hydroxy- 3-methoxybenzoic, *cis*-2-hydroxycinnamic, and 4-hydroxy-2-methoxycinnamic acid. The effects of π-electron interaction of the solutes with the resin matrix and of chelation with the ferric ion are seen from this elution order. Chelation seems to be the more important effect.

Otto et al.[95] used lanthanum ions as counter-ions on a sulfonated polystyrene resin and compared the retention of aromatic acids and neutral aromatic compounds on the resin with Na^+ and La^{3+} as counter-ions and dilute acetate buffers of pH 5 as eluents. Lanthanum salt had to be added to the buffer to prevent stripping of lanthanum from the column. The aromatic acids were retained much more strongly with La^{3+} as counter-ion, compared with Na^+, and were eluted in the order uric acid (first), benzoic, 4-methoxybenzoic, cinnamic, and mandelic acid (last). Separate experiments showed that La^{3+} associated with the organic acid anions to give doubly charged 1:1 complexes, LaA^{2+}, the most stable of which was the mandelate complex, with stability constant 130 ℓ mol^{-1}. The lanthanum column is selective for anions of aromatic carboxylic acids and hydroxy acids. (Mandelic acid is 2-hydroxyphenylacetic acid). Its utility in chemical analysis is clouded by the rather long elution times, up to 60 min, and the broad bands, related, as we have observed earlier, to the low water content of a resin carrying triply charged cations and the resulting slow diffusion.

Lanthanum-loaded resin absorbs aliphatic hydroxy-acids like malic, tartaric, and citric acids very strongly and may someday be used for their chromatography if suitable conditions are found.[37]

A bonded silica stationary phase carrying L-hydroxyproline groups and loaded with Cu(II) was described in Chapter 2 (Reference 48). Mandelic acid is resolved by it into its optically active D- and L-forms, using 10^{-4} *M* $CuSO_4$ as the mobile phase. In the same system mandelic acid was separated from hydroxymandelic, 2- and 3-phenyllactic and 2-hydroxycaproic acids.

Optical isomers were resolved, and hydroxy acids separated from one another, by exploiting outer-sphere complexation. Gaal and Inczedy[96] resolved the D- and L-forms of mandelic acid on a column of cation-exchange resin carrying optically active cobalt(III)-*tris*-ethylenediamine cations; aspartic acid was also resolved. The eluent was sodium sulfate solution. Extensive use was made of outer-sphere coordination by Carunchio and coworkers.[97,98] They prepared a chiral-bonded silica carrying $-NHCH_2CH_2NH_2Coen_2^{3+}$, where en = ethylene diamine; the cobalt ion is coordinated with six nitrogen atoms in this functional group. They used this bonded silica to separate the optical isomers of malic and tartaric acids from one another, as well as the D- and L-forms of the *tris*-oxalate complex anions of Co(III) and Cr(III). Nucleosides and cephalosporins were separated in the same system.

Outer-sphere coordination in the mobile phase was used by Halmos and Inczedy[99] to separate the anions of maleic and fumaric acids. The stationary phase was a strong-base anion exchanger, and the mobile phase contained the complex salt $Coen_3Cl_3$ or $Co(NH_3)_6Cl_3$, plus a buffer to give pH 8 to 10. The maleate ion with its *cis*-configuration associated much more strongly than the fumarate ion (*trans*-) and consequently, maleate was eluted before fumarate from the column. Mobile-phase coordination reinforces normal anion-exchange selectivity, which also leads maleate to be eluted first.

Aromatic acids were separated by ligand-exchange complexation in the mobile phase by Lee in 1970.[100] The stationary phase was a strong-base anion-exchange resin, 200-400 mesh; the mobile phase was 0.05 *M* ferric chloride in a water-alcohol mixture. The ferric ions formed cationic complexes with the anions of the acids, thus decreasing their retention by the resin. The anions forming the most stable complexes were eluted first. The complexes are strongly colored, allowing measurement by spectrophotometry as well as making them visible as colored bands in the glass column. Salicylic, sulfosalicylic, and nitrosalicylic acids were among the acids separated. The bands were broad and resolution was poor, but if the experiments were repeated with current techniques and current ion exchangers, they might well give useful results.

D. Miscellaneous Oxygen-Containing Ligands

A most interesting use of outer-sphere coordination is the recovery of nonionic surfactants from polluted water and the separation of homologous surfactants from one another.[101] The stationary phase was an iminodiacetate chelating resin carrying the cations $Co(NH_3)_6^{3+}$ or, better, $Co(NH_3)_5OH_2^{3+}$. Aqueous solutions 5×10^{-5} *M* in the polyoxyethylene surfactants, $C_9H_{19}{\cdot}C_6H_4{\cdot}O(CH_2CH_2O)_nH$, were passed through a small column of the chelating resin, which took up these compounds and also cationic surfactants, but not anionic surfactants. Stripping the column with alcoholic ammonia removed the nonionic surfactants, but not the cationic.

Metal-loaded macroporous resins are used in the analysis of crude oil; schemes were devised by Webster et al.[102,103] to absorb various compound classes, including aromatic acids and β-diketones, on resins loaded with Cu(II), Ni(II), and Fe(III). A sulfonated polystyrene resin loaded with Al(III) was used by Kothari[104] and Shankar[105] to fractionate DNA and RNA with various buffer solutions as eluents. Presumably there is metal-ligand interaction through the phosphate units of the nucleic acid chains.

Outer-sphere coordination was employed by Chow and Grushka[106] to separate nucleosides and nucleotides. The stationary phase was silica bonded with a *tris*-cobalt(III) ethylenediamine complex, similar to that used by Carunchio;[97] the eluent was a phosphate buffer containing magnesium ions.

REFERENCES

1. **Snyder, L. R.,** Nitrogen and oxygen compound types in petroleum, *Anal. Chem.*, 41, 314, 1969.
2. **Vogh, J. W. and Dooley, J. E.,** Separation of organic sulfides from aromatic concentrates by ligand-exchange chromatography, *Anal. Chem.*, 47, 816, 1975.
3. **Sluyterman, L. A. E. and Wijdenes, J.,** An agarose mercurial column for separation of mercaptopapain and nonmercaptopapain, *Biochem. Biophys. Acta,* 200, 593, 1970.
4. **Matuo, Y., Sano, R., Tosa, T., and Chibata, I.,** Purification of flavin-adenine dinucleotide and coenzyme A on *p*-acetoxymercurianiline-agarose columns, *Anal. Biochem.*, 68, 349, 1975.
5. **Hannestad, U., Lundqvist, P., and Sörbo, B.,** An agarose derivative containing an arsenical for affinity chromatography of thiols, *Anal. Biochem.*, 126, 200, 1982.
6. **Joshi, P. N., Shankar, V., Abraham, K. J., and Sreenivasan, K.,** Separation of chymopapain from papaya latex on Amberlite IR-120-Hg, *J. Chromatogr.*, 121, 65, 1976.
7. **Takayanagi, H., Hatano, O., Fujimura, K., and Ando, T.,** Ligand-exchange high-performance liquid chromatography of dialkyl sulfides, *Anal. Chem.*, 57, 1840, 1985.
8. **Skorokhod, O. R. and Varavva, A. G.,** Complexes of thiourea and cadmium in KU-2 sulfonic cation exchangers, *Zh. Fiz. Khim.*, 46, 1708, 1972.
9. **Honda, S.,** High-performance liquid chromatography of mono- and oligosaccharides, *Anal. Biochem.*, 140, 1, 1984
10. **Palmer, J. K.,** A versatile system for sugar analysis via liquid chromatography, *Anal. Lett.*, 8, 215, 1975.
11. **Schwarzenbach, R.,** A chemically bonded stationary phase for carbohydrate analysis, *J. Chromatogr.*, 117, 206, 1976.
12. **Rabel, F. M., Caputo, A. G., and Butts, E. T.,** Separation of carbohydrates on a new polar bonded phase, *J. Chromatogr.*, 126, 731, 1976.
13. **D'Amboise, M., Noel, D., and Hanai, T.,** Characterization of bonded amine packing for liquid chromatography of carbohydrates, *Carbohydr. Res.*, 79, 1, 1980.
14. **Verhaar, L. A. T. and Kuster, B. F. M.,** Liquid chromatography of sugars on silica-based stationary phases, *J. Chromatogr.*, 220, 313, 1981.
15. **Verhaar, L. A. T. and Kuster, B. F. M.,** Elucidation of mechanism of sugar retention on amine-bonded silica, *J. Chromatogr.*, 234, 57, 1982.
16. **Rooklin, R. D. and Phol, C. A.,** Determination of carbohydrates by anion-exchange chromatography with pulsed amperometric determination, *J. Liq. Chromatogr.*, 6, 1577, 1983.
17. **Baust, J. G., Lee, R. E., and James, H.,** Differential binding of sugars and polyhydric alcohols to ion-exchange resins; inappropriateness to quantitative HPLC, *J. Liq. Chromatogr.*, 5, 767, 1982.
18. **Samuelson, O.,** Partition chromatography of sugars, sugar alcohols and sugar derivatives, *Ion Exchange: A Series of Advances,* Vol. 2, Marcel Dekker, New York, 1969, chap. 5.
19. **Khym, J. X. and Zill, L. P.,** The separation of monosaccharides by ion exchange, *J. Am. Chem. Soc.*, 74, 2090, 1952.
20. **van Duyn, M., Peters, J. A., Kieboom, A. P. G., and van Bekkum, H.,** Studies on borate esters, *Tetrahedron,* 40, 2901, 1984.
21. **Paal, T. L.,** Study of glycol-boric acid complex formation equilibria, *Acta Chim. Acad. Sci. Hung.*, 103, 181, 1980; as cited in *Chem. Abstr.*, 93, 32552t, 1980.
22. **Davis, H. B. and Mott, C. J. B.,** Interaction of boric acid with carbohydrates and related substances, *J. Chem. Soc. Faraday Trans. I,* 76, 1991, 1980.
23. **Larsson, K. and Samuelson, O.,** Anion-exchange chromatography of alditols in borate medium, *Carbohydr. Res.*, 50, 1, 1976.
24. **Simatupang, M. H., Sinner, M., and Dietrichs, H. H.,** Simultaneous determination of reducing sugars and sugar alcohols, *J. Chromatogr.*, 155, 446, 1978.
25. **Simatupang, M. H.,** Ion-exchange chromatography of some neutral monosaccharides and uronic acids, *J. Chromatogr.*, 178, 588, 1979.
26. **Reimerdes, E. H., Rothkitt, K. D., and Schauer, R.,** Determination of carbohydrates by anion exchange of their borate complexes, *Fresenius Z. Anal. Chem.*, 318, 285, 1984.

27. **Honda, S., Matsuda, Y., Takahashi, M., Kakehi, K., and Ganno, S.,** Fluorimetric determination of reducing carbohydrates with 2-cyanoacetamide and application to automated analysis, *Anal. Chem.*, 52, 1079, 1980.
28. **Honda, S., Takahashi, M., Nishimura, Y., Kakehi, K., and Ganno, S.,** Sensitive UV monitoring of aldoses in automated borate complex anion-exchange chromatography, *Anal. Biochem.*, 118, 162, 1981.
29. **Honda, S., Takahashi, M., Kakehi, K., and Ganno, S.,** Rapid automated analysis of monosaccharides, *Anal. Biochem.*, 113, 130, 1981.
30. **Kennedy, J. F. and Fox, J. E.,** Fully automatic ion exchange and gel permeation chromatography of neutral mono- and oligosaccharides, *Carbohydr. Res.*, 54, 13, 1977.
31. **Mopper, K., Dawson, R., Lieberzeit, G., and Hansen, H. P.,** Borate complex ion-exchange chromatography with fluorimetric detection for determination of saccharides, *Anal. Chem.*, 52, 2018, 1980.
32. **Nordin, P.,** Monitoring of carbohydrates with periodate, *Anal. Biochem.*, 131, 492, 1983.
33. **Glad, M., Ohlson, S. Hansson, L., Maansson, M. O., and Mosbach, K.,** High-performance liquid affinity chromatography of nucleosides, nucleotides and carbohydrates with boronic acid substituted silica, *J. Chromatogr.*, 200, 254, 1980.
34. **Saunders, R. M.,** Separation of sugars on an ion-exchange resin, *Carbohyr. Res.*, 7, 76, 1968.
35. **Wong-Chong, J. and Martin, F. A.,** HPLC for the analysis of saccharides in sugar cane, *Sugar Azucar*, 75, 64, 1980; *Chem. Abstr.*, 93, 206460d, 1980.
36. **Pecina, R., Bonn, G., Burtscher, E., and Bobleter, O.,** HPLC elution behaviour of alcohols, aldehydes, ketones, organic acids and carbohydrates on a strong cation-exchange stationary phase, *J. Chromatogr.*, 287, 245, 1984.
37. **Walton, H. F.,** Counter-ion effects in partition chromatography, *J. Chromatogr.*, 332, 203, 1985.
38. **Hobbs, J. S. and Lawrence, J. G.,** Separation of carbohydrates on cation-exchange resin columns having organic counter-ions, *J. Chromatogr.*, 72, 311, 1972.
39. **Kuwamoto, T. and Okada, E.,** Separation of mono- and disaccharides by HPLC with a strong cation-exchange resin and an acetonitrile-rich eluent, *J. Chromatogr.*, 258, 284, 1983.
40. **Honda, S. and Suzuki, S.,** Common conditions for HPLC of aldoses, hexosamines and sialic acids in glycoproteins, *Anal. Biochem.*, 142, 167, 1984.
41. **Goulding, R. W.,** Liquid chromatography of sugars and related polyhydric alcohols on cation exchangers: effect of cation variation, *J. Chromatogr.*, 103, 229, 1975.
42. **Honda, S., Suzuki, S., and Kakehi, K.,** Analysis of aldose anomers by HPLC on cation-exchange columns, *J. Chromatogr.*, 291, 317, 1984.
43. **Jandera, P. and Churacek, J.,** Ion-exchange chromatography of aldehydes, ketones, ethers, alcohols, polyols and saccharides, *J. Chromatogr.*, 98, 55, 1974.
44. **Angyal, S. J. and Davies, K. P.,** Complexing of sugars with metal ions, *Chem. Commun.*, p. 500, 1971.
45. **Angyal, S. J.,** Complexing of polyols with cations, *Tetrahedron*, 30, 1695, 1974.
46. **Ekström, L. G. and Olin, A.,** Complex formation between Pb^{2+}, Ca^{2+} and some pentoses, *Acta Chem. Scand. Ser. A*, 31, 838, 1977.
47. **Lönnberg, H. and Vesala, A.,** Complexing of glycofuranosides with calcium ions, *Carbohydr. Res.*, 78, 53, 1980.
48. **Angyal, S. J. and Mills, J. A.,** Complexes of carbohydrates with metal ions: paper electrophoresis of polyols in solutions of calcium ions, *Aust. J. Chem.*, 32, 1993, 1979.
49. **Fitt, L. E., Hassler, W., and Just, D. E.,** A rapid method to determine the composition of corn syrup by liquid chromatography, *J. Chromatogr.*, 187, 381, 1980.
50. **Angyal, S. J., Bethell, G. S., and Beveridge, R. J.,** Separation of sugars and polyols on cation-exchange resins in the Ca form, *Carbohydr. Res.*, 73, 9, 1979.
51. **Verhaar, L. A. T. and Kuster, B. F. M.,** Improved column efficiency in chromatographic analysis of sugars on cation-exchange resins by use of water-triethylamine eluents, *J. Chromatogr.*, 210, 279, 1981.
52. **Vidalvalverde, C. and Martin-Villa, C.,** Improved separation of polyols and carbohydrates by HPLC, *J. Liq. Chromatogr.*, 5, 1941, 1982.
53. **Vidalvalverde, C., Olmedilla, B., and Martin-Villa, C.,** Reliable separation of xylitol from some carbohydrates and polyols by HPLC, *J. Liq. Chromatogr.*, 7, 2003, 1984.
54. **Duarte-Coelho, A. C., Dumoulin, E. D., and Guerain, J. T.,** HPLC determination of sucrose, glucose, fructose in complex products, *J. Liq. Chromatogr.*, 8, 59, 1985.
55. **Schlabach, T. D. and Robinson, J.,** Improvements in sensitivity with the cyanoacetamide reaction for determination of reducing sugars, *J. Chromatogr.*, 282, 169, 1983.
56. **Dokladova, J., Barton, A. Y., and Mackenzie, E. A.,** HPLC determination of sorbitol, *J. Assoc. Off. Anal. Chem.*, 63, 664, 1980.
57. **Baust, J. G., Lee, R. E., Rojas, R. R., Hendrix, D. L., Friday, D., and James, H.,** Separation of low-molecular-weight carbohydrates and polyols by HPLC: amine-modified silica vs. ion exchange, *J. Chromatogr.*, 261, 65, 1983.

58. **Schmidt, J., John, M., and Wandrey, C.,** Rapid separation of malto-, xylo- and cello-oligosaccharides, *J. Chromatogr.*, 213, 151, 1981.
59. **Scobell, H. D. and Probst, K. M.,** Rapid high-resolution separation of oligosaccharides on silver-form cation-exchange resins, *J. Chromatogr.*, 212, 51, 1981.
60. **Schwald, W., Concin, R., Bonn, G., and Bobleter, O.,** Analysis of oligomeric and monomeric carbohydrates, *Chromatographia,* 20, 35, 1985.
61. **Martinson, E. and Samuelson, O.,** Partition chromatography of sugars on ion-exchange resins, *J. Chromatogr.*, 50, 429, 1970.
62. **Ladisch, M. R., Huebner, A. L., and Tsao, G. T.,** High-speed LC of cellodextrins and other saccharide mixtures, *J. Chromatogr.*, 147, 185, 1978.
63. **Samarco, E. C. and Parente, E. S.,** Automated HPLC system for determination of mannitol, sorbitol and xylitol, *J. Assoc. Off. Anal. Chem.*, 65, 76, 1982.
64. **Guardiola, J. and Schultze, K. W.,** Thin-layer and liquid chromatography for quality control in manufacture of infusion solutions, *Fresenius Z. Anal. Chem.*, 318, 237, 1984.
65. **Scobell, H. D., Probst. K. M., and Steele, E. M.,** Automated LC system for analysis of carbohydrate mixtures, *Cereal Chem.*, 54, 905, 1977.
66. **McBee, G. C. and Maness, N. O.,** Determination of sucrose, glucose and fructose in plant tissue, *J. Chromatogr.*, 264, 474, 1985.
67. **Fitt, L. E.,** Convenient in-line purification of saccharide mixtures in automated HPLC, *J. Chromatogr.*, 152, 243, 1978.
68. **Barker, P. E. and Abusabah, E. K. E.,** Separation of mixtures of glucose and fructose using countercurrent chromatographic techniques, *Chromatographia,* 20, 9, 1985.
69. **Wortel, T. M. and van Bekkum, H.,** Carbohydrate separation by X-zeolites: cation and solvent effects, *Recl. Trav. Chim. Pays-Bas,* 97, 156, 1978.
70. **Petrus, L., Bilik, V., Kuniak, L., and Stankovic, L.,** Chromatographic separation of alditols on a cation-exchange resin in lanthanum form, *Chem. Zvesti,* 34, 530, 1980.
71. **Dua, V. K. and Bush, C. A.,** HPLC separation of amino sugars and peptides with metal-modified mobile phases, *J. Chromatogr.*, 244, 128, 1982.
72. **Bourne, E. J., Searle, F., and Weigel, H.,** Complexes between polyhydroxy compounds and copper(II) ions, *Carbohydr. Res.*, 16, 185, 1971.
73. **Leonard, J. L., Guyon, F., and Fabiani, P.,** HPLC of sugars on Cu(II)-modified silica gel, *Chromatographia,* 18, 600, 1984.
74. **Briggs, J., Finch, P., Matulewicz, M. C., and Weigel, H.,** Complexes of Cu(II), Ca and other metal ions with carbohydrates, *Carbohydr. Res.*, 97, 181, 1981.
75. **Zay, I., Gaizer, F., and Burger, K.** Iron(III) complex formation equilibria of sugar type ligands, *Inorg. Chem. Acta,* 80, L9, 1983.
76. **Shaw, V. and Walton, H. F.,** Chromatography of alcohols and sugars on Fe(III)-loaded resin, *J. Chromatogr.*, 68, 267, 1972.
77. **Martell, A. E. and Smith, R. M., Eds.,** *Critical Stability Constants,* Vol. 3, Plenum Press, New York, 1977.
78. **Milburn, R. M.,** The stability of iron(III)-phenol complexes, *J. Am. Chem. Soc.*, 77, 2064, 1955.
79. **McKay, J. F., Jewell, D. M., and Latham, D. R.,** The separation of acidic compound types isolated from high-boiling petroleum distillates, *Sep. Sci.*, 7, 361, 1972.
80. **Petronio, B. M., Lagana, A., and Russo, M. V.,** Some applications of ligand exchange: recovery of phenolic compounds from water, *Talanta,* 28, 215, 1981.
81. **Petronio, B. M., De Caris, E., and Ianuzzi, L.,** Separation of phenolic compounds, *Talanta,* 29, 691, 1982.
82. **Petronio, B. M., Lagana, A., and Andrea, G. D.,** Preparation and properties of phenol-formaldehyde-based resin in the iron(III) form, *Talanta,* 31, 357, 1984.
83. **Maslowska, J. and Pietek, W.,** Effect of complex formation in the separation of phenols by ion-exchange chromatography, *J. Chromatogr.*, 201, 293, 1980.
84. **Maslowska, J. and Pietek, W.,** Separation of chlorophenols on a cation exchanger in cerium(III) form, *Chromatographia,* 18, 704, 1984.
85. **Shahwan, G. J. and Jezorek, J. R.,** Liquid chromatography of phenols on an 8-quinolinol-silica gel-iron(III) stationary phase, *J. Chromatogr.*, 256, 39, 1983.
86. **Chanda, M., O'Driscoll, K. F., and Rempel, G. L.,** Removal of organic compounds by ligand sorption on polymer-anchored ferric ion, *React. Polym.*, 1, 183, 1983.
87. **Chanda, M., O'Driscoll, K. F., and Rempel, G. L.,** Removal and recovery of thiocyanate by ligand sorption on polymer-bound ferric ion, *React. Polym.*, 2, 175, 1984.
88. **Rawat, J. P. and Iqbal, M.,** Ligand exchange separation of phenols on alumina in Fe(III) form, *Chromatographia,* 17, 701, 1983.

89. **Bedetti, R., Carunchio, V., and Marino, A.,** Application of ligand exchange to separation of some aliphatic carboxylic acids, *J. Chromatogr.*, 95, 127, 1974.
90. **Funasaka, W., Hanai, T., Fujimura, K., and Ando, T.,** Complex chromatography between the metal ion of a cation-exchange resin and organic compounds, *J. Chromatogr.*, 78, 424, 1973.
91. **Fujimura, K., Koyama, T., Tanigawa, T., and Funasaka, W.,** Ligand-exchange chromatography: separation of hydroxybenzoic and hydroxynaphthoic isomers, *J. Chromatogr.*, 85, 101, 1973.
92. **Funasaka, W., Fujimura, K., and Kuriyana, S.,** Separation of aminobenzoic acid isomers by ligand-exchange chromatography, *Bunseki Kagaku,* 19, 104, 1970.
93. **Fukimura, K., Matsubara, M., and Funasaka, W.,** Separation of nitrosonaphthol isomers, *J. Chromatogr.*, 59, 383, 1971.
94. **Maslowska, J. and Pietek, W.,** Use of ligand exchange for the separation of aromatic acids, *Chromatographia,* 20, 46, 1985.
95. **Otto, J., de Hernandez, C. M., and Walton, H. F.,** Chromatography of aromatic acids on lanthanum-loaded ion-exchange resins, *J. Chromatogr.*, 247, 91, 1982.
96. **Gaal, J. and Inczedy, J.,** Chromatographic separation of optical isomers by outer-sphere complex formation, *Talanta,* 23, 78, 1976.
97. **Carunchio, V., Messina, A., Sinibaldi, M., and Corradini, D.,** Outer-sphere ligand-exchange chromatography on bonded chiral silica gel, *J. Liq. Chromatogr.*, 5, 819, 1982.
98. **Sinibaldi, M., Carunchio, V., Messina, A., and Corradini, D.,** Chromatography on bonded silica gel modified with amino complexes of cobalt(III), *Ann. Chim. (Rome),* 74, 175, 1984.
99. **Halmos, P. and Inczedy, J.,** Use of outer-sphere complex formation reactions in ion-exchange chromatography: separation of maleate and fumarate ions, *Talanta,* 29, 647, 1982.
100. **Lee, K. S., Lee, D. W., and Lee, E. K. Y.,** Anion-exchange chromatography of aromatic acids in ferric chloride-organic solvent medium, *Anal. Chem.*, 42, 554, 1970.
101. **Carunchio, V., Liberatori, A., Messina, A., and Petronio, B. M.,** Ligand-exchange techniques in analytical methods for surfactants, *Ann. Chim. (Rome),* 69, 165, 1979.
102. **Webster, P. V., Wilson, J. N., and Franks, M. C.,** Macro-reticular ion-exchange resins: some analytical applications to petroleum products, *Anal. Chim. Acta,* 38, 193, 1967.
103. **Webster, P. V., Wilson, J. N., and Franks, M. C.,** Applications of ion-exchange resins in nonaqueous media to separation and analysis of petroleum additives, *J. Inst. Petrol.*, 56, 50, 1970.
104. **Kothari, R. M.,** Aspects of fractionation of DNA on an IR-120-Al(III) column, *J. Chromatogr.*, 53, 580, (1970); **Kothari, R. M.,** 56, 151, 1971, **Kothari, R. M.,** *J. Chromatogr.*, 57, 83, 1971; **Kothari, R. M.,** *J. Chromatogr.*, 59, 194, 1971; and **Kothari, R. M.,** *J. Chromatogr.*, 64, 85, 1972.
105. **Shankar, V. and Joshi, P. N.,** Fractionation of RNA on a metal-ion equilibrated cation exchanger, *J. Chromatogr.*, 90, 99, 1974; and **Shankar, V. and Joshi, P. N.,** *J. Chromatogr.*, 95, 65, 1974.
106. **Chow, F. K. and Grushka, E.,** HPLC of nucleotides and nucleosides using outer-sphere and inner-sphere metal-solute complexes, *J. Chromatogr.*, 185, 361, 1979.

Chapter 7

METAL-OLEFIN COMPLEXES

J. D. Navratil

I. ARGENTATION CHROMATOGRAPHY

If LEC can be interpreted broadly enough to include all kinds of chromatography in which metal ions are held in the stationary phase and metal-ligand forces affect the retention of ligands, then argentation chromatography can be classified under LEC. This technique involves separating molecules having olefinic double bonds or aromatic ring systems which coordinate with the silver ions through π-electron interactions; because argentation chromatography has its own extensive literature and is sufficiently different from other LEC techniques, it will not be given the in-depth treatment as was done in other sections of this book.

The fact that ions of silver(I) form complexes with olefins was first used to advantage in chromatography by Bradford[1] in 1955. He used a solution of silver nitrate in glycerol as a gas chromatography (GC) stationary phase. Olefins were retarded by silver nitrate and separated from saturated hydrocarbons. Later, Bednas and Russell[2] separated *cis*- and *trans*-butene-2 in this manner, the *trans* isomer emerging first. The compounds 3- and 4-methyl-1-pentene were separated in the same way.[3] Using alumina impregnated with solid silver nitrate, Chapman and Kuemmel[4] separated isomers of dodecene.

Argentation chromatography was soon applied to liquid chromatography.[4] Paraffins were weakly sorbed on silver nitrate-treated alumina and were eluted by pentene. The olefin octadecene was held more strongly and was eluted, the *trans* isomer first, by pentene containing 1 to 3% ether. Oxygen-containing compounds were eluted later.

It was natural to use cation-exchange resins as supports for silver ions. This development brought "argentation chromatography" within the definition of LEC proposed by Muzzarelli, and, moreover, the first paper detailing the use of a silver-loaded resin[5] describes the use of another ligand as the displacing agent; a solution of butene in methanol was used to separate the esters of oleic and linoleic acid. A solution of water in methanol served the same purpose. Complex mixtures of isomeric *cis*- and *trans*-dienes, esters of naturally occurring unsaturated acids, were separated on macroporous resins loaded with a silver ion[6,7] and methyl *cis*-15-octadecenoate was isolated from mixtures.[8] Recently, the use of a silver-loaded bonded exchanger, Zipax SCX, to retain and separate heterocyclic aromatic compounds using 1% CH_3CN in hexone as eluent was reported;[9,10] the mechanism involves silver-to-nitrogen coordination besides the π-electron interactions.

Another obvious development was the use of "argentation chromatography" in thin layers. For this kind of chromatography, resins are generally unsuitable, and silica gel impregnated with silver nitrate is the logical choice for the stationary phase. Isomeric octadecanoates were separated on thin layers of silica gel impregnated with silver nitrate, and then brought to equilibrium with air of 30% relative humidity. The developing liquid was toluene at $-20°C$.[11] Chloroform-methanol mixtures were used to separate unsaturated acids derived from seaweed,[12] and hexane-ether mixtures to separate isomeric octadecadienoates.[13]

An interesting synergistic effect was found by Burns et al.[14] They used mixtures of benzene and cyclohexene to move triglycerides on thin-layer plates coated with silica gel and silver nitrate. Each solvent alone gave little displacement, but a 1:1 mixture gave maximum R_f values. Tristearin, with no double bonds, had $R_f = 0.98$. Displacement was less the more double bonds the compounds had.

Sterols[15] and phospholipids[16] were separated by thin-layer chromatography on alumina or silica gel containing silver nitrate. Again, the rate of movement is less the greater the degree of unsaturation.

Argentation chromatography, particularly argentation-TLC, has gained wide acceptance and popularity for lipid research and analysis.[17] Natural lipids consisting mainly of mixtures of a neutral lipid, phospholipid, and glycolipid types of molecules, have been separated from themselves and a variety of substances. These similar compounds only differ in chain length and degrees of unsaturation and/or substitution of their constituent ocyl or askyl chains.

Nichols[18] was the first to study silver complexes of lipid materials in 1952 and suggested using paper chromatography with silver nitrate in the liquid phase to separate the *cis-trans* isomers oleate and elaidate.

Argentation adsorption chromatography of lipids was first simultaneously reported by deVries[19] and Morris.[20] DeVries separated fatty acid methyl esters, triglycerides, and sterols on silver nitrate-impregnated columns whereas Morris used silver nitrate-impregnated thin layers to separate methyl esters of fatty acids and of epoxy and hydroxy fatty acids.

Mixtures of fatty acid methyl esters are separated according to the number of *cis* and/or *trans* double bonds of their constituents; numerous methods are used for isolation, detection and characterization of lipids. Furthermore, intact lipids (neutral glycerides, phospholipids, and glycolipids) are analyzed more readily by argentation than by GC.

II. MISCELLANEOUS SEPARATIONS ON SUPPORTED METAL SALTS

In this section we shall consider some separations of a kind similar to argentation chromatography, but performed with salts of metals other than silver. We shall start with thin-layer chromatography. Sterols have been separated on silica gel impregnated with lead and mercury salts,[21] and nitrotoluenes on silica gel mixed with zinc dust.[22] In the second instance, the active stationary phase was probably zinc oxide, and the nitro compounds presumably were reduced to the corresponding amines.

Shimomura[23] made an exhaustive study of the thin-layer chromatography of aromatic amines on support containing metal ions. She first tried finely ground inorganic exchangers, such as zirconium phosphate and "molecular sieves", and found them to give bad "tailing", bad resolution, or both. She got very good results with silica gel and activated alumina impregnated with zinc, cadmium, and nickel nitrates. Impregnation was done by mixing the dry adsorbent with twice its weight of 10% metal nitrate solution, spreading the slurry on glass plates and drying at 110°C. Comparison plates were prepared which had only the adsorbent, silica or alumina, without metal salt. Detection was done with ninhydrin, or by the quenching of fluorescence of Rhodamine B. Twenty amines were tested and five different solvent mixtures, all consisting of a nonpolar component (benzene, carbon, tetrachloride, chloroform) and a polar component (methanol, acetone, methyl ethyl ketone). Selected R_f values are given in Table 1. The plates that contained metal salts gave much better discrimination between amines than those without metal salt, and in nearly every case the metal slowed down the rate of migration.

Turning to the thin-layer chromatography of compounds that do not contain nitrogen, we note the separation of anthocyanins, anthocyanidins, and sugars on silica gel impregnated with basic lead acetate.[24] The solvents were butanol-methanol-water mixtures. Without the lead salt, silica gel alone did not affect separation. Alumina, however, does lead to the separation of anthocyanins[25] using 1% hydrochloric acid in methanol, and this attributed to the complexes formed between anthocyanins and aluminum ions.

Ferric chloride adsorbed in kaolin is effective in separating nonbasic nitrogen compounds from petroleum. These compounds are adsorbed on a column and then eluted with 1,2-

Table 1
THIN-LAYER CHROMATOGRAPHY ON ADSORBENTS IMPREGNATED WITH METAL SALTS (R_f VALUES)

Solvent:	Benzene-methanol, 5:1 Silica gel		Benzene-methyl ethyl ketone, 3:1 Silica gel		Benzene-methanol, 20:1 Alumina	
Compound **Absorbent:**	**Alone**	**With Zn**	**Alone**	**With Cd**	**Alone**	**With Zn**
Aniline	0.53	0.28	0.47	0.21	0.83	0.40
o-Toluidine	0.60	0.45	0.51	0.29	0.86	0.70
m-Toluidine	0.55	0.27	0.47	0.15	0.83	0.46
p-Toluidine	0.50	0.17	0.43	0.10	0.82	0.22
2,4-Xylidine	0.58	0.31	0.48	0.25	0.86	0.48
N-Methylaniline	0.68	0.56	0.62	0.57	—	—
N,N'-dimethylaniline	0.74	0.67	0.69	0.61	—	—
N,N'-diethylaniline	0.78	0.28	0.73	0.35	—	—
α-Naphthylamine	0.60	0.59	0.52	0.44	0.85	0.79
β-Naphthylamine	0.55	0.44	0.47	0.21	0.85	0.55
N-Methylnaphthylamine	0.72	0.82	0.67	0.72	—	—
p-Phenetidine	0.49	0.11	0.31	0.06	0.70	0.12
Methyl *p*-aminobenzoate	0.49	0.50	0.46	0.45	0.70	0.82

Adapted from Shimomura, K. and Walton, H. F., *Sep. Sci.*, 3, 493, 1968.

dichloroethane.[26] They are eluted as iron(III) complexes that are then passed through a strong-base macroporous resin to adsorb the iron and release the organic nitrogen compound. It is doubtful that this sequence of operations should be called "ligand exchange", but the compounds are not adsorbed by cation-exchange alone; adsorption depends on the formation of iron(III) complexes.

In his original publications on ligand exchange, Helfferich suggested that the technique should be applicable to GC. GC by "argentation" was described in the previous section. Pecsok[27] has shown that the metal phthalocyanins dispersed in silicone oil are useful as stationary phases in gas-liquid chromatography. Amines, alcohols, and nitriles are retarded selectively by phthalocyanins of copper(II), zinc, iron(II), cobalt(II), nickel(II), aluminum, and chromium(III), while hydrocarbons are no more retarded than they would be by silicone oil alone.

In 1959, Barber et al.[28] used liquid films of metal stearates supported on Celite® as stationary phases in gas chromatography. An ammonia-hydrogen mixture was used as the carrier gas. Comparing the metal stearates with Apiezon oil, they retarded amines very considerably and also retarded alcohols. Interestingly, straight-chain primary aliphatic amines were retarded more than branched-chain amines; this is the same effect that we found in liquid chromatography with metal-loaded ion-exchange resins.

In the laboratory of H. F. Walton, attempts were made to use macroporous resins as stationary phases in gas chromatography; experiments were also made with the liquid ion exchanger, di-(2-ethylhexyl) phosphate, loaded with metal ions and supported on Chromosorb®. Amines were retarded and, in fact, could hardly be displaced if helium were used as carrier gas. Ammonia and ammonia-helium mixtures displaced the amines and gave reasonable retention times, but tailing was very bad.

REFERENCES

1. **Bradford, B. W.,** The chromatographic analysis of hydrocarbon mixtures, *J. Inst. Petrol.*, 41, 80, 1955.
2. **Bednas, M. E. and Russell, D. S.,** A study of silver nitrate solutions in gas chromatography, *Can. J. Chem.*, 36, 1272, 1958.
3. **Smith, B. and Ohlson, R.,** Gas chromatographic separation of 3- and 4-methyl-1-pentene, *Acta Chem. Scand.*, 13, 1253, 1959.
4. **Chapman, L. R. and Kuennel, D. F.,** Liquid-solid and capillary gas-liquid chromatography of internal olefin isomers, *Anal. Chem.*, 37, 1598, 1965.
5. **Wurster, C. F., Copenhaven, J. H., and Shafer, P. R.,** Separation of the methyl esters of oleic, linoleic, and linolenic acids by column chromatography using cation exchange resin containing silver ion, *J. Am. Oil Chem. Soc.*, 40, 513, 1963.
7. **Emken, E. A., Scholfield, C. R., Davison, V. L., and Frankel, E. N.,** Separation of conjugated methyl octadecadienoate and thenoate geometric isomers by silver-resin column and preparative gas-liquid chromatography, *J. Am. Oil Chem. Soc.*, 44, 373, 1967.
8. **Schofield, C. R. and Emken, E. A.,** Isolation of methyl *cis*-15-octadecanoate by chromatography on a silver-treated macroreticular exchange resin, *Lipids*, 1, 235, 1966.
9. **Vivilecchia, R., Thiebaud, M., and Frei, R. W.,** Separation of polynuclear aza-heterocyclics by high-performance liquid chromatography, using a silver-impregnated adsorbent, *J. Chromatogr. Sci.*, 10, 411, 1972.
10. **Frei, R. W., Beall, K., and Cassidy, R. M.,** Determination of aromatic nitrogen heterocycles in air samples by high-speed liquid chromatography, *Mikrochim. Acta*, p. 859, 1974.
11. **Morris, L. J., Wharry, D. M., and Hammond, E. M.,** Chromatographic behavior or isomeric long-chain aliphatic compounds II — argentation thin-layer chromatography of isomeric octadecenoates, *J. Chromatogr.*, 31, 69, 1967.
12. **Dutta, S. P. and Barnta, A. K.,** Separation of *cis*- and *trans*- isomers of α, β-unsaturated acids by thin-layer chromatography, *J. Chromatogr.*, 29, 263, 1967.
13. **Christie, W. W.,** Chromatography of isomeric methylene interrupted methyl *cis, cis*-octadecadienoates, *J. Chromatogr.*, 31, 69, 1967.
14. **Burns, D. T., Stretton, R. J., Shepherd, G. F., and Dallas, M. S. J.,** Synergistic complexation effects in thin-layer chromatography of certain triglycerides on silica impregnated with silver nitrate, *J. Chromatogr.*, 44, 399, 1969.
15. **Kammerek, R., Lee, W.-H., Paliokas, A., and Schroepfer, G. J.,** Thin-layer chromatography of sterols on neutral alumina impregnated with silver nitrate, *J. Lipid Res.*, 8, 282, 1967.
16. **Hoevet, S. P., Viswanathan, C. V., and Lundberg, W. O.,** Fractionation of a natural mixture of alkenyl acyl and diacyl ethanolamine phosphatides by argentation adsorption thin-layer chromatography, *J. Chromatogr.*, 34, 195, 1968.
17. **Morris, L. J. and Nichols, B. W.,** Argentation thin layer chromatography of lipids, *Prog. Thin-Layer Chromatogr. Relat. Methods*, 1, 74, 1972.
18. **Nichols, P. L.,** Coordination of silver ion with methyl esters of oleic and elaidic acids, *J. Am. Chem. Soc.*, 74, 1091, 1952.
19. **de Vries, B.,** Paper presented at Sixth Congr. Int. Soc. Fat Research, London, April 1962; as cited in *Chem. Ind. (London)*, p. 1049, 1962.
20. **Morris, L. J.,** Paper presented at Sixth Congr. Int. Soc. Fat Research, London, April 1962; as cited in *Chem. Ind. (London)*, p. 1238, 1962.
21. **Stahl, E.,** *Dunnschicht-Chromatographie*, 2nd ed., Springer-Verlag, Berlin, 1967, 534.
22. **Yasuda, S. K.,** Separation and identification of tetryl and related compounds by two-dimensional thin-layer chromatography, *J. Chromatogr.*, 50, 453, 1970.
23. **Shimomura, K. and Walton, H. F.,** Thin-layer chromatography of amines by ligand exchange, *Sep. Sci.*, 3, 493, 1968.
24. **Pifferi, P. G.,** Thin-layer chromatography of sugars, anthocyanins and anthocyanidines on Kieselgel G impregnated with basic lead acetate, *J. Chromatogr.*, 43, 530, 1969.
25. **Birkhofer, L., Kaiser, C., and Donikem, M.,** Trennung von Anthocyangemischen Durch Komplexbildung an Aluminiumoxid, *J. Chromatogr.*, 22, 303, 1966.
26. **Jewell, D. M. and Snyder, R. E.,** Selective separation of "nonbasic" nitrogen compounds from petroleum by anion exchange of ferric chloride complexes, *J. Chromatogr.*, 38, 351, 1968.
27. **Pescok, R. L. and Vary, E. M.,** Gas-solid-liquid chromatography: dispersed metal phthalocyanines as a substrate, *Anal. Chem.*, 39, 287, 1967.
28. **Barber, D. W., Phillips, C. S. G., Tusa, F. F., and Berdin, A.,** The chromatography of gases and vapours. VI. Use of the stearates of bivalent Mn, Co, Ni, Cu and Zn as column liquids in gas-liquid chromatography, *J. Chem. Soc.*, p. 18, 1959.

Chapter 8

CONCLUSION

J. D. Navratil

We have interpreted ligand-exchange chromatography broadly to include many techniques for separating organic compounds in which coordination with metal ions plays a part. The method is very versatile and distinguished by high selectivity, especially stereoselectivity. Experimental conditions are flexible and selectivity orders can be manipulated almost at will. Practical difficulties remain, but they are on their way to solution. The main difficulties are explained below.

Slow mass transfer — It leads to broad bands and slow flow rates. This is a problem with conventional ion-exchange resins and appears to have two components, slow diffusion and slow chemical reactions. Slow exchange of ligands around metal ions held in resins is being avoided by having some of the ligand exchange take place in the mobile phase. Slow diffusion is becoming less important with the use of ion-exchange resins of small particle size and the use of bonded phases of various kinds, ionic and nonionic.

Softness and chemical instability of stationary phases — One of the best resins for LEC has been the acrylic polymer, Bio-Rex® 70. Unfortunately the granules are soft and will not stand high flow rates and pressure gradients. Bonded phases based on silica are rigid and do stand high pressure gradients, but silica is attacked by aqueous solutions of pH greater than 8, and alkaline eluents are much used in LEC. A possible solution to both these problems may be in the development of porous polymers with anionic functional groups that combine chemical and mechanical stability with good chromatographic behavior, that is, homogeneous active sites and fast mass transfer. The nonionic polymers available today, mainly macroporous polystyrene, show good chromatographic behavior, but those carrying ionic groups are less satisfactory. A good porous polymer with functional carboxyl groups would have obvious applications in LEC.

Detection — Eluents used in LEC contain metal ions and displacing ligands which interfere seriously with many of the methods used in liquid chromatography detection. Post-column derivatization has solved some of the problems, but not all. Variable-wavelength UV and fluorescence detectors are very useful. The area of liquid chromatography detectors is an active field of research today, and we may expect developments that will help ligand-exchange chromatography to come into its own.

INDEX

D

I

K

L

M

N

O

P

Q

R

S

T

U

V

W

X

Z